Quantum Mechanics

Quantum Mechanics

Professor Suresh Chandra
M.Sc., Ph.D., F.R.A.S. (London),
AvH-Fellow (Germany), F.M.A.S. (Pune)
Director, School of Physical Sciences,
S.R.T.M. University, Nanded - 431 606 (India)

CBS Publishers & Distributors Pvt. Ltd.
New Delhi • Bengaluru • Chennai • Kochi • Kolkata • Mumbai
Hyderabad • Uttarakhand • Nagpur • Patna • Pune • Jharkhand

ISBN: 978-81-239-1609-5

First Edition: 2008
Reprint: 2014, 2019

Published by **Satish Kumar Jain** and produced by **Varun Jain** for
CBS Publishers & Distributors Pvt. Ltd.,
4819/XI Prahlad Street, 24 Ansari Road, Daryaganj, New Delhi - 110002
delhi@cbspd.com, cbspubs@airtelmail.in • www.cbspd.com
Ph.: 23289259, 23266861, 23266867 • Fax: 011-23243014

Corporate Office: 204 FIE, Industrial Area, Patparganj, Delhi - 110 092
Ph: 49344934 • Fax: 011-49344935
E-mail: publishing@cbspd.com • publicity@cbspd.com

Branches:
- ***Bengaluru:*** 2975, 17th Cross, K.R. Road, Bansankari 2nd Stage, Bengaluru - 70 • Ph: +91-80-26771678/79 • Fax: +91-80-26771680 E-mail: cbsbng@gmail.com, bangalore@cbspd.com
- ***Chennai:*** No. 7, Subbaraya Street, Shenoy Nagar, Chennai - 600030 Ph: +91-44-26681266, 26680620 • Fax: +91-44-42032115 E-mail: chennai@cbspd.com
- ***Kochi:*** Ashana House, 39/1904, A.M. Thomas Road, Valanjambalam, Ernakulum, Kochi • Ph: +91-484-4059061-65 Fax: +91-484-4059065 • E-mail: cochin@cbspd.com
- ***Kolkata:*** 6-B, Ground Floor, Rameshwar Shaw Road, Kolkata - 700014 Ph: +91-33-22891126/7/8 • E-mail: kolkata@cbspd.com
- ***Mumbai:*** 83-C, Dr. E. Moses Road, Worli, Mumbai - 400018 Ph: +91-9833017933, 022-24902340/41 • E-mail: mumbai@cbspd.com

Representatives:

- Bhubaneswar 0-9911037372 • Hyderabad 0-9885175004 • Jharkhand 0-9811541605
- Nagpur 0-9021734563 • Patna 0-9334159340 • Pune 0-9623451994
- Uttarakhand 0-9716462459 • Dhaka (Bangladesh) 01912-003485

Printed at:
India Binding House, Noida, UP (India)

Preface

Quantum mechanics is one of the fundamental courses for graduate and postgraduate students of Physics as well as Chemistry. A number of books have been written on these topics and every one claims his/her presentation to be lucid and student-friendly. However, this topic has always been little difficult for a common student. We also made this attempt to present the topic in a student-friendly manner using simple language. In the presentation we tried to give exercises which help in order to make the topic as simple as possible. We hope, this attempt would be appreciated by the students and teachers. We shall appreciate receiving comments/suggestions, which can be sent at the following email also:

suresh492000@yahoo.co.in

While preparing the manuscript, I have been helped, advised, and encouraged by my seniors, friends and colleagues working in various institutions in India as well as abroad, and by my friends in personal life. I am heartily thankful to all of them. I am grateful to Prof. J.V. Narlikar, Prof. Dr. W.H. Kegel, Prof. S.P. Khare, for their encouragement. For preparation of the manuscript, I have been helped by my Ph.D. students, Dr. P.G. Musrif, Mr. S.V. Shinde, Mr. R.M. Dharmkare and Mr. Amit Kumar. My son, Mr. Mohit Kumar Sharma also helped in verifying some derivations. I am thankful to all of them. Special thanks are due to my wife Mrs. Purnima Sharma for her valuable cooperation in my life and for sharing a major part of the responsibility of family affairs, so that I could spend my time for this book. I am deeply grateful to my family members who always have been a source of encouragement and happiness for me. In the last but not the least, I am highly thankful to the publisher for bringing out this book nicely and in time.

Nanded (Maharashtra) **Suresh Chandra**

Contents

I

CHAPTER

Advent of Quantum Theory

By the end of the nineteenth century, scientists could explain all the observed phenomena. The theories employed for explanation of those phenomena are now called the classical theories. The light was considered as a wave. This wave theory of light could successfully explain the phenomena, such as reflection, refraction, diffraction, polarization, etc., known at that time. Since discovery is a continuously going on natural process, the scientists went on searching new phenomena. For example, the scientists discovered the black-body radiation spectrum, line spectra from various elements (atoms), photoelectric effect, Compton effect, variation of specific heat of solids at low temperature. With the discoveries of these phenomena, the available classical theories failed to provide any explanation for them. In the present chapter, we shall discuss the limitations of these classical theories and then will proceed for further developments needed for the explanation of the observed phenomena.

1. Limitations of classical theory

Here, we shall discuss how classical theories were found inadequate for providing any explanation for some specific phenomena.

1.1 Black-body radiation

Energy density u_λ of radiation from a black-body was found to depend on the wavelength λ of the radiation and the temperature T of the cavity. Typical variations of u_λ versus λ at $T = 100$ K and 120 K are shown in Figure 1. The black-body spectrum was found to have the following features.

1. Energy density u_λ for a given wavelength λ of radiation increases with the increase of the temperature T of the body.

2. For a given value of T, u_λ becomes maximum at some wavelength, denoted by λ_m.

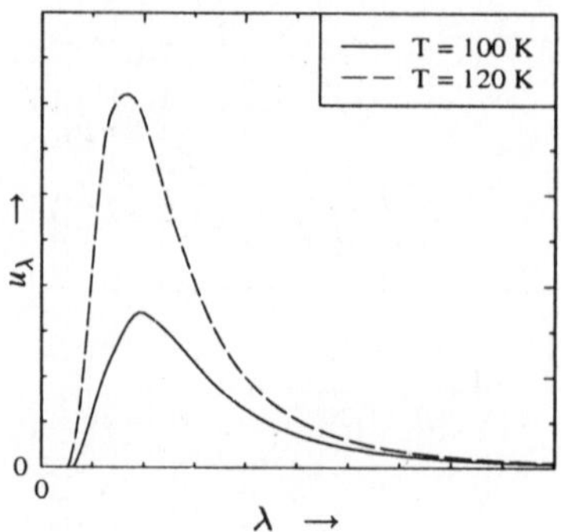

Figure 1: Typical variation of u_λ versus λ at $T = 100$ K and 120 K.

3. With the increase of T, λ_m is found to decrease such that $\lambda_m\, T =$ constant.

Attempts for explanation

Based on the thermodynamic arguments, Wien attempted to explain the black-body radiation spectrum and found the energy density u_ν for a frequency range from ν to $\nu + \mathrm{d}\nu$ at temperature T as

$$u_\nu \mathrm{d}\nu = c_1 \nu^3 \exp\Big(-\frac{c_2 \nu}{T}\Big)\, \mathrm{d}\nu \tag{1.1}$$

where c_1 and c_2 are positive constants, which could be varied freely. Taking $c_1 = 8\pi h/c^3$ and $c_2 = h/k$, the variation of u_ν versus ν (equation 1.1) for $T = 50$ K, along with the black-body radiation at $T = 50$ K is shown in Figure 2. Here, c is the speed of light, h the Planck constant and k the Boltzmann constant. This relation (1.1) could obviously explain the black-body spectrum in the low wavelength (high frequency) region, but failed miserably in the high wavelength (low frequency) region.

Rayleigh and Jeans made another attempt to explain the black-body spectrum, where they considered the radiation inside a black-body as the standing electromagnetic waves, and found the energy density u_ν for a frequency range from ν to $\nu + \mathrm{d}\nu$ at temperature T as

$$u_\nu \mathrm{d}\nu = \frac{8\pi k T \nu^2}{c^3}\, \mathrm{d}\nu \tag{1.2}$$

Variation of u_ν versus ν (Equation 1.2) for $T = 50$ K, along with the black-body radiation at $T = 50$ K is shown in Figure 3. This relation could explain the black-body spectrum in the high wavelength (low frequency) region, but failed completely in the low wavelength (high frequency) region.

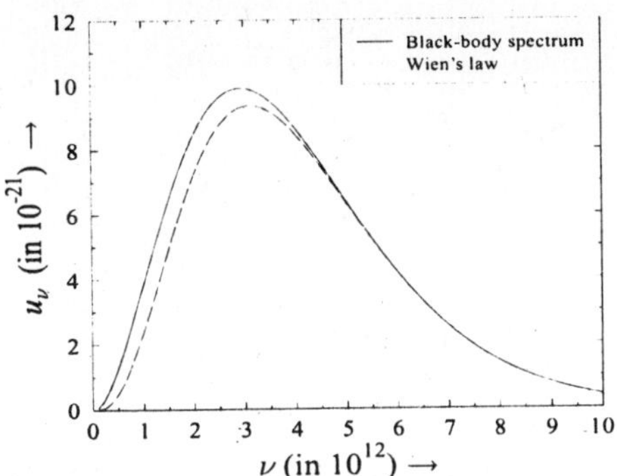

Figure 2: Black-body radiation at $T = 50$ K is shown by a continuous line. Variation of u_ν versus ν (equation 1.1) at $T = 50$ K is shown by a dotted line.

It shows that neither of the classical theories proposed by Wien as well as by Rayleigh and Jeans could explain the observed black-body radiation spectrum for all wavelengths.

1.2 Optical spectra of hydrogen atom

On heating, each element in a gaseous state was found to emit characteristic line spectrum. The simplest observed spectrum was from the hydrogen atom. Balmer in 1885 could successively fit the wavelengths λ of these spectral lines in the form of the following empirical relation.

$$\frac{1}{\lambda} = R\left(\frac{1}{2^2} - \frac{1}{n^2}\right) \tag{1.3}$$

where $n = 3, 4, 5, \ldots$ and R was a constant, which is now known as the Rydberg constant. Classical theory failed to give any explanation for these spectral lines as well as for the variation of intensity of the lines with wavelength.

1.3 Photoelectric effect

It was found that when some light was incident on a metal surface, electrons were found released from the surface. These released electrons are known as the photoelectrons and the phenomenon is known as the photoelectric effect. Some important conclusions drawn regarding the photoelectric effect are as the following.

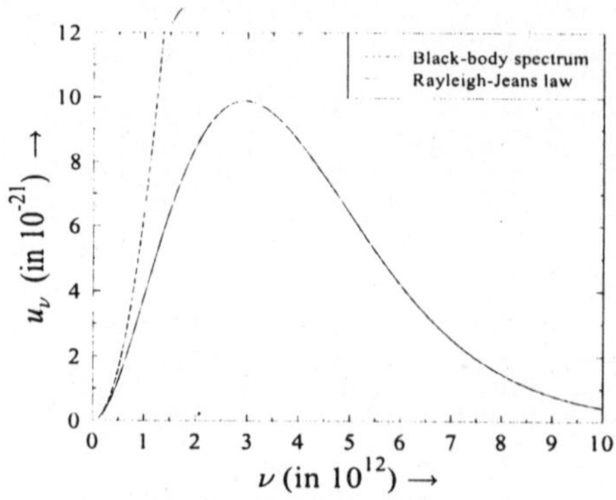

Figure 3: Black-body radiation at $T = 50$ K is shown by a continuous line. Variation of u_ν versus ν (equation 1.2) at $T = 50$ K is shown by a dashed line.

1. Energy distribution of photoelectrons was independent of the intensity of the incident radiation.

2. The maximum kinetic energy of photoelectrons from a given metal was proportional to the frequency of incident radiation.

3. For a given metal, photoelectrons were not emitted when frequency of incident light was less than a certain value, irrespective of the intensity of light.

4. For a given frequency, the number of photoelectrons emitted were directly proportional to the intensity of the incident light.

5. Emission of photoelectrons was found to be instantaneous with the incident of radiation.

Classical theory failed to give any explanation for the photoelectric effect.

1.4 Compton effect

The spectrum of x-rays scattered from a graphite block was found to have intensity peaks at two wavelengths in each direction, one at the same wavelength λ as the incident radiation and the other at a longer wavelength λ'. The wavelength λ' was found to depend on the direction of scattering whereas the wavelength λ was independent of the direction. This effect known as the Compton effect, could not be explained by the classical theory of radiation.

1.5 Specific heat of solids

Variation of the measured specific heat of a solid with temperature was found as shown in Figure 4. Classical theory for the specific heat can be understood in the following manner. In a solid, atoms are arranged in a lattice and each of them, vibrates about its equilibrium position. Following the law of the equipartition of energy, average energy per degree of freedom is kT, where k is the Boltzmann constant and T the absolute temperature of the solid. Since an atom is free to vibrate in three mutually perpendicular directions, the average energy per atom is $3kT$. The energy per gram atom is $E = 3NkT \equiv 3RT$, where N is the Avogadro number and $R(\equiv Nk)$ the gas constant. Thus, the specific heat c_V is

$$c_V = \frac{\partial E}{\partial T} = 3R$$

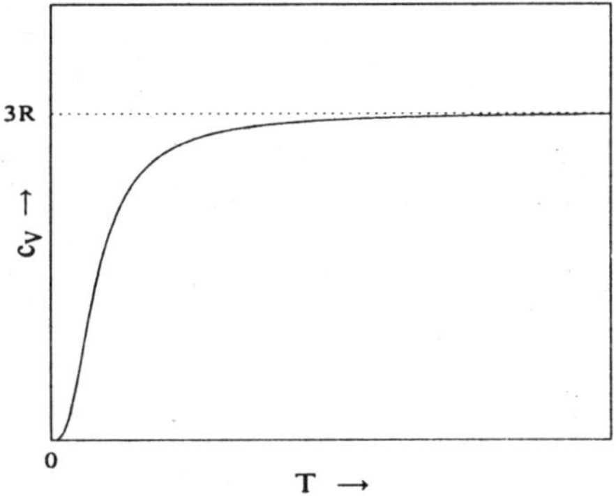

Figure 4: Variation of the measured specific heat c_V versus T for a metal.

It shows that the specific heat of solids is constant with a value $3R$. This is known as the Dulong and Petit's law. As Figure 4 shows, at the ordinary temperatures, solids generally follow this relation. But, at low temperatures, the specific heat decreases with the decrease of temperature and becomes zero at zero Kelvin temperature. This phenomenon also could not be explained with the help of the classical theory.

Remark

Besides the above, a number of other observed phenomena could not be explained with the help of the classical theories. By the end of the nineteenth century, scientist were convinced that classical theories were insufficient and there was a dire need of some revolutionary idea that could help in the explanation of these phenomena.

2. Nature of light

The first attempt to explain the nature of light was made by Newton in 1675. He considered the light as consists of tiny, light and perfectly elastic particles, called corpuscules, traveling in all directions with the speed of light. This theory, named as the 'corpuscular theory', could explain successfully some optical phenomenon, such as reflection and refraction. But, it completely failed to explain other known phenomena, such as the interference, diffraction and polarization. Moreover, this theory predicted that the speed of light would increase with the increase of the density of the medium of propagation, which was against the observations.

In 1678, Huygens proposed a wave theory where light was considered as a wave propagating through a hypothetical elastic medium, called ether.[1] This theory could explain successfully, the reflection, refraction, interference, diffraction, polarization and even the fact that the light travels with different speed in different media. In the wave nature of light, distribution of light-energy in a medium is continuous.

In the second half of nineteenth century, Maxwell proposed an entirely new idea that light consists of electromagnetic waves whose electric and magnetic fields lie in a plane, perpendicular to the direction of its propagation. The electric and magnetic fields associated with the wave were expressed by Maxwell in the form of four mathematical equations:

$$\nabla\times \vec{B} = \frac{4\pi}{c}\vec{j} + \frac{1}{c}\frac{\partial \vec{E}}{\partial t} \qquad \nabla . \vec{E} = 4\pi\rho$$

$$\nabla\times \vec{E} = -\frac{1}{c}\frac{\partial \vec{B}}{\partial t} \qquad \nabla . \vec{B} = 0$$

where $\vec{E}$, $\vec{B}$, $\vec{j}$ and ρ are respectively the electric field, magnetic field, electric current density and the electric charge density. The electromagnetic wave nature of light was later on confirmed experimentally

[1]This was name of the medium and has nothing to do with the chemical, ether in chemistry.

by Hertz. This electromagnetic theory, likewise the wave theory, could explain successfully, the reflection, refraction, interference, diffraction, polarization and the variation of the speed of light in different media. Towards the end of the nineteenth century, all the observed phenomena were explained with the help of the available theories.

Later on, some new phenomena, such as the black-body spectrum, photoelectric effect, Compton effect, optical spectral lines, variation of specific heat of solids with temperature, were observed. Further, some experimental discoveries such as x-rays (1895), radioactivity (1896) and electron (1897) also could not be explained on the basis of the classical theory.

Scientists were therefore looking for a new theory which could explain newly discovered phenomena.

3. Planck hypothesis

In 1900, Max Planck made a theoretical attempt to explain the black-body spectrum. He modified Wien's formula in such a way that the experimental curves could be fitted precisely. Then he started looking for a sound basis for his modifications.

Planck proposed a hypothesis where he considered that atoms of the wall of a cavity behave like oscillators, each of them oscillating with a characteristic frequency. Further, he assumed that these oscillators exchange electromagnetic energy with the cavity, and for these atomic oscillators, he made the following postulates:

1. An oscillator with a characteristic frequency ν, can have energies

$$E_n = nh\nu \tag{1.4}$$

   where h is known as the Planck constant and the integer n can have the values 0, 1, 2, 3, ...

2. An oscillator can exchange energy with the cavity only in discrete units

$$\Delta E_n = \Delta n\ h\nu$$

   where Δn = 0, 1, 2, 3, ... Thus, the exchange of energy is in the form of the packets of energy, known as quanta.

In the Planck's hypothesis, the average energy E of an oscillator is[2]

$$E = \frac{\sum_{n=0}^{\infty} E_n \exp(-E_n/kT)}{\sum_{n=0}^{\infty} \exp(-E_n/kT)}$$

$$= \frac{\sum_{n=0}^{\infty} nh\nu \exp(-nh\nu/kT)}{\sum_{n=0}^{\infty} \exp(-nh\nu/kT)}$$

$$= \frac{h\nu}{\exp(h\nu/kT) - 1}$$

In this hypothesis, the quantization condition (equation 1.4) invalidates the theorem of the equipartition of energy, which is based on the classical picture.

We know that the number of oscillators per unit volume in the frequency range from ν to $\nu + \mathrm{d}\nu$ is $(8\pi\nu^2/c^3)\mathrm{d}\nu$. Thus, the spectral energy density u_ν is

$$u_\nu \, \mathrm{d}\nu = \frac{8\pi\nu^2}{c^3} \, \mathrm{d}\nu \, E$$

$$= \frac{8\pi h\nu^3}{c^3} \frac{1}{\exp(h\nu/kT) - 1} \, \mathrm{d}\nu \qquad (1.5)$$

This relation is known as the Planck's radiation formula, and it could explain successfully the black-body spectrum for all the wavelengths.

This relation reduces to the Wien's formula when $h\nu >> kT$ and to the Rayleigh and Jeans formula when $h\nu << kT$. That is

$$u_\nu = \frac{8\pi h\nu^3}{c^3} \exp\left(-\frac{h\nu}{kT}\right) \qquad \text{for} \qquad h\nu >> kT$$

$$= \frac{8\pi kT\nu^2}{c^3} \qquad \text{for} \qquad h\nu << kT$$

[2]Using $x = \exp(-h\nu/kT)$, we have

$$E = \frac{h\nu \sum_{n=0}^{\infty} nx^n}{\sum_{n=0}^{\infty} x^n} = \frac{h\nu[x + 2x^2 + 3x^3 + 4x^4 + \ldots]}{1 + x + x^2 + x^3 + x^4 + \ldots}$$

$$= \frac{h\nu x[1 + 2x + 3x^2 + 4x^3 + \ldots]}{1 + x + x^2 + x^3 + x^4 + \ldots} = \frac{h\nu x(1-x)^{-2}}{(1-x)^{-1}}$$

$$= \frac{h\nu x}{1 - x}$$

Using the expression for x, we have

$$E == \frac{h\nu \exp(-h\nu/kT)}{1 - \exp(-h\nu/kT)} = \frac{h\nu}{\exp(h\nu/kT) - 1}$$

Planck's radiation formula in terms of wavelengths can be obtained by using the relation $\nu = c/\lambda$ and $d\nu = -c\, d\lambda/\lambda^2$ in equation (1.5) to get

$$u_\lambda\, d\lambda = \frac{8\pi ch}{\lambda^5} \frac{1}{\exp(hc/\lambda kT) - 1}\, d\lambda$$

Note:

Planck's explanation of black-body radiation provided a crucial step in the development of quantum ideas in physics, that the energy of an electromagnetic radiation is not continuous, but discrete in the form of packets. This hypothesis of quantization of energy was put on a firm basis by the pioneering work of Einstein, Bohr and others later on. The concept of the quantization of energy was however so radical that Planck himself was reluctant to accept it.

4. Einstein theory for photoelectric effect

Einstein used the idea of the quantization of energy of electromagnetic radiation for explanation of the photoelectric effect. According to this idea, the electromagnetic radiation consists of the packets of energy, known as photons. The energy of a photon of frequency ν is $h\nu$, where h is the Planck constant. When a photon of energy $h\nu$ falls on a metal surface, a part $h\nu_0$ of the energy of the photon is used to liberate an electron from the metal surface and the rest part appears in the form of the kinetic energy $\frac{1}{2}mv^2$ of the photo-electron. The incident photon is destroyed. From the conservation of energy, we have

$$h\nu = h\nu_0 + \frac{1}{2}mv^2 \tag{1.6}$$

Here, $h\nu_0$ is often known as the work-function of the metal. The frequency ν_0 is known as the threshold frequency, *i.e.*, no photo electron will be emitted by a photon whose frequency is less than the threshold frequency. The threshold frequency as well as the work-function are the characteristics of the metal and not of the incident radiation. This equation (1.6) explains all the observed facts about the photoelectric effect.

Exercise 1. Work functions of barium and tungsten are 2.5 eV and 4.2 eV, respectively. Find out if these metals are useful in a photocell operating in the visible light.

Solution: Wavelength (λ) in the visible light has a range from 4000 Å to 7000 Å.

$$\text{Energy of 4000 Å photon} = \frac{hc}{\lambda} = \frac{6.63 \times 10^{-34} \times 3 \times 10^8}{4000 \times 10^{-10} \times 1.602 \times 10^{-19}} = 3.10 \text{ e}$$

$$\text{Energy of 7000 Å photon} = \frac{hc}{\lambda} = \frac{6.63 \times 10^{-34} \times 3 \times 10^8}{7000 \times 10^{-10} \times 1.602 \times 10^{-19}} = 1.77 \text{ e}$$

Since work-function of tungsten is more than the energy range of the visible light, it cannot be used in the photocell operating in the visible light. As the work-function of barium lies in the energy range of the visible light, it can be used for the radiation whose energy is more than 2.5 eV.

Exercise 2. Calculate the potential difference applied to stop the fastest photoelectrons emitted by a surface when a photon of frequency 1.7 $\times 10^{15}$ Hz is allowed to fall on it. Work function of the surface is 5 eV.

Solution: Energy of the incident photon is

$$h\nu = 6.63 \times 10^{-34} \times 1.7 \times 10^{15}\ J$$

$$= \frac{6.63 \times 10^{-34} \times 1.7 \times 10^{15}}{1.602 \times 10^{-19}} = 7.04 \text{ eV}$$

Energy of the fastest electron = 7.04 − 5 = 2.04 eV
Hence, the potential difference required to stop the fastest electron is 2.04 V.

Exercise 3. Electromagnetic radiation of wavelengths 3100 Å and 3600 Å cause emission of electrons from a sodium surface with maximum kinetic energies 2.12 eV and 1.59 eV, respectively. Calculate Planck constant and the work-function of sodium.

Solution: Einstein relation for photoelectric effect is

$$\frac{hc}{\lambda} = \frac{hc}{\lambda_0} + \text{kinetic energy}$$

Using the given data, we have

$$\frac{hc}{3100 \times 10^{-10}} = \frac{hc}{\lambda_0} + 2.12 \times 1.602 \times 10^{-19}$$

$$\frac{hc}{3600 \times 10^{-10}} = \frac{hc}{\lambda_0} + 1.59 \times 1.602 \times 10^{-19}$$

$$\frac{hc}{10^{-10}}\left(\frac{1}{3100} - \frac{1}{3600}\right) = 0.53 \times 1.602 \times 10^{-19}$$

Hence,

$$h = \frac{0.53 \times 1.602 \times 10^{-19} \times 10^{-10} \times 3100 \times 3600}{500 \times 3 \times 10^{8}} = 6.32 \times 10^{-34}\,\text{Js}$$

From the first equation, the work-function is

$$\frac{hc}{\lambda_0} = \frac{hc}{3100 \times 10^{-10}} - 2.12 \times 1.602 \times 10^{-19}$$

$$= \frac{6.32 \times 10^{-34} \times 3 \times 10^{8}}{3100 \times 10^{-10}} - 2.12 \times 1.602 \times 10^{-19}$$

$$= 2.72 \times 10^{-19}\,\text{J} = 1.70\ \text{eV}$$

5. Compton effect

Let a photon of frequency ν incident on an electron at rest. For the photon having frequency ν, the energy is $h\nu$ and momentum is $h\nu/c$. Let the electron be placed at the origin of the coordinate system and the photon is moving along the x-axis. Suppose, after collision the frequency of photon is ν' while scattered in a direction making an angle θ with the x-axis and the electron recoils with velocity $\vec{v}$ in a direction making an angle ϕ with the x-axis (Figure 5). For the scattered photon having frequency ν', the energy is $h\nu'$ and momentum is $h\nu'/c$.

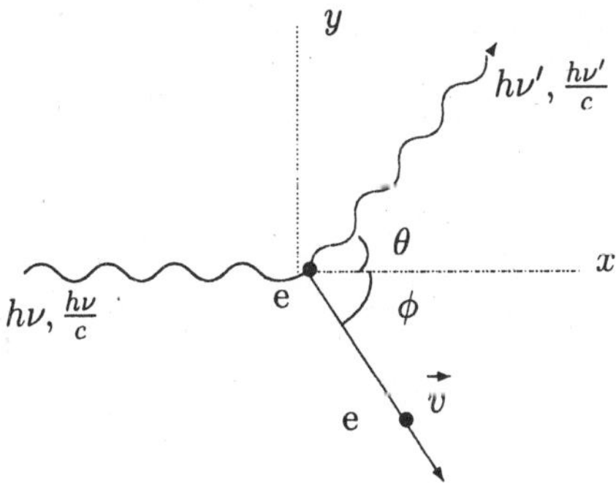

Figure 5: Photon of frequency ν moving along the x-axis incident on an electron at rest placed at the origin of the coordinate system. After collision, the frequency of photon is ν' while scattered in a direction θ with the x-axis and the electron recoils with velocity $\vec{v}$ in a direction ϕ with the x-axis.

Before collision, total energy and momentum of electron at rest are m_0c^2 and zero, respectively. Here, m_0 is rest mass of electron and c the speed of light. It is a case of the relativistic motion and therefore mass of the electron varies with its velocity according to the relation

$$m = \frac{m_0}{\sqrt{1 - v^2/c^2}}$$

Using the principle of the conservation of energy, we have

$$h\nu + m_0c^2 = h\nu' + mc^2$$

From the principle of the conservation of momentum along the x-axis and perpendicular to it, we have

$$\frac{h\nu}{c} = \frac{h\nu'}{c} \cos\theta + mv \cos\phi$$

$$0 = \frac{h\nu'}{c} \sin\theta - mv \sin\phi$$

These equations can be rearranged in the following form.

$$\frac{\nu - \nu'}{\nu\nu'} = \frac{h}{m_0c^2}(1 - \cos\theta)$$

Using $\nu = c/\lambda$ and $\nu' = c/\lambda'$, we have

$$\Delta\lambda \equiv \lambda' - \lambda = \frac{h}{m_0c}(1 - \cos\theta) \qquad (1.7)$$

The factor h/m_0c is known as the Compton wavelength. Since ϕ varies from zero to π radian, the wavelength λ' of the scattered x-ray is longer than the wavelength λ of the incident x-ray, and it obviously depends on the direction of the scattering.

When instead of one free electron, the metal block participates in the scattering of x-rays, in equation (1.7), m_0 is replaced by the mass M of the block and thus, we have $\Delta\lambda = 0$, *i.e.*, $\lambda' = \lambda$. Hence, the wavelength of the scattered x-ray is the same as that of the incident one. This explains the two wavelengths in the intensity peaks of the scattered x-rays.

Exercise 4. X-ray with $\lambda = 1$ Å are scattered from a free electron. The scattered radiation are observed at 90° to the incident direction. Calculate the Compton shift.

Solution: The Compton shift is

$$\Delta\lambda = \frac{h}{m_0c}(1 - \cos\phi)$$

$$= \frac{6.63 \times 10^{-34}}{9.11 \times 10^{-31} \times 3 \times 10^8}(1 - \cos 90)$$

$$= 2.43 \times 10^{-12}\text{m} = 0.024 \text{ Å}$$

6. Quantum theory of specific heat

Using the idea of the quantization of energy, Einstein explained the anomaly in the specific heat of solids at low temperatures. For $3N$ degrees of freedom of N atoms in a solid, he considered $3N$ oscillators, all of them having the same frequency ν and obtained the vibrational specific heat of solids as

$$c_V = 3R\left(\frac{h\nu}{kT}\right)^2 \frac{\exp(h\nu/kT)}{[\exp(h\nu/kT) - 1]^2}$$

$$= 3R\left(\frac{\theta}{T}\right)^2 \frac{\exp(\theta/T)}{[\exp(\theta/T) - 1]^2}$$

where R is a gas constant and $\theta = h\nu/k$ is known as the Einstein temperature. At ordinary temperatures, $T >> \theta$ and we have

$$c_V = 3R\left(\frac{\theta}{T}\right)^2 \frac{1 + (\theta/T) + \dots}{[1 + (\theta/T) + (\theta^2/2T^2) + \dots - 1]^2}$$

$$= 3R\left(\frac{\theta}{T}\right)^2 \frac{1 + (\theta/T) + \dots}{(\theta/T)^2[1 + (\theta/2T) + \dots]^2}$$

$$= 3R\frac{1 + (\theta/T) + \dots}{[1 + (\theta/2T) + \quad]^2} = 3R$$

At low temperatures, $T << \theta$ and we have

$$c_V = 3R\left(\frac{\theta}{T}\right)^2 \frac{\exp(\theta/T)}{[\exp(\theta/T)]^2} = 3R\left(\frac{\theta}{T}\right)^2 \exp\left(-\frac{\theta}{T}\right)$$

The specific heat reduces to zero as the temperature goes to zero Kelvin. However, it is not in agreement with the experimentally observed T^3 dependence of the specific heat.

Debye improved the Einstein model by assuming that the frequencies of vibrations for $3N$ oscillators were not the same, but distributed as in the case of the black-body spectrum. The expression derived by Debye is

$$c_V = 9R\left[4\left(\frac{T}{\theta}\right)^3 \int_0^{\theta/T} \frac{x^3}{e^x - 1}\,dx - \left(\frac{\theta}{T}\right)\frac{1}{e^{\theta/T} - 1}\right] \tag{1.8}$$

(i) When T is large, we have

$$\left(\frac{\theta}{T}\right)\frac{1}{\mathrm{e}^{\theta/T}-1} \approx \left(\frac{\theta}{T}\right)\frac{1}{[1+\theta/T\ldots-1]} \approx 1$$

For large T, $x \equiv h\nu/kT$ is small and we have

$$\frac{x^3}{\mathrm{e}^x-1} \approx \frac{x^3}{1+x\ldots-1} \approx x^2$$

and

$$\int_0^{\theta/T}\frac{x^3}{\mathrm{e}^x-1}\,\mathrm{d}x \approx \int_0^{\theta/T} x^2\,\mathrm{d}x \approx \frac{1}{3}\left(\frac{\theta}{T}\right)^3$$

We therefore have

$$c_V = 9R\left[4\left(\frac{T}{\theta}\right)^3\frac{1}{3}\left(\frac{\theta}{T}\right)^3 - 1\right] = 3R$$

(ii) For very low temperature, $\theta/T \to \infty$, so that second term on the right side of equation (1.8) is negligible and in the first term, the upper limit in the integral can be taken as ∞. Thus, we have

$$\int_0^{\infty}\frac{x^3}{\mathrm{e}^x-1}\,\mathrm{d}x = \frac{\pi^4}{15}$$

and

$$c_V = 9R\left[4\left(\frac{T}{\theta}\right)^3\frac{\pi^4}{15}\right] = \frac{12}{5}\frac{\pi^4 R}{\theta^3}\,T^3$$

This expression explains the T^3 dependence of the specific heat at low temperatures.

7. Bohr theory of hydrogen atom

Rutherford, on the basis of his experiment on scattering of α particles by a thin gold foil, proposed a nuclear model of atom. In this model, the positive charge is assumed to be confined to a very small sphere, called the nucleus, and the electrons are considered revolving around the nucleus. According to the classical electrodynamics, an accelerated charge radiates energy and therefore, this model of revolving electrons was highly unstable, as the radius of the orbit of an electron would go on decreasing continuously and the electron will spiral into the nucleus. Moreover, that model failed to explain the observed optical spectrum emitted by various elements. As we stated earlier, the optical spectral lines of hydrogen atom could be expressed by an empirical relation (1.3).

Theoretical explanation of hydrogen spectrum, based on the quantum ideas, was given by Bohr in 1913. His model for hydrogen atom is based on the following postulates.

1. The atom has a positively charged nucleus and the electron revolves around it in a circular orbit.

2. Out of a large number of circular orbits, the electron revolves in those orbits only for which the angular momentum is equal to an integral multiple of $h/2\pi$. That is

$$mvr = nh/2\pi$$

   where n is an integer, which can assume the values 1, 2, 3, ...

3. Bohr assumed that as long as the electron does not move from one orbit to another, it does not emit or absorb any amount of energy. These orbits are therefore known as the stationary states. Hence, the emission or absorption of energy occurs only when the electron goes from one stationary state to another. When electron goes from an outer orbit n (with energy E_n) to an inner orbit m (with energy E_m), the frequency ν of the emitted photon is

$$h\nu = E_n - E_m$$

Based on these assumptions, an expression for energy of electron in the n-th orbit is obtained as

$$E_n = -\frac{me^4}{8h^2\epsilon_0^2}\frac{1}{n^2} = -E_1\frac{1}{n^2}$$

where $E_1(\equiv me^4/8h^2\epsilon_0^2)$ is the amount of energy of electron in the first Bohr-orbit of hydrogen atom. Thus, the total energy in a state is inversely proportional to n^2. When an electron in an initial state n (with energy E_n) jumps to a lower state m (with energy E_m), a photon of energy $h\nu$ is emitted. Thus, we have

$$h\nu = E_n - E_m = -\frac{E_1}{n^2} + \frac{E_1}{m^2}$$

$$\nu = \frac{E_1}{h}\left(\frac{1}{m^2} - \frac{1}{n^2}\right)$$

Using the relation $\nu = c/\lambda$, where c is the speed of light and λ the wavelength of the photon, we get

$$\frac{1}{\lambda} = R_\infty\left(\frac{1}{m^2} - \frac{1}{n^2}\right)$$

where $R_\infty(\equiv E_1/hc)$ is known as the Rydberg constant. Thus, the optical spectra of hydrogen corresponds to $m = 2$.

The value of R_∞ is 10973731.1 m^{-1}. Various series observed in the spectrum of hydrogen spectrum may be reproduced as the following.

Series	m	n
Lyman series	1	2, 3, 4, ...
Balmer series	2	3, 4, 5, ...
Paschen series	3	4, 5, 6, ...
Brackett series	4	5, 6, 7, ...
Pfund series	5	6, 7, 8, ...

Wavelengths of first few lines of various series are given in the following table. The Lyman series lies in the UV region, Balmer series lies in the visible region whereas Paschen, Brackett and Pfund series lie in the infrared region.

Wavelengths are given in Å

Lyman	Balmer	Paschen	Brackett	Pfund
UV	Visible	IR	IR	IR
1216	6563	18751	40500	74560
1026	4861	12818	26300	46520
973	4340	10938	21660	37400
950	4102	10050	19450	
938	3970	9546		
931	3889			
926				

8. Franck and Hertz experiment

Existence of discrete energy states in atoms was demonstrated experimentally by Franck and Hertz in 1914. Outline of their experiment is shown in Figure 6.

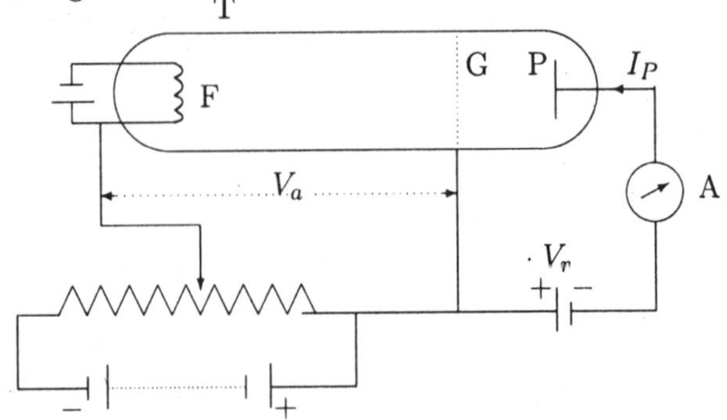

Figure 6: Outline of the Franck and Hertz experiment.

Vapour of element (atoms) under investigation is filled in the tube T at low pressure. The electrons emitted from the heated filament F are accelerated towards the grid G by a potential difference V_a applied between G and F. The electrons reach at the plate P after facing a small retarding potential difference V_r between G and P, and the current I_p is measured by the ammeter A.

If we assume that the electrons emitted from F have zero value for the initial velocity. Then the maximum velocity v_1 of the electrons reaching at G is

$$\frac{1}{2}\,mv_1^2 = eV_a$$

where m and e are the mass and charge of the electron, respectively. Further, only those electrons can reach from G to P whose velocity at G is greater than v_2, given by

$$\frac{1}{2}\,mv_2^2 = eV_r$$

Obviously, by changing the value of V_a, the velocities of electrons reaching G can be changed, and thus only those electrons whose velocity at G is greater than v_2 will reach at P and will contribute to I_p. A qualitative behaviour of variation of I_p versus V_a is shown in Figure 7.

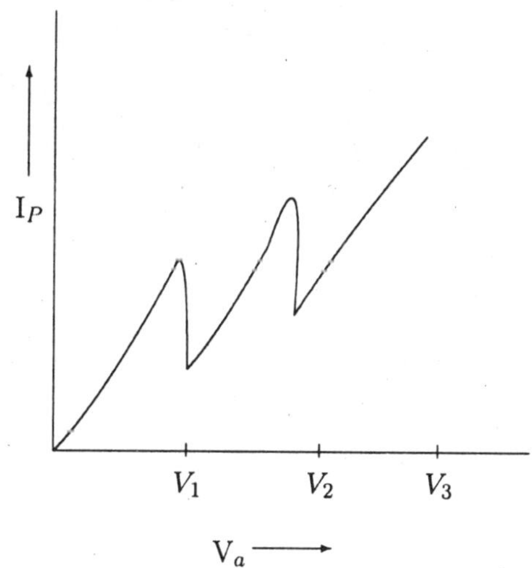

Figure 7: Variation of I_p versus V_a.

Figure 7 can be understood in the following manner. At low values of V_a ($V_a < V_1$), energy of electron is so small that it is not sufficient for excitation of an atom. The current I_p increases with V_a. At $V_a = V_1$, the electron has enough energy to excite an atom to the first excited state. After loosing the energy for excitation of an atom, the electron does not have enough energy to reach at P after facing the retarding potential V_r. Consequently a sudden drop in the current is observed.

For $V_a > V_1$, but smaller than V_2, electron looses energy for excitation of an atom to the first excited state and the remaining energy is sufficient enough to reach at P. The current I_p again increases with V_a. For $V_a = V_2$, the electron has enough energy to excite an atom to the second excited state. After loosing the energy for excitation of an atom to the second excited state, the electron does not have enough energy to reach at P after facing the retarding potential V_r. Consequently a sudden drop in the current is observed.

For $V_a > V_2$, but smaller than V_3, electron looses energy for excitation of atom to the second excited state and the remaining energy is sufficient enough to reach P. The current I_p again increases with V_a. We further get a sudden drop in current at $V = V_3$, $V = V_4$, etc. Since all electrons do not behave in the same manner and there is distribution of energy among the electrons, therefore, at $V_a = V_1, V_2, V_3, \ldots$, the value of I_p is not zero, but has a finite value.

It is obvious that the amount of energy transferred to an atom by an electron is not continuous, but has certain discrete values. This experiment provides an excellent proof for the existence of discrete energy states in an atom.

9. Inadequacy of quantum theory

As mentioned above, the quantum theory developed by Planck, Einstein and Bohr could explain some observed phenomena such as black-body spectrum, photoelectric effect, optical spectra of hydrogen atom, specific heat of solids at low temperatures, Compton effect, etc. However, it was still found inadequate to explain a number of other phenomena. Some of them are as the following.

1. The general quantization rule is valid only for a periodic system. Hence, this quantum theory cannot be applied to a non-periodic system.

2. Even for a hydrogen atom, the quantum theory could explain the broad features only and it failed to explain the fine structure of the spectral lines.

3. Application of quantum theory to helium atom as well as to the hydrogen molecule gave the results contrary to those obtained from

the experiments.

4. The quantum theory does not accept half-odd integer quantum numbers.

5. This theory has no provision to evaluate the transition probabilities and the intensities of the spectral lines.

Such shortcomings of the quantum theory show that the theory is not yet sufficient as per the requirement. Therefore, there was a need to look for a more basic theory which could account for all the phenomena.

10. Problems and questions

1. Write in brief about some observed phenomena which could not be explained with the help of a classical theory.

2. Describe the Planck's quantum hypothesis and obtain an expression for energy density of radiation emitted from a black-body. Show that Wien expression as well as Rayleigh and Jeans expression are special cases of it.

3. Describe the photoelectric effect. Explain the Einstein theory for photoelectric effect.

4. Monochromatic radiation of wavelength 1500 Å falls on a metallic surface of work-function 4 eV. Calculate the kinetic energy of the fastest photoelectrons emitted from the surface. Also calculate the stopping potential for radiation and threshold wavelength for the metal.

5. Calculate the work-function of a metal for which the threshold wavelength is 5600 Å. If a monochromatic radiation of wavelength 2000 Å falls on the metal, what is the required stopping potential.

6. Describe the quantum theory of specific heat of solids.

7. What is the Compton effect? Derive an expression for the change in wavelength of a scattered photon.

8. Describe the Bohr theory of hydrogen atom.

9. A photon of wavelength 3 Å is scattered by an electron at rest in a direction 60° to its original direction. Calculate the wavelength of photon after scattering.

10. Write notes on the following

    (i) Dual nature of light
    (ii) Planck's quantum hypothesis
    (iii) Einstein theory for photoelectric effect
    (iv) Compton effect
    (v) Quantum theory of specific heat
    (vi) Bohr theory of hydrogen atom
    (vii) Specific heat of solids
    (viii) Franck and Hertz experiment and its utility
    (ix) Inadequacy of quantum theory

Wave Mechanical Concepts

In the last chapter, we have seen that some optical phenomena like reflection, refraction, diffraction, polarization, can be explained on the basis of the wave nature of light. However, some other phenomena like black-body spectrum, photoelectric effect, Compton effect, optical spectra, can be explained on the basis of the particle (quantum) nature of light. Hence, it is difficult to decide between the wave and particle natures of light. Therefore, dual nature of light has been unanimously accepted by the scientific community world wide.

1. Wave nature of particles

Dual nature of light prompted de Broglie to extend this concept of duality to the material particles, as he was convinced with the well known facts:

(i) Our universe is composed of radiation and matter.

(ii) The nature has a number of symmetries.

Hence, the radiation and matter should have a similar nature. That is, when radiation (light) shows a dual nature, the matter also should have a dual nature. Hence, the matter should show a wave nature also. Finally, de Broglie in 1924 proposed that a particle in motion has an associated wave, called the matter wave, which may be regarded localized with the particle and the wavelength λ of the matter wave is

$$\lambda = \frac{h}{mv} \tag{2.1}$$

where h is the Planck constant, and m and v are, respectively, the mass and velocity of the particle. This equation (2.1), known as the de Broglie relation, relates the wave as well as particle natures. The relation shows that the wavelength would be the longest for an electron - the lightest available particle. Therefore, experiments were performed to see the wave nature of electrons.

Exercise 1. Write down two equations which show a relation between the particle and wave natures.

Solution: The two equations showing a relation between the particle and wave natures are

$$E = h\nu \qquad\qquad \lambda = \frac{h}{p}$$

Here, the wavelength λ and frequency ν are the characteristics of a wave, whereas the energy E and momentum p are the characteristics of a particle. The Planck constant h obviously behaves like a bridge between the two natures.

Exercise 2. Calculate de Broglie wave length of an electron moving with a kinetic energy of 1500 eV. Compare this wavelength with that of an X-ray having the same energy. (Mass of electron = 9.11×10^{-31} kg; 1 eV = 1.602×10^{-19} J)

Solution: Kinetic energy of 1500 eV corresponds to a non-relativistic motion. The momentum p of electron is

$$p = \sqrt{2mE} = \sqrt{2 \times (9.11 \times 10^{-31}) \times 1500 \times (1.602 \times 10^{-19})}$$

$$= 2.09 \times 10^{-23} \text{ kg m/s}$$

de Broglie wave length is

$$\lambda = \frac{h}{p} = \frac{6.63 \times 10^{-34}}{2.09 \times 10^{-23}} = 3.17 \times 10^{-11}\text{m} = 0.317 \text{ Å}$$

Wavelength for x-ray of energy 1500 eV is

$$\lambda = \frac{hc}{E} = \frac{(6.63 \times 10^{-34}) \times (3 \times 10^{8}))}{1500 \times (1.602 \times 10^{-19})} = 8.277 \times 10^{-10}\text{m} = 8.277 \text{ Å}$$

Hence, the ratio of two wave lengths is

$$\frac{\text{X-ray wave length}}{\text{de Broglie wave length}} = \frac{8.277}{0.317} = 26.11$$

1.1 Electron diffraction experiment

In order to show wave nature of electrons, experiments were performed independently by Davisson and Germer and by G.P. Thompson. Outlines of the experimental arrangement for diffraction of electrons is shown in Figure 1(a). The electrons emitted from the heated filament C are accelerated towards the tube A by a potential difference ranging from 10,000 to 50,000 volts. The electrons collimated by the tube A fall on a thin gold film F of thickness of the order of 10^{-6} cm. The discharge tube T is evacuated to avoid collisions of electrons with the gas particles. The diffracted beam is allowed to fall on a fluorescent screen or on a photographic plate S. The diffracted beam is found to have a pattern of concentric rings, as shown in Figure 1(b). Diffraction is a characteristic of a wave. On measuring radii of the rings and the distance between the film F and the fluorescent screen or photographic plate S, the angle of diffraction θ can be calculated. For the known distance d between atomic planes, the wavelength of the diffracted beam can be obtained. As the diffraction can be shown by a wave, this experiment clearly demonstrates the wave nature of electrons.

Before falling on the film F, the electrons behave like a particle whereas in the diffraction process, they behave like a wave. Here, the electrons show particle nature in one part of the experiment whereas in the second part of the experiment, they show the wave nature. Thus, for a complete description of the physical phenomena, both particle as well as wave aspects of the material are required. It proves the existence of a wave associated with the moving electrons.

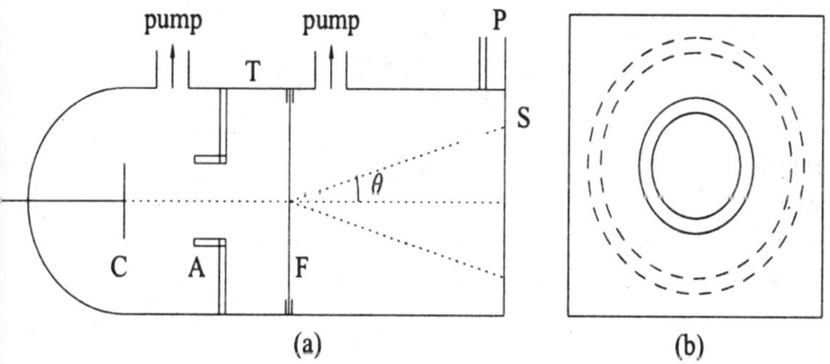

Figure 1: (a) Outline of the experiment for diffraction of electrons. (b) Circles indicate diffraction pattern which is a characteristic of a wave.

Exercise 3. Determine the de Broglie wavelength of an electron which is accelerated through a potential difference of 200 V. (Mass of electron = 9.11×10^{-31} kg; 1 eV = 1.602×10^{-19} J)

Solution: After going through a potential difference of 200 V, the energy gained by the electron is 200 eV. This energy corresponds to a non-relativistic motion. The momentum p of electron is

$$p = \sqrt{2mE} = \sqrt{2 \times (9.11 \times 10^{-31}) \times 200 \times (1.602 \times 10^{-19})}$$

$$= 7.64 \times 10^{-24} \text{kg m/s}$$

de Broglie wave length is

$$\lambda = \frac{h}{p} = \frac{6.63 \times 10^{-34}}{7.64 \times 10^{-24}} = 8.68 \times 10^{-11} \text{m} = 0.868 \text{ Å}$$

1.2 Standing waves of an electron in an orbit

Since an electron revolving around a nucleus in a circular path of radius r is bound, the motion can be represented by a standing wave. That is, the circumference of the orbit must be equal to an integer multiple of the wavelength of the wave. That is

$$2\pi r = n\lambda \tag{2.2}$$

Using equation (2.1) in (2.2), we have

$$2\pi r = n\frac{h}{mv}$$

$$mvr = n\frac{h}{2\pi}$$

This relation is the Bohr quantization rule. Thus, the quantization rule (postulate 2 of the Bohr theory in the preceding chapter), which was introduced by Bohr in an ad-hoc manner, has an explanation in terms of the matter waves.

2. Uncertainty principle

According to the classical theory, value of a dynamical variable can be determined to any desired degree of accuracy. This principle of determinism is a backbone of the classical theory. However, the accuracy of measurements will depend on the limitation of the measuring equipments. But, the quantum mechanics does not support this idea of determinism. Quantum mechanics says about the uncertainty in the measurements of the physical quantities. Further, it relates the uncertainties between two canonical conjugates variables in the following manner.

2.1 Position-momentum uncertainty

The idea of wave associated with a particle came in existence after the de Broglie hypothesis. If we consider a plane wave, the position is completely indeterminate, as a plane wave extends to an infinite distance. When we consider a wave associated with a particle, indeterminacy comes in the picture automatically, as a wave of a definite frequency is not localized. Heisenberg realized this situation of indeterminacy and after an intelligent thought, he formulated a principle, called the uncertainty principle. This principle states that no two canonically conjugate quantities can be measured accurately, simultaneously. For example, for the two canonically conjugate variables x and p_x, the uncertainty principle can be mathematically expressed as

$$\Delta x \, \Delta p_x \approx h$$

where Δx and Δp_x are uncertainties in determination of x and p_x, respectively.

Experimental verification of this uncertainty relation can be obtained with the help of a single-slit experiment, discussed in the following section.

2.2 Single-slit experiment

Consider a beam of monochromatic electrons moving with a speed u along the y-direction, taken horizontal in the figure (Figure 2). Let us try to measure uncertainty Δx in the position along the x-direction, taken vertical in the figure, and in the velocity component v (in the vertical direction) of an electron at the slit S_1. The slit-width Δx obviously gives the uncertainty in the x-component of the position. The uncertainty Δx can be decreased by making the slit narrow. When an electron passes through the slit S_1, the x-component of velocity at the slit is zero.

According to the hypothesis of matter waves, an electron has a wave nature and therefore, it should show the diffraction phenomenon at the slit producing a pattern as shown in Figure 2. The formation of diffraction pattern shows that after passing through the slit, the electron has developed the x-component (vertical component) of velocity. Further, for the first minimum, the theory of diffraction gives

$$\sin \theta = n\lambda$$

where λ is wavelength of the diffracted wave. For the first minimum, we have $n = 1$. Further, for the small angle θ, we have $\sin \theta \approx \theta$. Thus, we have

$$\theta = \frac{\lambda}{\Delta x} \tag{2.3}$$

Let t be the time of transit from the position O to the position A and v be the x-component (vertical component) of velocity of electron reaching at the position B. Thus, we have

$$u\,t = \text{OA} \qquad \text{and} \qquad v\,t = \text{AB}$$

$$\tan\theta \approx \theta = \frac{\text{AB}}{\text{OA}} = \frac{vt}{ut} = \frac{v}{u} \tag{2.4}$$

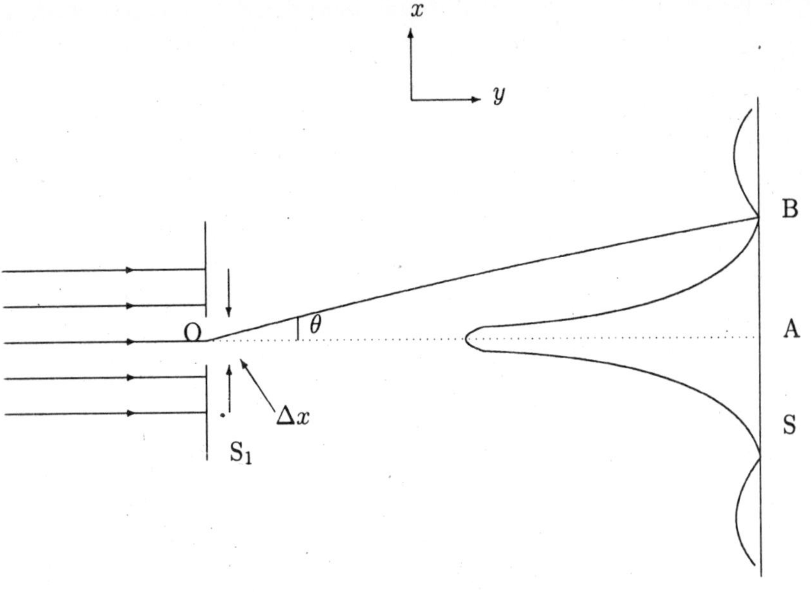

Figure 2: Single-slit experiment for illustration of Heisenberg uncertainty principle.

From equations (2.3) and (2.4), we have

$$\frac{\lambda}{\Delta x} = \frac{v}{u}$$

Using $\lambda = h/mu$ and taking v as a change in the vertical component of velocity Δv, we have

$$\frac{h}{mu\Delta x} \approx \frac{\Delta v}{u} \qquad\qquad \Delta x\,\Delta p_x \approx h$$

where $\Delta p_x = m\Delta v$. This is the desired relation for uncertainty in x and p_x. Accurate measurements give $\Delta x \; \Delta p_x \approx \hbar/2$, where $\hbar$ is the h divided by 2π. The value of $\hbar$ is $1.054 \cdot \times 10^{-34}$ J s.However, it is customary to use $\Delta x \; \Delta p_x \approx \hbar$.

In the same way, we can get

$$\Delta y \; \Delta p_y \approx \hbar \qquad\qquad \Delta z \; \Delta p_z \approx \hbar$$

As the product of two uncertainties is a universal constant h, the more precisely we determine one variable, the less accurate is the determination of the second variable. Though this result is based on the said experiment, it is a very general result, as it is independent of the particle mass and the parameters of the apparatus used. Heisenberg considered this principle to be a fundamental law of the nature.

Exercise 4. An electron is moving with a speed of 400 m/s with an accuracy of 0.005%. Calculate the centainty with which we can locate the position of the electron. (Mass of electron = 9.11×10^{-31} kg)

Solution: Momentum of electron is

$$p = mv = (9.11 \times 10^{-31}) \times 400 = 3.64 \times 10^{-28} \text{ kg m/s}$$

$$\text{percentage accuracy} = \frac{\Delta p}{p} \times 100 = 0.005$$

Thus, we have

$$\Delta p = \frac{0.005 \times (3.64 \times 10^{-28})}{100} = 1.82 \times 10^{-32} \text{ kg m/s}$$

Thus, using the uncertainty principle, the uncertainty in the position is

$$\Delta x = \frac{\hbar}{\Delta p} = \frac{1.054 \times 10^{-34}}{1.82 \times 10^{-32}} = 0.0058 \text{ m}$$

It shows that the position of the electron can be measured to an accuracy less than 0.0058 m.

2.3 Uncertainty relations for other variables

Uncertainty relations can also be obtained for other pairs of canonically conjugate variables.

(i) For a free particle moving in the x direction, the energy E is

$$E = \frac{p_x^2}{2m} \qquad \text{or} \qquad \Delta E = \frac{p_x}{m} \, \Delta p_x = v_x \, \Delta p_x$$

where $v_x (\equiv p_x/m)$ is the velocity of the particle moving along the x-direction, which can be expressed as the rate of change of displacement.

$$v_x = \frac{\Delta x}{\Delta t}$$

Thus, we have

$$\Delta E = \frac{\Delta x}{\Delta t}\,\Delta p_x \qquad \text{or} \qquad \Delta E\,\Delta t = \Delta x\,\Delta p_x$$

Therefore,

$$\Delta E\,\Delta t \approx \hbar$$

This equation relates energy and time, saying that if a system remains in a particular state for a time Δt, its energy is uncertain at least by $\Delta E = h/\Delta t$.

(ii) The uncertainty relation between angular momentum and angular position can be obtained in the following manner. The energy E, angular momentum L and moment of inertia I of a system are related through the equation

$$E = \frac{L^2}{2I}$$

On differentiating, we have

$$\Delta E = \frac{L}{I}\,\Delta L \qquad \text{or} \qquad \Delta E = \omega\,\Delta L$$

where $\omega (\equiv L/I)$ is the angular velocity, which can be expressed as the rate of change of angle.

$$\omega = \frac{\Delta \theta}{\Delta t}$$

Thus, we have

$$\Delta E = \frac{\Delta \theta}{\Delta t}\,\Delta L \qquad \text{or} \qquad \Delta L\,\Delta \theta = \Delta E\,\Delta t$$

Therefore,

$$\Delta L\,\Delta \theta \approx \hbar$$

2.4 Applications of uncertainty relation

The above stated uncertainty relations are useful in explanation of a number of observed phenomena, where the classical theory fails. Some of the such phenomena are as the following.

(i) Natural width of a spectral line

Each line in a spectrum has a finite width due to some reasons, such as the Doppler broadening, pressure broadening and natural broadening. The Doppler broadening (width) is due to the motion of the source of the line whereas the pressure broadening is due to the gas pressure. The natural broadening can only be understood with the help of the uncertainty principle. When pressure of a gas is very low and the speed of the source relative to the observer is zero, the line broadening can only be natural, which can be understood in the following manner. Before making a transition in the downward direction, an atom remains in an excited state for a finite time τ, called the life-time of the state. Hence, an uncertainty in the time is τ and therefore the uncertainty in energy ΔE, according to the uncertainty principle, is

$$\tau\,\Delta E \approx \hbar \qquad \text{or} \qquad \Delta\nu \approx \frac{1}{2\pi\tau}$$

For a large number of excited states, the life time $\tau \approx 10^{-8}$ s. Therefore, $\Delta\nu \approx 10^{8}$ Hz. Widths of spectral lines of this order have been experimentally observed when the pressure of gas is very low.

Exercise 5. In an atom, average life-time of an excited state is 10^{-8} s. If the mean wavelength of the spectral line generated due to deexcitation from this level is 6000 Å, calculate the width of the line.

Solution: Life time of the excited level gives the uncertainty in time Δt. Thus, $\Delta t = 10^{-8}$ s. From the expression

$$E = \frac{hc}{\lambda} \qquad \text{we have} \qquad \Delta E = \frac{hc}{\lambda^2}\Delta\lambda$$

Using the uncertainty principle $\Delta E \Delta t = \hbar$, we have

$$\Delta\lambda = \frac{\lambda^2}{2\pi c\,\Delta t} = \frac{(6000\times 10^{-10})^2}{(2\times 3.14\times 3\times 10^{8})(10^{-8})} = 1.91\times 10^{-14}\ \text{m}$$

(ii) Non-existence of electron in a nucleus

Let us imagine that an electron can exist in a nucleus. The radius of nucleus is of the order of 10^{-15} m. Thus, the maximum uncertainty in the position of electron in a nucleus can be equal to the diameter of the nucleus, *i.e.*, 2×10^{-15} m. Following the uncertainty principle, the minimum uncertainty in momentum of the electron is

$$\Delta p = \frac{\hbar}{\Delta x} = \frac{1.054 \times 10^{-34}}{2 \times 10^{-15}} = 5.27 \times 10^{-20} \text{ kg m/s}$$

So much momentum of an electron can only be possible when its velocity is comparable to the speed of light. Therefore, the total energy E of the electron is given by the relativistic formula

$$E^2 = p^2 c^2 + m_0^2 c^4$$

For the relativistic electrons, $p^2c^2 >> m_0^2c^4$, and we have

$$E = pc \qquad \text{or} \qquad \Delta E = \Delta p\, c$$

Thus, the minimum uncertainty in energy of an electron to exit inside a nucleus is

$$\Delta E = 5.27 \times 10^{-20} \times 3 \times 10^8 = 1.58 \times 10^{-11} \text{J}$$
$$= \frac{1.58 \times 10^{-11}}{1.602 \times 10^{-19} \times 10^6} = 98.6 \text{ MeV}$$

Thus, for an electron to exist in a nucleus, the minimum uncertainty in energy is about 98.6 MeV. The minimum energy of an electron in a nucleus must be at least of the order of 98.6 MeV. It is several orders of magnitude larger than the energy of a β-particle emitted from the nucleus. Thus, an electron cannot exist in a nucleus.

3. Principle of superposition

When two or more waves travel in a common space simultaneously, total displacement at a point in the space is simply the vector sum of the displacements produced by the individual waves. This phenomenon is known as the 'principle of superposition', which plays a magnificent role in various phenomena. Superposition principle allows us to analyze a complicated wave motion, in the form of a combination of a number of simple harmonic waves. Though a linear combination of waves is often used in optics, sound etc., it is not used in the classical theory of particles. Since now the wave motion is assigned to each moving particle, application of the principle of superposition needs to be extended to the matter waves also. This concept of superposition can help in the construction of wave packets.

Exercise 6. Two periodic and perpendicular forces

$$x = a\sin(\omega t) \qquad \text{and} \qquad y = b\sin(\omega t + \phi)$$

having equal frequency ω, amplitudes a and b, and a constant phase difference ϕ are acting simultaneously on a particle. Discuss the resultant motion of the particle.

Solution: Suppose the two given forces are acting in the xy plane perpendicular to each other. We have

$$\frac{x}{a} = \sin(\omega t) \tag{2.5}$$

and

$$\frac{y}{b} = \sin(\omega t + \phi) = \sin(\omega t)\ \cos\phi + \cos(\omega t)\ \sin\phi \tag{2.6}$$

Using equation (2.5) in (2.6), we have

$$\frac{y}{b} = \frac{x}{a}\ \cos\phi + \sqrt{1 - \frac{x^2}{a^2}}\ \sin\phi$$

$$\frac{y}{b} - \frac{x}{a}\ \cos\phi = \sqrt{1 - \frac{x^2}{a^2}}\ \sin\phi$$

Squaring both sides, we have

$$\frac{y^2}{b^2} + \frac{x^2}{a^2}\ \cos^2\phi - \frac{2xy}{ab}\ \cos\phi = \left(1 - \frac{x^2}{a^2}\right)\ \sin^2\phi$$

$$\frac{x^2}{a^2} + \frac{y^2}{b^2} = \sin^2\phi + \frac{2xy}{ab}\ \cos\phi \tag{2.7}$$

This is the general equation for the orbit of the particle. Depending on the value of the phase difference ϕ, some known forms of the orbit can be obtained as the following.

(i) When $\phi = 0$, equation (2.7) reduces to

$$y = \frac{b}{a}\ x$$

That is, when the two forces are in phase, the particle moves in a straight line.

(ii) When $\phi = \pm\pi/2$, equation (2.7) reduces to

$$\frac{x^2}{a^2} + \frac{y^2}{b^2} = 1$$

That is, the motion of the particle is along the ellipse when the phase difference is $\pm\pi/2$. Further, when $a = b$, the particle moves in a circular path.

Exercise 7. Two periodic and parallel forces

$$y_1 = a\sin(\omega t) \qquad \text{and} \qquad y_2 = b\sin(\omega t + \phi)$$

having equal frequency ω, amplitudes a and b, and a constant phase difference ϕ are acting simultaneously on a particle. Discuss the resultant motion of the particle.

Solution: The two parallel forces are working simultaneously on a particle, the resultant displacement y of the particle is

$$\begin{aligned} y &= y_1 + y_2 \\ &= a\sin(\omega t) + b\sin(\omega t + \phi) \\ &= a\sin(\omega t) + b\sin(\omega t)\ \cos\phi + b\cos(\omega t)\ \sin\phi \\ &= \sin(\omega t)\ (a + b\cos\phi) + \cos(\omega t)\ b\ \sin\phi \end{aligned}$$

Since a, b and ϕ are constant. Let us write $A\ \cos\alpha = a + b\cos\phi$ and $A\ \sin\alpha = b\ \sin\phi$, so that we have

$$y = A\sin(\omega t + \alpha)$$

where

$$A = \sqrt{a^2 + b^2 + 2ab\cos\phi} \qquad \text{and} \qquad \tan\alpha = \frac{b\ \sin\phi}{a + b\ \cos\phi}$$

This is the general equation for the motion of the particle. Depending on the value of the phase difference ϕ, some known forms of the orbit can be obtained as the following.

(i) When $\phi = 0$, we have $A = a + b$, $\alpha = 0$ and

$$y = (a + b)\sin(\omega t)$$

That is, the case of the constructive interference.

(ii) When $\phi = \pm\pi/2$, we have $A = \sqrt{a^2 + b^2}$, $\tan\alpha = \pm b/a$ and

$$y = \sqrt{(a + b)}\sin\left(\omega t \pm \tan^{-1}\frac{b}{a}\right)$$

(ii) When $\phi = \pi$, we have $A = a - b$, $\alpha = 0$ and

$$y = (a - b)\sin(\omega t)$$

That is, the case of the destructive interference.

4. Wave packet

Following the de Broglie hypothesis, a particle behaves like a wave under suitable conditions. The Heisenberg's uncertainty principle states that when the momentum of a particle is well defined, the associated wave could be of infinite extent. Hence, a free particle, with a well defined momentum, moving in the positive x-direction can be described by an infinite plane wave $\Psi(x,t)$ expressed as

$$\Psi(x,t) = A_0 \exp\Big[i(kx - \omega t)\Big]$$

where A_0 is amplitude, $k = 2\pi/\lambda$ is the wave vector and ω the angular frequency. The quantity $\Psi(x,t)$, known as the wave function, is generally used to express a matter wave associated with a moving particle. The basic requirement here is that the wave function $\Psi(x,t)$ must be large in the region where the particle is likely to be found and be small in the region where the particle is less likely to be found. The wave function of the matter wave which is confined to a small region is known as the 'wave packet' or 'wave group' (Figure 3).

Mathematically, a wave packet can be constructed by superposition of an infinite number of plane waves with different values of k.

$$\Psi(x,t) = \int_{-\infty}^{\infty} A(k) \exp[i(kx - \omega(k)t)]\, dk \tag{2.8}$$

As the particle is localized, we want to have a superposition which gives a wave packet, traveling without change in shape. This can be obtained when $A(k)$ is zero everywhere except for a small region for the values of k lying in the range:

$$\left(k_0 - \frac{\Delta k}{2}\right) < k < \left(k_0 + \frac{\Delta k}{2}\right) \tag{2.9}$$

where $\Delta k << k_0$. Since k lies in a very small interval, $\omega(k)$ can be expressed as a power series in $(k - k_0)$ about k_0 as

$$\omega(k) = \omega[k_0 + (k - k_0)] = \omega(k_0) + (k - k_0)\left(\frac{d\omega}{dk}\right)_{k=k_0} + \ldots$$

On neglecting higher order terms and writing

$$\omega(k_0) = \omega_0 \qquad \text{and} \qquad \left(\frac{d\omega}{dk}\right)_{k=k_0} = \frac{d\omega}{dk}$$

can be expressed as

$$\Psi(x,t) = \int_{-\infty}^{\infty} A(k) \exp\Big[ikx - i\omega_0 t - i(k - k_0)\Big(\frac{d\omega}{dk}\Big)t\Big]\, dk \tag{2.10}$$

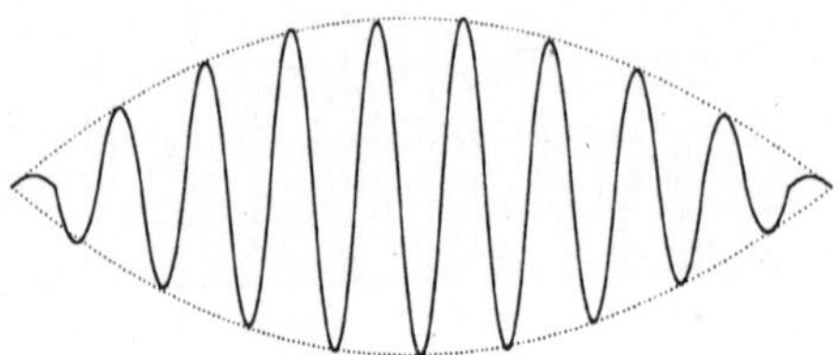

Figure 3: Representation of a wave packet.

According to the requirement (equation 2.9), the limits of the integral can be changed to $(k_0 - \Delta k/2)$ and $(k_0 + \Delta k/2)$, as the particle should not exist beyond these limits. On adding and subtracting ik_0x to the exponential, equation (2.10) can be rearranged as

$$\Psi(x,t) = \int_{k_0-\Delta k/2}^{k_0+\Delta k/2} A(k)\, \exp\left[ix(k-k_0) + ik_0x - i\omega_0 t - i(k-k_0)\left(\frac{\mathrm{d}\omega}{\mathrm{d}k}\right)t\right]\, \mathrm{d}k$$

$$= \exp[i(k_0x - \omega_0 t)] \int_{k_0-\Delta k/2}^{k_0+\Delta k/2} A(k)\, \exp\left[i(k-k_0)\left(x - \frac{\mathrm{d}\omega}{\mathrm{d}k}\, t\right)\right]\, \mathrm{d}k$$

$$= F(x,t)\, \exp\left[i(k_0x - \omega_0 t)\right] \qquad (2.11)$$

where

$$F(x,t) = \int_{k_0-\Delta k/2}^{k_0+\Delta k/2} A(k)\, \exp\left[i(k-k_0)\left(x - \frac{\mathrm{d}\omega}{\mathrm{d}k}\, t\right)\right]\, \mathrm{d}k$$

Equation (2.11) represents a plane wave with propagation vector k_0, angular frequency ω_0 and amplitude $F(x,t)$ which depends on the position x and time t through $[x - (\mathrm{d}\omega/\mathrm{d}k)t]$. It shows that the wave packet moves with the 'group velocity'

$$v_g = \frac{\mathrm{d}\omega}{\mathrm{d}k}$$

While the wave packet (dotted line in Figure 3) moves with the group velocity v_g, each of the waves of the packet travels with velocity v_p, known as the 'phase velocity' or 'wave velocity'. However, it is the wave packet that carries the energy and what we measure experimentally is the group velocity.

Exercise 8. Show that the group velocity v_g of a wave-packet is equal to the velocity of a material particle. Derive the relation $v_g v_p = c^2$, where v_p is the phase velocity of a wave in the wave-packet and c the speed of light.

Solution: For the relations

$$E = \hbar\,\omega \qquad \text{and} \qquad p = \hbar k$$

The group velocity

$$v_g = \frac{d\omega}{dk} = \frac{dE}{dp}$$

(i) For a free nonrelativistic particle, potential energy is zero and we have

$$E = \frac{p^2}{2m} \qquad \text{or} \qquad \frac{dE}{dp} = \frac{p}{m} = v$$

(ii) For a relativistic particle, we have

$$E^2 = c^2\,p^2 + m_0^2 c^4$$

and thus,

$$\frac{dE}{dp} = \frac{c^2 p}{E} = \frac{c^2 mv}{mc^2} = v$$

This shows that velocity of a particle and the group velocity of the corresponding wave packet are the same.

The phase velocity given by

$$v_p = \frac{\omega}{k} = \frac{E}{p} = \frac{mc^2}{mv} = \frac{c^2}{v}$$

has no physical significance and it is not a measurable quantity.

The product of the group velocity and phase velocity is

$$v_g v_p = v\,\frac{c^2}{v} = c^2$$

5. Time-dependent Schrödinger equation

We have introduced the wave function $\Psi(x,t)$ in the preceding section. For the detailed study of a system, we however require an equation of motion for $\Psi(x,t)$. That is, we require an equation of motion for the system whose solution is the required wave function $\Psi(x,t)$. This equation of motion was formulated by Schrödinger in 1926, and it plays key role in the quantum mechanics.

5.1 One-dimensional equation of motion for a free particle

Wave equation of a localized free particle is given by equation (2.8). For a free particle, classical expression for energy E is

$$E = \frac{p_x^2}{2m} \tag{2.12}$$

On replacing p_x by $\hbar k$ and E by $\hbar\omega$, we have

$$\omega = \frac{\hbar k^2}{2m} \tag{2.13}$$

Using equation (2.13) in (2.8), we have

$$\Psi(x,t) = \int_{-\infty}^{\infty} A(k) \exp\Big[i\Big(kx - \frac{\hbar k^2}{2m}t\Big)\Big]\, \mathrm{d}k \tag{2.14}$$

On differentiating equation (2.14) with respect to t, we get

$$\frac{\partial\Psi}{\partial t} = -\frac{i\hbar}{2m}\int_{-\infty}^{\infty} k^2 A(k) \exp\Big[i\Big(kx - \frac{\hbar k^2}{2m}t\Big)\Big]\, \mathrm{d}k \tag{2.15}$$

Now, differentiating equation (2.14) twice with respect to x, we get

$$\frac{\partial^2\Psi}{\partial x^2} = -\int_{-\infty}^{\infty} k^2 A(k) \exp\Big[i\Big(kx - \frac{\hbar k^2}{2m}t\Big)\Big]\, \mathrm{d}k \tag{2.16}$$

From equations (2.15) and (2.16), we get

$$\frac{\partial\Psi}{\partial t} = \frac{i\hbar}{2m}\frac{\partial^2\Psi}{\partial x^2}$$

$$i\hbar\,\frac{\partial\Psi}{\partial t} = -\frac{\hbar^2}{2m}\frac{\partial^2\Psi}{\partial x^2} \tag{2.17}$$

This is one-dimensional Schrödinger equation for a free particle.

5.2 Energy and momentum operators

For getting energy and momentum operators, equation (2.17) can be expressed as

$$\Big(i\hbar\,\frac{\partial}{\partial t}\Big)\Psi = \frac{1}{2m}\Big(-i\hbar\,\frac{\partial}{\partial x}\Big)\Big(-i\hbar\,\frac{\partial}{\partial x}\Big)\Psi \tag{2.18}$$

Comparison of equations (2.12) and (2.18) tells that the energy E and momentum p_x can be represented as the following differential operators.

$$E \to i\hbar\,\frac{\partial}{\partial t} \qquad \text{and} \qquad p_x \to -i\hbar\,\frac{\partial}{\partial x}$$

Note that equation (2.17) can also be obtained even if we take

$$p_x \rightarrow i\hbar \frac{\partial}{\partial x}$$

The choice of negative sign in the operator for p_x is very significant and would be explained in section 7.

5.3 Three-dimensional equation of motion for a free particle

The one-dimensional treatment discussed above can easily be extended to the three-dimensional case. The three-dimensional wave packet can be expressed as

$$\Psi(\vec{r},t) = \int_{-\infty}^{\infty} A(\vec{k}) \exp\left[i\left(\vec{k}\,.\,\vec{r} - \frac{\hbar k^2}{2m}t\right)\right] dk_x\, dk_y\, dk_z$$

$$= \int_{-\infty}^{\infty} A(\vec{k}) \exp\left[i\left(\{k_x x + k_y y + k_z z\} - \frac{\hbar k^2}{2m}t\right)\right] dk_x\, dk_y\, dk_z \quad (2.19)$$

On differentiating equation (2.19) with respect to t, we get

$$\frac{\partial \Psi}{\partial t} = -\frac{i\hbar}{2m} \int_{-\infty}^{\infty} k^2 A(\vec{k}) \exp\left[i\left(\vec{k}\,.\,\vec{r} - \frac{\hbar k^2}{2m}t\right)\right] dk_x\, dk_y\, dk_z \quad (2.20)$$

Now, differentiating equation (2.19) twice with respect to x, we get

$$\frac{\partial^2 \Psi}{\partial x^2} = -\int_{-\infty}^{\infty} k_x^2\, A(\vec{k}) \exp\left[i\left(\vec{k}\,.\,\vec{r} - \frac{\hbar k^2}{2m}t\right)\right] dk_x\, dk_y\, dk_z \quad (2.21)$$

Now, differentiating equation (2.19) twice with respect to y, we get

$$\frac{\partial^2 \Psi}{\partial y^2} = -\int_{-\infty}^{\infty} k_y^2\, A(\vec{k}) \exp\left[i\left(\vec{k}\,.\,\vec{r} - \frac{\hbar k^2}{2m}t\right)\right] dk_x\, dk_y\, dk_z \quad (2.22)$$

Now, differentiating equation (2.19) twice with respect to z, we get

$$\frac{\partial^2 \Psi}{\partial z^2} = -\int_{-\infty}^{\infty} k_z^2\, A(\vec{k}) \exp\left[i\left(\vec{k}\,.\,\vec{r} - \frac{\hbar k^2}{2m}t\right)\right] dk_x\, dk_y\, dk_z \quad (2.23)$$

On addition of equations (2.21), (2.22) and (2.23), we get

$$\nabla^2 \Psi = -\int_{-\infty}^{\infty} k^2 A(\vec{k}) \exp\left[i\left(\vec{k}\,.\,\vec{r} - \frac{\hbar k^2}{2m}t\right)\right] dk_x\, dk_y\, dk_z \quad (2.24)$$

From equations (2.20) and (2.24), we get

$$i\hbar \frac{\partial \Psi}{\partial t} = -\frac{\hbar^2}{2m} \nabla^2 \Psi \quad (2.25)$$

This is three-dimensional Schrödinger equation for a free particle.

5.4 Energy and momentum operators

For getting the energy and momentum operators, equation (2.25) can be expressed as

$$\left(i\hbar\,\frac{\partial}{\partial t}\right)\Psi = \frac{1}{2m}\left(-\,i\hbar\,\nabla\right)\left(-\,i\hbar\,\nabla\right)\Psi$$

This gives that the energy E and momentum $\vec{p}$ can be considered as the following differential operators.

$$E \rightarrow i\hbar\,\frac{\partial}{\partial t} \qquad \text{and} \qquad \vec{p} \rightarrow -i\hbar\,\nabla$$

Note that equation (2.25) can also be obtained even if we take

$$\vec{p} \rightarrow i\hbar\,\nabla$$

The choice of negative sign in the operator for $\vec{p}$ is very significant and would be explained in section 7.

5.5 Inclusion of potential energy

Classical expression for energy E of a particle of mass m moving in a potential field $V(\vec{r}, t)$ is

$$E = \frac{p^2}{2m} + V(\vec{r}, t) \tag{2.26}$$

Schrödinger then considered the operators for the position $\vec{r}$ and time t as

$$\vec{r} \rightarrow \vec{r} \qquad \text{and} \qquad t \rightarrow t$$

Replacing E, $\vec{p}$, $\vec{r}$ and t in equation (2.26) by the respective operators and allowing the so obtained operator equation to operate on a wave function $\Psi(\vec{r}, t)$, we have

$$i\hbar\,\frac{\partial\Psi}{\partial t} = \left[-\,\frac{\hbar^2}{2m}\,\nabla^2 + V(\vec{r}, t)\right]\Psi \tag{2.27}$$

This equation (2.27) is the time-dependent Schrödinger equation for a particle of mass m moving under a potential $V(\vec{r}, t)$. The quantity in the square bracket of equation (2.27) is the operator for Hamiltonian of the system. Since a factor i appears in the equation, its solution is in general a complex quantity.

Exercise 9. Wave function of a particle of mass m moving in a potential field $V(x)$ is

$$\Psi(x,t) = A\,\exp\Big(-\,ikt - \frac{km}{\hbar}\,x^2\Big)$$

where A and k are constants. Find the explicit expression for $V(x)$.

Solution: We have the wave function

$$\Psi(x,t) = A\,\exp\Big(-\,ikt - \frac{km}{\hbar}\,x^2\Big)$$

Differentiating with respect to t, we get

$$\frac{\mathrm{d}}{\mathrm{d}t}\Psi(x,t) = -ik\;\Psi(x,t)$$

Differentiating with respect to x, we get

$$\frac{\mathrm{d}}{\mathrm{d}x}\Psi(x,t) = -\frac{2kmx}{\hbar}\;\Psi(x,t)$$

Differentiating again with respect to x, we get

$$\frac{\mathrm{d}^2}{\mathrm{d}x^2}\Psi(x,t) = \Big(-\,\frac{2km}{\hbar} + \frac{4k^2m^2x^2}{\hbar^2}\Big)\;\Psi(x,t)$$

Using these expressions in the time-dependent Schrödinger equation

$$i\hbar\,\frac{\partial}{\partial t}\Psi(x,t) = -\frac{\hbar^2}{2m}\frac{\mathrm{d}^2}{\mathrm{d}x^2}\Psi(x,t) + V(x)\Psi(x,t)$$

we get

$$k\hbar = -\frac{\hbar^2}{2m}\Big(-\,\frac{2km}{\hbar} + \frac{4k^2m^2x^2}{\hbar^2}\Big) + V(x)$$

On solving this equation, the potential energy $V(x)$ is equal to $2k^2mx^2$

6. Interpretation of wave function

As mentioned earlier, the solution of the Schrödinger equation (2.27), *i.e.*, $\Psi(\vec{r},t)$, is a complex quantity. Hence, $\Psi(\vec{r},t)$ cannot have any physical meaning. Further, since a probability is always a real and positive quantity, $\Psi(\vec{r},t)$ cannot be directly related to the probability of existence of the particle at the position $\vec{r}$ at time t. However, the wave function $\Psi(\vec{r},t)$ should be in some way related to the presence of the particle. Born in 1926 suggested a statistical interpretation for the wave function, which was later on accepted unanimously. According to this interpretation, a product of $\Psi(\vec{r},t)$ and its complex conjugate $\Psi^*(\vec{r},t)$ is the position probability density $P(\vec{r},t)$.

$$P(\vec{r},t) = \Psi(\vec{r},t)\ \Psi^*(\vec{r},t) = |\Psi(\vec{r},t)|^2$$

The quantity $|\Psi(\vec{r},t)|^2\ d\tau$ is therefore the probability of finding the particle at time t in a small volume element $d\tau$ surrounding the point $\vec{r}$. Obviously, on integrating $|\Psi(\vec{r},t)|^2\ d\tau$ over the entire space, the probability must be unity, and we have[1]

$$\int_{-\infty}^{\infty} |\Psi(\vec{r},t)|^2\ d\tau = 1 \tag{2.28}$$

Here, $\Psi(\vec{r},t)$ must tend to zero sufficiently rapidly as r tends to $\pm\infty$. In case the wave function $\Psi(\vec{r},t)$ does not satisfy equation (2.28), one can multiply the wave function by a constant, say N, so that $N\Psi(\vec{r},t)$ satisfies the condition (2.28), and we have

$$|N|^2 \int_{-\infty}^{\infty} |\Psi(\vec{r},t)|^2\ d\tau = 1 \tag{2.29}$$

This constant N is known as the normalization constant and the equation (2.29) is called the normalization condition. Since the Schrödinger equation is a differential equation of one degree with solution $\Psi(\vec{r},t)$, then $N\Psi(\vec{r},t)$ also is a solution of the equation.

6.1 Probability current density

The probability of finding a particle in a volume v is

$$\int_v \Psi(\vec{r},t)\ \Psi^*(\vec{r},t)\ d\tau$$

and this changes as the wave function varies with time. Now, consider the Schrödinger wave equation and its complex conjugate form as the following.

$$i\hbar\ \frac{\partial\Psi(\vec{r},t)}{\partial t} = \Big[-\frac{\hbar^2}{2m}\ \nabla^2 + V(\vec{r},t)\Big]\Psi(\vec{r},t) \tag{2.30}$$

$$-i\hbar\ \frac{\partial\Psi^*(\vec{r},t)}{\partial t} = \Big[-\frac{\hbar^2}{2m}\ \nabla^2 + V(\vec{r},t)\Big]\Psi^*(\vec{r},t) \tag{2.31}$$

[1]In fact there are three integrals, one for each coordinate of space. Here, the limits from $-\infty$ to ∞ indicate that the integrals over various coordinates are taken in such a manner that the entire space is covered.

Here, the potential energy $V(\vec{r},t)$ is considered as a real quantity. Multiplying equation (2.30) by $\Psi^*(\vec{r},t)$ and equation (2.31) by $\Psi(\vec{r},t)$ from left and subtracting the former from the later, we get

$$i\hbar\left(\Psi^*\frac{\partial\Psi(\vec{r},t)}{\partial t}+\Psi\frac{\partial\Psi^*(\vec{r},t)}{\partial t}\right)=-\frac{\hbar^2}{2m}\left[\Psi^*\nabla^2\Psi-\Psi\nabla^2\Psi^*\right]$$

This equation can be expressed as

$$\frac{\partial(\Psi^*\Psi)}{\partial t}=\frac{i\hbar}{2m}\left[\nabla.(\Psi^*\nabla\Psi-\Psi\nabla\Psi^*)\right] \tag{2.32}$$

On integrating equation (2.32) with respect to space coordinates, we have

$$\begin{aligned}\frac{\partial}{\partial t}\int_{-\infty}^{\infty}\Psi^*\Psi\ d\tau &= \frac{i\hbar}{2m}\int_{-\infty}^{\infty}\nabla.(\Psi^*\nabla\Psi-\Psi\nabla\Psi^*)\ d\tau\\ &= \frac{i\hbar}{2m}\left[\Psi^*\nabla\Psi-\Psi\nabla\Psi^*\right]_{-\infty}^{\infty}\end{aligned} \tag{2.33}$$

For a localized wave packet, Ψ and Ψ^* must vanish as r tends to $\pm\infty$ and thus the right side of equation (2.33) becomes zero. Hence, we have

$$\frac{\partial}{\partial t}\int_{-\infty}^{\infty}\Psi^*\Psi\ d\tau=0 \qquad \text{or} \qquad \int_{-\infty}^{\infty}\Psi^*\Psi\ d\tau=\text{constant in time}$$

It shows that the normalization-integral does not depend on time.

By defining the probability current density $\vec{j}(\vec{r},t)$ as

$$\vec{j}(\vec{r},t)=\frac{i\hbar}{2m}(\Psi\nabla\Psi^*-\Psi^*\nabla\Psi)$$

equation (2.32) can be expressed as

$$\frac{\partial}{\partial t}P(\vec{r},t)+\nabla.\ \vec{j}(\vec{r},t)=0$$

This equation is known as the equation of continuity. It may be noted that when Ψ is a real quantity the probability current density $\vec{j}(\vec{r},t)$ vanishes.

6.2 Expectation value

For the known wave function for a given state of a system, the expectation value $\langle v\rangle$ of a variable v is defined as

$$\langle v\rangle=\int\Psi^*v_{op}\Psi d\tau$$

where v_{op} is an operator corresponding to the variable v and Ψ is a wave function representing the state. Some physical quantities and their operators are given in the following table.

Physical quantity	Operator
Position x	x
Position $\vec{r}$	$\vec{r}$
Energy E	$i\hbar \frac{\partial}{\partial t}$
Momentum p_x	$-i\hbar \frac{\partial}{\partial x}$
Momentum $\vec{p}$	$-i\hbar \nabla$
Potential energy V	V
Time t	t

When the wave function Ψ is not normalized, the expectation value $\langle v \rangle$ of a variable v is defined as

$$\langle v \rangle = \frac{\int \Psi^* v_{op} \Psi d\tau}{\int \Psi^* \Psi d\tau}$$

On multiplying equation (2.30) by $\Psi*$ from left and integrating over the space, we get

$$\int_{-\infty}^{\infty} \Psi^* \left(i\hbar \frac{\partial}{\partial t}\right) \Psi \, d\tau = \int_{-\infty}^{\infty} \Psi^* \left(-\frac{\hbar^2}{2m} \nabla^2\right) \Psi \, d\tau + \int_{-\infty}^{\infty} \Psi^* V \Psi \, d\tau$$

In the form of the expectation values, we have

$$\left\langle i\hbar \frac{\partial}{\partial t} \right\rangle = \left\langle -\frac{\hbar^2}{2m} \nabla^2 \right\rangle + \left\langle V \right\rangle$$

Thus, we have

$$\langle E \rangle = \left\langle \frac{p^2}{2m} \right\rangle + \langle V \rangle$$

Obviously, this is a quantum analogue of the energy expression in the classical mechanics.

Exercise 10. Normalize the wave function $\psi(x) = A \sin(n\pi x/L)$ of a particle over the domain $0 \leq x \leq L$. Here, A is a constant and n is an integer. Calculate the expectation value of the momentum of the particle.

Solution: For the normalization condition

$$\int_0^L \psi^*(x)\psi(x) \, dx = 1$$

we have

$$A^2 \int_0^L \sin^2\left(\frac{n\pi x}{L}\right) \, dx = 1$$

The integral can be solved as

$$I = \int_0^L \sin^2\left(\frac{n\pi x}{L}\right) \, dx = \int_0^L \frac{1}{2}\left[1 - \cos\left(\frac{2n\pi x}{L}\right)\right] \, dx = \frac{L}{2}$$

Thus, we have

$$A = \sqrt{\frac{2}{L}}$$

Therefore, the normalized wave function is

$$\psi(x) = \sqrt{\frac{2}{L}} \, \sin\left(\frac{n\pi x}{L}\right)$$

The expectation value of p_x is

$$\begin{aligned}
\langle p_x \rangle &= \int_0^L \Psi^* \left(- i\hbar \frac{d}{dx}\right) \Psi \, dx \\
&= -i\hbar \frac{2}{L} \frac{n\pi}{L} \int_0^L \sin\left(\frac{n\pi x}{L}\right) \cos\left(\frac{n\pi x}{L}\right) dx \\
&= -\frac{i\hbar n\pi}{L^2} \int_0^L \sin\left(\frac{2n\pi x}{L}\right) \, dx = 0
\end{aligned}$$

7. Ehrenfest theorem

Ehrenfest in 1927 proved a correspondence between the motion of a classical particle and the motion of a wave packet. In the limit, when a wave packet associated with a particle reduces to a point, we expect the wave packet to behave like a classical particle. Let us consider a particle of mass m moving along the positive x-direction. The expectation value of position x of a particle is

$$\langle x \rangle = \int_{-\infty}^{\infty} \Psi^* \, x\Psi \, dx$$

The time derivative of $\langle x \rangle$ is

$$\frac{d}{dt}\langle x \rangle = \frac{\partial}{\partial t} \int_{-\infty}^{\infty} \Psi^* \, x\Psi dx = \int_{-\infty}^{\infty} \left(\frac{\partial \Psi^*}{\partial t} \, x\Psi + \Psi^* x \, \frac{\partial \Psi}{\partial t}\right) \, dx$$

Substituting the values of $\partial\Psi/\partial t$ and $\partial\Psi^*/\partial t$ from equations (2.30) and (2.31), we have

$$\frac{\mathrm{d}}{\mathrm{d}t}\langle x\rangle = \int_{-\infty}^{\infty}\Big\{-\frac{x\Psi}{i\hbar}\Big[-\frac{\hbar^2}{2m}\frac{\partial^2\Psi^*}{\partial x^2}+V\Psi^*\Big]+\frac{x\Psi^*}{i\hbar}\Big[-\frac{\hbar^2}{2m}\frac{\partial^2\Psi}{\partial x^2}+V\Psi\Big]\Big\}\,\mathrm{d}x$$

$$= \frac{i\hbar}{2m}\int_{-\infty}^{\infty}\left(x\Psi^*\frac{\partial^2\Psi}{\partial x^2}-x\Psi\frac{\partial^2\Psi^*}{\partial x^2}\right)\mathrm{d}x \tag{2.34}$$

On integrating second term on the right side of equation (2.34), we have

$$\int_{-\infty}^{\infty}x\Psi\frac{\partial^2\Psi^*}{\partial x^2}\,\mathrm{d}x = \left[x\Psi\frac{\partial\Psi^*}{\partial x}\right]_{-\infty}^{\infty}-\int_{-\infty}^{\infty}\frac{\partial(x\Psi)}{\partial x}\,\frac{\partial\Psi^*}{\partial x}\,\mathrm{d}x$$

Using the fact that Ψ vanishes as x tends to ∞, we have

$$\int_{-\infty}^{\infty}x\Psi\frac{\partial^2\Psi^*}{\partial x^2}\,\mathrm{d}x = -\int_{-\infty}^{\infty}\frac{\partial(x\Psi)}{\partial x}\,\frac{\partial\Psi^*}{\partial x}\,\mathrm{d}x$$

Integrating by parts the right side of this equation, we have

$$\int_{-\infty}^{\infty}x\Psi\frac{\partial^2\Psi^*}{\partial x^2}\,\mathrm{d}x = \left[\frac{\partial(x\Psi)}{\partial x}\Psi^*\right]_{-\infty}^{\infty}+\int_{-\infty}^{\infty}\frac{\partial^2(x\Psi)}{\partial x^2}\,\Psi^*\,\mathrm{d}x$$

Using the fact that Ψ^* vanishes as x tends to ∞, we have

$$\int_{-\infty}^{\infty}x\Psi\frac{\partial^2\Psi^*}{\partial x^2}\,\mathrm{d}x = \int_{-\infty}^{\infty}\frac{\partial^2(x\Psi)}{\partial x^2}\,\Psi^*\,\mathrm{d}x = \int_{-\infty}^{\infty}\Psi^*\left[\frac{\partial}{\partial x}\left(x\,\frac{\partial\Psi}{\partial x}+\Psi\right)\right]\mathrm{d}x$$

$$= \int_{-\infty}^{\infty}\Psi^*\left[x\,\frac{\partial^2\Psi}{\partial x^2}+\frac{\partial\Psi}{\partial x}+\frac{\partial\Psi}{\partial x}\right]\mathrm{d}x = \int_{-\infty}^{\infty}\Psi^*\left[x\,\frac{\partial^2\Psi}{\partial x^2}+2\frac{\partial\Psi}{\partial x}\right]\mathrm{d}x$$

Using this equation in (2.34), we have

$$\frac{\mathrm{d}}{\mathrm{d}t}\langle x\rangle = \frac{i\hbar}{2m}\int_{-\infty}^{\infty}\left(x\Psi^*\frac{\partial^2\Psi}{\partial x^2}-2\Psi^*\frac{\partial\Psi}{\partial x}-\Psi^*x\,\frac{\partial^2\Psi}{\partial x^2}\right)\mathrm{d}x$$

$$= -\frac{i\hbar}{m}\int_{-\infty}^{\infty}\Psi^*\,\frac{\partial\Psi}{\partial x}\,\mathrm{d}x = \frac{1}{m}\int_{-\infty}^{\infty}\Psi^*\left(-i\hbar\frac{\partial}{\partial x}\right)\Psi\,\mathrm{d}x = \frac{\langle p_x\rangle}{m}$$

Hence,

$$m\,\frac{\mathrm{d}}{\mathrm{d}t}\langle x\rangle = \langle p_x\rangle$$

Notice that here we have taken operator of p_x as $-i\hbar\frac{\partial}{\partial x}$. it is not possible to take it as $i\hbar\frac{\partial}{\partial x}$. This clarifies the reason for choosing negative sign in the operator of the momentum: Equivalence of this relation in classical mechanics is that the momentum of a particle is the product of its mass and velocity.

The expectation value of momentum p_x of a particle is

$$\langle p_x \rangle = \int_{-\infty}^{\infty} \Psi^* \Big(-i\hbar \frac{\partial}{\partial x}\Big) \Psi \; \mathrm{d}x$$

The time derivative of $\langle p_x \rangle$ is

$$\frac{\mathrm{d}}{\mathrm{d}t}\langle p_x \rangle = \frac{\partial}{\partial t}\int_{-\infty}^{\infty} \Psi^* \Big(-i\hbar \frac{\partial}{\partial x}\Big) \Psi \mathrm{d}x$$

$$= -i\hbar \int_{-\infty}^{\infty} \Big[\frac{\partial \Psi^*}{\partial t}\frac{\partial \Psi}{\partial x} + \Psi^* \frac{\partial}{\partial t}\Big(\frac{\partial \Psi}{\partial x}\Big)\Big] \mathrm{d}x$$

$$= -i\hbar \int_{-\infty}^{\infty} \Big[\frac{\partial \Psi^*}{\partial t}\frac{\partial \Psi}{\partial x} + \Psi^* \frac{\partial}{\partial x}\Big(\frac{\partial \Psi}{\partial t}\Big)\Big] \; \mathrm{d}x$$

Substituting the values of $\partial\Psi/\partial t$ and $\partial\Psi^*/\partial t$ from equations (2.30) and (2.31), we have

$$\frac{\mathrm{d}}{\mathrm{d}t}\langle p_x \rangle = \int_{-\infty}^{\infty} \Big[\frac{\partial \Psi}{\partial x}\Big(-\frac{\hbar^2}{2m}\frac{\partial^2 \Psi^*}{\partial x^2} + V\Psi^*\Big) - \Psi^* \frac{\partial}{\partial x}\Big(-\frac{\hbar^2}{2m}\frac{\partial^2 \Psi}{\partial x^2} + V\Psi\Big)\Big] \; \mathrm{d}x$$

$$= \frac{\hbar^2}{2m}\int_{-\infty}^{\infty} \Big[\Psi^* \frac{\partial}{\partial x}\Big(\frac{\partial^2 \Psi}{\partial x^2}\Big) - \frac{\partial \Psi}{\partial x}\frac{\partial^2 \Psi^*}{\partial x^2}\Big] \mathrm{d}x + \int_{-\infty}^{\infty} \Big[V\Psi^* \frac{\partial \Psi}{\partial x} - \Psi^* \frac{\partial V\Psi}{\partial x}\Big] \mathrm{d}x$$

$$= \frac{\hbar^2}{2m}\int_{-\infty}^{\infty} \Big[\frac{\partial}{\partial x}\Big(\Psi^* \frac{\partial^2 \Psi}{\partial x^2}\Big) - \frac{\partial \Psi^*}{\partial x}\frac{\partial^2 \Psi}{\partial x^2} - \frac{\partial \Psi}{\partial x2}\frac{\partial^2 \Psi^*}{\partial x^2}\Big] \; \mathrm{d}x$$

$$+ \int_{-\infty}^{\infty} \Big[V\Psi^* \frac{\partial \Psi}{\partial x} - \Psi^* V \frac{\partial \Psi}{\partial x} - \Psi\Psi^* \frac{\partial V}{\partial x}\Big] \; \mathrm{d}x$$

$$= \frac{\hbar^2}{2m}\int_{-\infty}^{\infty} \Big[\frac{\partial}{\partial x}\Big(\Psi^* \frac{\partial^2 \Psi}{\partial x^2}\Big) - \frac{\partial}{\partial x}\Big(\frac{\partial \Psi^*}{\partial x}\frac{\partial \Psi}{\partial x}\Big)\Big] \mathrm{d}x + \int_{-\infty}^{\infty} \Psi^* \Big(-\frac{\partial V}{\partial x}\Big) \Psi \mathrm{d}x$$

$$= \frac{\hbar^2}{2m}\Big[\Psi^* \frac{\partial^2 \Psi}{\partial x^2} - \frac{\partial \Psi^*}{\partial x}\frac{\partial \Psi}{\partial x}\Big]_{-\infty}^{\infty} + \Big\langle -\frac{\partial V}{\partial x}\Big\rangle$$

As Ψ^* as well as $\partial\Psi/\partial x$ vanish as x tends to $\pm\infty$, we have

$$\frac{\mathrm{d}}{\mathrm{d}t}\langle p_x \rangle = -\Big\langle \frac{\partial V}{\partial x}\Big\rangle = \big\langle F_x \big\rangle$$

This is equivalent to the Newton's second law of motion in the classical mechanics.

8. Time-independent Schrödinger equation

The solution $\Psi(\vec{r},t)$ of time-dependent Schrödinger equation

$$i\hbar\frac{\partial\Psi(\vec{r},t)}{\partial t} = \Big[-\frac{\hbar^2}{2m}\nabla^2 + V(\vec{r})\Big]\Psi(\vec{r},t) \tag{2.35}$$

describes evolution of a quantum system. However, there no time-dependence in the operators. When the Hamiltonian operator does not depend on time, the wave function $\Psi(\vec{r},t)$ can be expressed as product of two functions $\psi(\vec{r})$ and $\phi(t)$, the first depending on $\vec{r}$ and the other on t. Using $\Psi(\vec{r},t) \equiv \psi(\vec{r})\ \phi(t)$ in equation (2.35), we have

$$i\hbar\psi(\vec{r})\frac{\partial\phi(t)}{\partial t} = \phi(t)\Big[-\frac{\hbar^2}{2m}\nabla^2 + V(\vec{r})\Big]\psi(\vec{r})$$

On dividing this equation by $\psi(\vec{r})\ \phi(t)$, we get

$$i\hbar\frac{1}{\phi(t)}\frac{\partial\phi(t)}{\partial t} = \frac{1}{\psi(\vec{r})}\Big[-\frac{\hbar^2}{2m}\nabla^2 + V(\vec{r})\Big]\psi(\vec{r}) \tag{2.36}$$

In equation (2.36), the left side depends on time t whereas the right side depends on the space coordinates $\vec{r}$. Such situation can only be possible when each side is equal to some constant, say E. Thus, we have

$$i\hbar\frac{1}{\phi(t)}\frac{\mathrm{d}\phi(t)}{\mathrm{d}t} = E \tag{2.37}$$

$$\frac{1}{\psi(\vec{r})}\Big[-\frac{\hbar^2}{2m}\nabla^2 + V(\vec{r})\Big]\psi(\vec{r}) = E \tag{2.38}$$

Equation (2.37) cam be rearranged as

$$\frac{\mathrm{d}\phi(t)}{\phi(t)} = -\frac{iE}{\hbar}\mathrm{d}t$$

having a solution

$$\phi(t) = C\ \exp\Big(-\frac{iEt}{\hbar}\Big)$$

where C is a constant. The equation (2.38) can be written as

$$\Big[-\frac{\hbar^2}{2m}\nabla^2 + V(\vec{r})\Big]\psi(\vec{r}) = E\psi(\vec{r}) \tag{2.39}$$

This equation (2.39) is known as the time-independent Schrödinger equation.

8.1 Significance of the constant E

Significance of the constant E can be understood in the following manner. The wave function $\Psi(\vec{r},t)$ can be expressed as

$$\Psi(\vec{r},t) = C\,\psi(\vec{r})\,\exp\Big(-\frac{iEt}{\hbar}\Big)$$

Differentiating this equation with respect to t and multiplying by $i\hbar$, we get

$$i\hbar\frac{\partial\Psi(\vec{r},t)}{\partial t} = E\Psi(\vec{r},t)$$

Multiplying this equation by $\Psi^*(\vec{r},t)$ from left and integrating over the space, we have

$$\int_{-\infty}^{\infty}\Psi^*(\vec{r},t)\Big(i\hbar\frac{\partial}{\partial t}\Big)\Psi(\vec{r},t)\ d\tau = \int_{-\infty}^{\infty}\Psi^*(\vec{r},t)E\Psi(\vec{r},t)\ d\tau$$

$$= E\int_{-\infty}^{\infty}\Psi^*(\vec{r},t)\Psi(\vec{r},t)\ d\tau = E$$

As on the left side of this equation is the energy operator, the constant E on the right side is the energy of the system.

Exercise 11. Normalize the wave function $\psi(x) = A\exp(-ax^2)$, where A and a are constants over the domain $-\infty \le x \le \infty$.

Solution: For the normalization condition

$$\int_{-\infty}^{\infty}\psi^*(x)\psi(x)\ dx = 1$$

we have

$$A^2\int_{-\infty}^{\infty}\exp(-2ax^2)\ dx = 1$$

Using the integral

$$\int_{-\infty}^{\infty}\exp(-2ax^2)\ dx = 2\int_{0}^{\infty}\exp(-2ax^2)\ dx = \sqrt{\frac{\pi}{2a}}$$

we get

$$A = \Big(\frac{2a}{\pi}\Big)^{1/4}$$

Therefore, the normalized wave function is

$$\psi(x) = \Big(\frac{2a}{\pi}\Big)^{1/4}\exp(-ax^2)$$

Exercise 12. Find energy of the particle of mass m moving in a potential field $V(x) = 2\hbar^2 b^2 x^2/m$ for which the time-independent wave function is

$$\psi(x) = \exp\left(-bx^2\right)$$

Here, b is a constant.

Solution: We have the wave function

$$\psi(x) = \exp\left(-bx^2\right)$$

Differentiating with respect to x, we get

$$\frac{\mathrm{d}}{\mathrm{d}x}\psi(x) = -2bx\ \psi(x)$$

Differentiating again with respect to x, we get

$$\frac{\mathrm{d}^2}{\mathrm{d}x^2}\psi(x) = (-2b + 4b^2x^2)\psi(x)$$

Using these expressions in the time-independent Schrödinger equation

$$-\frac{\hbar^2}{2m}\frac{\mathrm{d}^2}{\mathrm{d}x^2}\psi(x) + V(x)\psi(x) = E\psi(x)$$

we get

$$-\frac{\hbar^2}{2m}(-2b + 4b^2x^2)\psi(x) + \frac{2\hbar^2b^2}{m}x^2\psi(x) = E\psi(x)$$

Thus, the energy E is

$$E = \frac{\hbar^2 b}{m}$$

9. Problems and questions

1. Describe the de Broglie hypothesis. Give an experimental proof for the matter waves.
2. Calculate de Broglie wave length of an electron which is accelerated from rest by a potential difference of 200 V.
3. Calculate the ratio of the de Broglie wave length of electron to that of proton when (a) both have the same kinetic energy and (b) the electron kinetic energy is 2000 eV and that of proton is 200 V.

4. Calculate the ratio of the kinetic energy of electron to that of proton when their de Broglie wave lengths are equal.

5. Describe the uncertainty principle. How it can be illustrated with the help of a single-slit experiment.

6. Describe the uncertainty principle and its two applications.

7. Describe the uncertainty principle with the help of single-slit experiment.

8. Speed of a particle of mass 0.02 kg moving with a velocity of 400 m/s is measured to an accuracy of 0.02%. Calculate the uncertainty in its position.

9. Considering a standing wave of an electron revolving in an orbit, obtain the Bohr's quantization rule.

10. Describe a wave-packet. How is it represented analytically and diagrammatically.

11. Show that the velocity of a particle and that of the corresponding wave packet are the same.

12. Derive time-dependent Schrödinger equation for one-dimensional motion of a free-particle.

13. Describe the phase velocity v_p and the group velocity v_g. Show that $v_p v_g = c^2$.

14. Derive time-dependent Schrödinger equation for three-dimensional motion of a free-particle.

15. Under what situation, the time-dependent Schrödinger equation can reduce to the time-independent Schrödinger equation. Obtain time-independent Schrödinger equation.

16. Find energy of the particle of mass m moving in a potential field $V(x) = ax^2$ for which the time-independent wave function is

$$\psi(x) = \exp\left(-\sqrt{\frac{ma}{2\hbar^2}}\, x^2\right)$$

Here, a is a constant.

17. Drive time-dependent Schrödinger equation when a particle of mass m moves in a potential field $V(\vec{r}, t)$. Provide an interpretation for the wave function.

18. Define probability current density. What is its value when the wave function is a real quantity.

19. State and prove Ehrenfest theorem.

20. Prove the following relations:

$$\frac{\mathrm{d}}{\mathrm{d}t}\langle x\rangle = \langle p_x\rangle \qquad \text{and} \qquad \frac{\mathrm{d}}{\mathrm{d}t}\langle p_x\rangle = -\left\langle\frac{\partial V}{\partial x}\right\rangle$$

21. Write notes on the following

    (i) Dual nature of light
    (ii) de Broglie hypothesis
    (iii) Matter waves
    (iv) Electron diffraction experiment
    (v) Standing waves of an electron in an orbit
    (vi) Uncertainty principle between position and momentum
    (vii) Uncertainty principle between energy and time
    (viii) Uncertainty principle between angular momentum and angle
    (ix) Single-slit experiment
    (x) Natural width of a spectral line
    (xi) Non-existence of electron in a nucleus
    (xii) Principle of superposition
    (xiii) Wave packet
    (xiv) Interpretation of a wave function
    (xv) Equation of continuity
    (xvi) Expectation value

Basis of Quantum Mechanics

In the preceding chapter, following the wave nature of matter, we obtained a Schrödinger equation, which plays key role in quantum mechanics. Then we discussed the physical significance of the wave function and the uncertainty principle. In the present chapter, we shall discuss about the mathematical formulation of quantum mechanics along with the basic postulates used in its formulation.

1. Linear vector space

Let us begin the discussion with the usual three-dimensional (Euclidean) space and then we shall extend the discussion to a space having a large number of dimensions.

1.1 Vectors in a three-dimensional space

In a three-dimensional space, a vector quantity $\vec{a}$ is expressed as

$$\vec{a} = a_1\, \hat{e}_1 + a_2\, \hat{e}_2 + a_3\, \hat{e}_3$$

where $\hat{e}_1$, $\hat{e}_2$, $\hat{e}_3$ are unit vectors along the positive directions of three mutually perpendicular axes of the system of Cartesian coordinates and a_1, a_2, a_3 are scalars, generally called the components of the vector along the respective axes. The unit vectors $\hat{e}_1$, $\hat{e}_2$, $\hat{e}_3$ form a basis for vectors in the three dimensional space. Since these unit vectors are perpendicular to each other, they form an orthogonal basis. The vector $\vec{a}$ can also be expressed by a column matrix

$$\vec{a} \equiv \begin{bmatrix} a_1 \\ a_2 \\ a_3 \end{bmatrix}$$

where the elements are the components of the vector along the three axes. The unit vectors $\hat{e}_1$, $\hat{e}_2$, $\hat{e}_3$ can be expressed as

$$\hat{e}_1 \equiv \begin{bmatrix} 1 \\ 0 \\ 0 \end{bmatrix} \qquad \hat{e}_2 \equiv \begin{bmatrix} 0 \\ 1 \\ 0 \end{bmatrix} \qquad \hat{e}_3 \equiv \begin{bmatrix} 0 \\ 0 \\ 1 \end{bmatrix}$$

For two vectors

$$\vec{a} = \begin{bmatrix} a_1 \\ a_2 \\ a_3 \end{bmatrix} \qquad \text{and} \qquad \vec{b} = \begin{bmatrix} b_1 \\ b_2 \\ b_3 \end{bmatrix}$$

in the three-dimensional space, the inner product (or scalar product) of $\vec{a}$ and $\vec{b}$, denoted by $(\vec{a}, \vec{b})$ is expressed as

$$(\vec{a}, \vec{b}) = \sum_{i=1}^{3} a_i^* b_i$$

where a_i^* denotes the complex conjugate of a_i. When the components of the vectors are real, we have

$$(\vec{a}, \vec{b}) = \sum_{i=1}^{3} a_i b_i$$

1.2 Vectors in an n-dimensional space

The above concept of the three-dimensional space can be generalized to an n-dimensional space where $\hat{e}_1$, $\hat{e}_2$, $\hat{e}_3 \ldots \hat{e}_n$ form an orthogonal basis. Now, a vector $\vec{a}$ can be expressed as

$$\vec{a} = \sum_{i=1}^{n} a_i \hat{e}_i$$

The inner product of two vectors

$$\vec{a} = \sum_{i=1}^{n} a_i \hat{e}_i \qquad \text{and} \qquad \vec{b} = \sum_{i=1}^{n} b_i \hat{e}_i$$

is expressed as

$$(\vec{a}, \vec{b}) - \sum_{i=1}^{n} a_i^* b_i$$

The inner product of a vector with itself

$$(\vec{a}, \vec{a}) = \sum_{i=1}^{n} a_i^* a_i = \sum_{i=1}^{n} |a_i|^2$$

is always a real quantity. The norm N (or the magnitude) of a vector $\vec{a}$ is defined as $N = (\vec{a}, \vec{a})^{1/2}$. A vector quantity whose norm is unity is said to be normalized.

Two vectors $\vec{a}$ and $\vec{b}$ are said to be orthogonal when

$$(\vec{a}, \vec{b}) = 0$$

The concept of inner product can be used to define an orthonormal set of the vectors $\vec{a}_1, \vec{a}_2, \vec{a}_3, \ldots, \vec{a}_n$ as

$$(\vec{a}_i, \vec{a}_j) = \delta_{ij} \qquad\qquad i, j = 1, 2, 3, \ldots, n$$

where the Kronecker delta δ_{ij} is defined as

$$\delta_{ij} = \begin{cases} 1 & \text{when } i = j \\ 0 & \text{when } i \neq j \end{cases}$$

1.3 Linear independence of vectors

In an n-dimensional space, n vectors $\vec{a}_1, \vec{a}_2, \vec{a}_3, \ldots \vec{a}_n$ are said to be linearly dependent when they have a relation

$$\sum_{i=1}^{n} c_i \, \vec{a}_i = 0 \tag{3.1}$$

where all the scalars $c_1, c_2, c_3, \ldots, c_n$ are not zero. Hence, the n vectors $\vec{a}_1, \vec{a}_2, \vec{a}_3, \ldots \vec{a}_n$ are said to be linearly independent when no relation, such as equation (3.1) exists unless all the c's are zero, *i.e.*, $c_1 = c_2 = c_3 = \ldots = c_n = 0$. In an n-dimensional space, we can have maximum n linearly independent vectors. For n linearly independent vectors $\vec{a}_1, \vec{a}_2, \vec{a}_3, \ldots, \vec{a}_n$, any vector $\vec{v}$ can be expressed as

$$\vec{v} = \sum_{i=1}^{n} d_i \, \vec{a}_i$$

where $d_1, d_2, d_3, \ldots, d_n$ are scalars.

1.4 Hilbert space

In the vector space, the unit vectors $\hat{e}_1, \hat{e}_2, \hat{e}_3, \ldots, \hat{e}_n$ form an orthonormal basis. We can now consider an n-dimensional space in which a set of functions $\phi_1(x), \phi_2(x), \phi_3(x), \ldots, \phi_n(x)$ behaves as an orthonormal basis. Such space is known as the 'Hilbert space'. We can now consider some facts about the basis of the Hilbert space.

(i) Any two different functions $\phi_i(x)$ and $\phi_j(x)$ in an interval $a \le x \le b$ are orthogonal to each other, as their inner product, denoted by (ϕ_i, ϕ_j), is zero. That is

$$(\phi_i, \phi_j) = \int_a^b \phi_i^*(x)\ \phi_j(x)\ \mathrm{d}x = 0$$

where $\phi_i^*(x)$ is the complex conjugate of $\phi_i(x)$.

(ii) Any function $\phi_i(x)$ in an interval $a \le x \le b$ is normalized as its norm N, defined as the square root of the inner product of the function with itself, is unity. That is

$$N^2 = (\phi_i, \phi_i) = \int_a^b \phi_i^*(x)\ \phi_i(x)\ \mathrm{d}x = \int_a^b |\phi_i(x)|^2\ \mathrm{d}x = 1$$

(iii) A set of n orthonormal functions $\phi_1(x)$, $\phi_2(x)$, $\phi_3(x)$, ..., $\phi_n(x)$ obviously satisfy the relation

$$(\phi_i, \phi_j) = \int_a^b \phi_i^*(x)\ \phi_j(x)\ \mathrm{d}x = \delta_{ij} \qquad i, j = 1, 2, 3, \ldots, n$$

where the Kronecker delta δ_{ij} is defined as

$$\delta_{ij} = \begin{cases} 1 & \text{when } i = j \\ 0 & \text{when } i \ne j \end{cases}$$

(iv) The functions $\phi_1(x)$, $\phi_2(x)$, $\phi_3(x)$, ..., $\phi_n(x)$ are linearly independent as they satisfy the relation

$$\sum_{i=1}^{n} c_i\ \phi_i(x) = 0$$

only when all the scalars c_1, c_2, c_3, ..., c_n are zero.

(v) Any other function $F(x)$ can be expressed in term of the basis as

$$F(x) = \sum_{i=1}^{n} d_i\ \phi_i(x)$$

where d_1, d_2, d_3, ..., d_n are scalars, given by[1]

$$d_i = (\phi_i, F)$$

[1]For the function

$$F(x) = d_1\ \phi_1(x) + d_2\ \phi_2(x) + \ldots + d_k\ \phi_k(x) + \ldots + d_n\ \phi_n(x)$$

2. Linear operator

An operator is a rule by which a new function is obtained from a given function. For example, differentiation of sin x with respect to x gives cos x.

$$\frac{\mathrm{d}}{\mathrm{d}x}\sin x = \cos x$$

Here, differentiation is the operator which gives a new function cos x from the given function sin x.

An operator $\hat{A}$ is said to be linear when it satisfies the relation

$$\hat{A}\Big[c_1\ f_1(x) + c_2\ f_2(x) + \ldots + c_n\ f_n(x)\Big] = c_1\ \hat{A}f_1(x) + c_2\ \hat{A}f_2(x) + \ldots + c_n\ \hat{A}f_n(x)$$

where c_1, c_2, c_3, ..., c_n are constants. For example, the relation

$$\frac{\mathrm{d}}{\mathrm{d}x}\Big[c_1\ f_1(x) + c_2\ f_2(x)\Big] = c_1\ \frac{\mathrm{d}}{\mathrm{d}x}f_1(x) + c_2\ \frac{\mathrm{d}}{\mathrm{d}x}f_2(x)$$

shows that the differentiation is a linear operator.

Let us check whether an operator that squares a function is linear or not. For simplicity, we consider a combination of two functions. Now, we have

$$\begin{aligned} g(x) &= [c_1\ f_1(x) + c_2\ f_2(x)]^2 \\ &= c_1^2\ [f_1(x)]^2 + c_2^2\ [f_2(x)]^2 + c_1\ c_2\ f_1(x)\ f_2(x) \end{aligned}$$

Since $g(x)$ does not satisfy the relation

$$g(x) = c_1\ [f_1(x)]^2 + c_2\ f_2(x)]^2$$

we find that squaring is not a linear operation.

Exercise 1. A and B are two operators defined by

$$A\ \psi(x) = \psi(x) + a\ x \qquad \text{and} \qquad B\ \psi(x) = \frac{\mathrm{d}\psi(x)}{\mathrm{d}x} + b\ \psi(x)$$

---

we have

$$(\phi_k, F) = d_1 \int_{-\infty}^{\infty} \phi_k^*\ \phi_1\ \mathrm{d}x + d_2 \int_{-\infty}^{\infty} \phi_k^*\ \phi_2\ \mathrm{d}x + \ldots + d_k \int_{-\infty}^{\infty} \phi_k^*\ \phi_k\ \mathrm{d}x + \ldots + d_n \int_{-\infty}^{\infty} \phi_k^*\ \phi_n\ \mathrm{d}x = d_k$$

as all the integrals except one for the index k will be zero and we have

$$(\phi_k, F) = d_k$$

where a and b are constants. Decide if these operators are linear or not.

Solution: For an operator O to be linear, we must have

$$O[c_1 f_1(x) + c_2 f_2(x)] = c_1 O f_1(x) + c_2 O f_2(x)$$

For the given operator A, we have

$$A[c_1 f_1(x) + c_2 f_2(x)] = c_1 f_1(x) + c_2 f_2(x) + a\ x$$

Now, we have

$$A f_1(x) = f_1(x) + a\ x \qquad \text{and} \qquad A f_2(x) = f_2(x) + a\ x$$

and therefore,

$$\begin{aligned} c_1 A f_1(x) + c_2 A f_2(x) &= c_1 f_1(x) + c_1 a\ x + c_2 f_2(x) + c_2 a\ x \\ &= c_1 f_1(x) + c_2 f_2(x) + (c_1 + c_2) a\ x \end{aligned}$$

which is not equal to $A[c_1 f_1(x) + c_2 f_2(x)]$. It shows that A is not a linear operator.

For the given operator B, we have

$$B[c_1 f_1(x) + c_2 f_2(x)] = \frac{\mathrm{d}}{\mathrm{d}x}[c_1 f_1(x) + c_2 f_2(x)] + b[c_1 f_1(x) + c_2 f_2(x)]$$

Now, we have

$$B f_1(x) = \frac{\mathrm{d}}{\mathrm{d}x}[f_1(x)] + b\ f_1(x) \qquad \text{and} \qquad B f_2(x) = \frac{\mathrm{d}}{\mathrm{d}x}[f_2(x)] + b\ f_2(x)$$

and therefore,

$$\begin{aligned} c_1 B f_1(x) + c_2 B f_2(x) &= c_1 \frac{\mathrm{d}}{\mathrm{d}x}[f_1(x)] + c_1 b f_1(x) + c_2 \frac{\mathrm{d}}{\mathrm{d}x}[f_2(x)] + c_2 b f_2(x) \\ &= \frac{\mathrm{d}}{\mathrm{d}x}[c_1 f_1(x) + c_2 f_2(x)] + b[c_1 f_1(x) + c_2 f_2(x)] \end{aligned}$$

which is equal to $B[c_1 f_1(x) + c_2 f_2(x)]$. It shows that B is a linear operator.

2.1 Properties of operators

(i) Sum and difference of two operators $\hat{A}$ and $\hat{B}$ is expressed as

$$(\hat{A} \pm \hat{B}) f(x) = \hat{A} f(x) \pm \hat{B} f(x)$$

Addition of operators satisfies the commutation property as

$$\hat{A} + \hat{B} = \hat{B} + \hat{A}$$

Addition of operators satisfies the association property as

$$(\hat{A} + \hat{B}) + \hat{C} = \hat{A} + (\hat{B} + \hat{C})$$

(ii) Product of two operators $\hat{A}$ and $\hat{B}$ is expressed as

$$(\hat{A}\,\hat{B})f(x) = \hat{A}[\hat{B}f(x)]$$

Product of operators is generally neither commutative nor associative. It however satisfies the distribution property as

$$\hat{A}(\hat{B} + \hat{C})f(x) = (\hat{A}\,\hat{B} + \hat{A}\,\hat{C})f(x)$$

(iii) Commutator of two operators $\hat{A}$ and $\hat{B}$, denoted by $[\hat{A},\,\hat{B}]$, is expressed as

$$[\hat{A}, \hat{B}] = \hat{A}\,\hat{B} - \hat{B}\,\hat{A}$$

Note that the operators are enclosed in a pair of square brackets. Since

$$[\hat{B}, \hat{A}] = \hat{B}\,\hat{A} - \hat{A}\,\hat{B}$$

we have

$$[\hat{A}, \hat{B}] = -[\hat{B}, \hat{A}]$$

(a) When $\hat{A}\,\hat{B} = \hat{B}\,\hat{A}$, *i.e.*, $[\hat{A}, \hat{B}] = 0$, the operators $\hat{A}$ and $\hat{B}$ are said to commute with each other. In this case, the product of the operators is commutative.

(b) When $\hat{A}\,\hat{B} = -\hat{B}\,\hat{A}$, *i.e.*, $\hat{A}\,\hat{B} + \hat{B}\,\hat{A} = 0$, the operators $\hat{A}$ and $\hat{B}$ are said to anticommute with each other.

(c) An operator can be applied in succession on a function. This can be expressed as

$$\hat{A}\,\hat{A}\,\hat{A}\,\hat{A}\ldots\hat{A}f(x) = \hat{A}^n f(x)$$

Exercise 2. show that various powers of the same operator commute.

Solution: We have

$$[\hat{A}^m, \hat{A}^n] = \hat{A}^m\,\hat{A}^n - \hat{A}^n\,\hat{A}^m = \hat{A}^{m+n} - \hat{A}^{n+m} = 0$$

It shows that various powers of the same operator commute.

(d) An operator $\hat{A}^{-1}$, which is the inverse of $\hat{A}$, is expressed as

$$\hat{A}\,\hat{A}^{-1} = \hat{A}^{-1}\,\hat{A} = I$$

where I is the identity operation such that $If(x) = f(x)$.

Exercise 3. show that an operator always commutes with its inverse.

Solution: We have

$$[\hat{A}^{-1}, \hat{A}] = \hat{A}^{-1}\,\hat{A} - \hat{A}\,\hat{A}^{-1} = I - I = 0$$

It shows that an operator always commutes with its inverse.

Note: For convenience, in the further discussion, we shall denote an operator associated with a dynamical operator without a cap on the same letter. For example, the operator $\hat{A}$ will written without cap as A.

Exercise 4. Show that

$$\left[x, \frac{\mathrm{d}}{\mathrm{d}x}\right] = -1 \qquad \text{and} \qquad \left[\frac{\mathrm{d}}{\mathrm{d}x}, x\right] = 1$$

Solution: We have

$$\left[x, \frac{\mathrm{d}}{\mathrm{d}x}\right] f = x\,\frac{\mathrm{d}f}{\mathrm{d}x} - \frac{\mathrm{d}(xf)}{\mathrm{d}x}$$

$$= x\,\frac{\mathrm{d}f}{\mathrm{d}x} - x\,\frac{\mathrm{d}f}{\mathrm{d}x} - f = -f$$

Hence,

$$\left[x, \frac{\mathrm{d}}{\mathrm{d}x}\right] = -1$$

Now, we have

$$\left[\frac{\mathrm{d}}{\mathrm{d}x}, x\right] f = \frac{\mathrm{d}(xf)}{\mathrm{d}x} - x\,\frac{\mathrm{d}f}{\mathrm{d}x}$$

$$= x\,\frac{\mathrm{d}f}{\mathrm{d}x} + f - x\,\frac{\mathrm{d}f}{\mathrm{d}x} = f$$

Hence,

$$\left[\frac{\mathrm{d}}{\mathrm{d}x}, x\right] = 1$$

Exercise 5. Show that

$$\left[[A, B], C\right] + \left[[B, C], A\right] + \left[[C, A], B\right] = 0$$

Solution: We have

$$[A, B] = AB - BA$$

Then we have[2]

$$\Big[[A,B],C\Big] = [AB - BA, C] = [AB, C] - [BA, C]$$
$$= A[B,C] + [A,C]B - B[A,C] - [B,C]A$$
$$= ABC - ACB + ACB - CAB - BAC + BCA - BCA + CBA$$
$$= ABC - CAB - BAC + CBA$$

Similarly, we have

$$\Big[[B,C],A\Big] = BCA - ABC - CBA + ACB$$

and

$$\Big[[C,A],B\Big] = CAB - BCA - ACB + BAC$$

Hence,

$$\Big[[A,B],C\Big] + \Big[[B,C],A\Big] + \Big[[C,A],B\Big] = ABC - CAB - BAC + CBA$$
$$+ BCA - ABC - CBA + ACB + CAB - BCA - ACB + BAC = 0$$

Exercise 6. Show that $[x, p_x] = i\hbar$.

Solution: We have

$$[x, p_x]f = x\Big(-i\hbar\frac{\partial}{\partial x}\Big)f - \Big(-i\hbar\frac{\partial}{\partial x}\Big)(xf)$$
$$= -i\hbar x\frac{\partial f}{\partial x} + i\hbar x\frac{\partial f}{\partial x} + i\hbar f = i\hbar f$$

Thus, we have $[x, p_x] = i\hbar$.

Exercise 7. If x and p_x are position and momentum operators, show that $[x, p_x^n] = ni\hbar p_x^{n-1}$.

Solution: We have

$$[x, p_x^n] = [x, p_x p_x^{n-1}] = [x, p_x]p_x^{n-1} + p_x[x, p_x^{n-1}]$$
$$= i\hbar p_x^{n-1} + p_x\Big([x, p_x]p_x^{n-2} + p_x[x, p_x^{n-2}]\Big)$$

---

[2]We have

$$[\alpha\beta, \gamma] = \alpha\beta\gamma - \gamma\alpha\beta$$
$$= \alpha\beta\gamma - \alpha\gamma\beta + \alpha\gamma\beta - \gamma\alpha\beta = \alpha[\beta, \gamma] + [\alpha, \gamma]\beta$$

$$\begin{aligned}
&= i\hbar p_x^{n-1} + p_x i\hbar p_x^{n-2} + p_x^2[x, p_x^{n-2}] \\
&= 2i\hbar p_x^{n-1} + p_x^2\left([x, p_x]p_x^{n-3} + p_x[x, p_x^{n-3}]\right) \\
&= 2i\hbar p_x^{n-1} + p_x^2 i\hbar p_x^{n-3} + p_x^3[x, p_x^{n-3}] \\
&= 3i\hbar p_x^{n-1} + p_x^3[x, p_x^{n-3}]
\end{aligned}$$

This procedure continues and finally we get

$$[x, p_x^n] = ni\hbar p_x^{n-1}$$

3. Eigenfunctions and eigenvalues

When an operator A and a function $f(x)$ satisfy the relation

$$Af(x) = af(x)$$

where a is a scalar, then $f(x)$ is known as the eigenfunction of the operator A and a is the corresponding eigenvalue. For example, in the relation

$$\frac{\mathrm{d}}{\mathrm{d}x} \exp(kx) = k \exp(kx)$$

$\exp(kx)$ is an eigenfunction of the operator $\mathrm{d}/\mathrm{d}x$ and the corresponding eigenvalue is k.

3.1 Hermitian operators

An operator A is said to be Hermitian when, for two arbitrary functions $\psi_m(x)$ and $\psi_n(x)$, it satisfies the relations

$$\int_{-\infty}^{\infty} \psi_m^* \, A\psi_n \, \mathrm{d}x = \int_{-\infty}^{\infty} (A\,\psi_m)^* \, \psi_n \, \mathrm{d}x = \left[\int_{-\infty}^{\infty} \psi_n^* \, A\psi_m \, \mathrm{d}x\right]^*$$

These relations in the bracket notation[3] can be expressed as

$$(\psi_m, A\psi_n) = (A\psi_m, \psi_n) = (\psi_n, A\psi_m)^*$$

Two important theorems regarding the Hermitian operators, which are used in quantum mechanics, are as the following.

Theorem 1. Prove that the eigenvalues of a Hermitian operator are real.

[3]This notation will be introduced soon.

Proof: For a Hermitian operator A, an eigenvalue a_n and the corresponding eigenfunction ψ_n are expressed as

$$A\ \psi_n = a_n\ \psi_n$$

Taking inner product with ψ_n, we have

$$(\psi_n, A\psi_n) = a_n\ (\psi_n, \psi_n) = a_n$$

Since A is a Hermitian operator, we have

$$(\psi_n, A\psi_n) = (A\psi_n, \psi_n) = a_n^*\ (\psi_n, \psi_n) = a_n^*$$

These relations give that $a_n = a_n^*$. This shows that a_n is real. That is, the eigenvalue of a Hermitian operator is real. The same will be true for other eigen values as well.

Theorem 2. Prove that two eigenfunctions of a Hermitian operator corresponding to two different eigenvalues are orthogonal.

Proof: Let ψ_m and ψ_n be two eigenfunctions of a Hermitian operator A corresponding to two different eigenvalues a_m and a_n, respectively. Then, we have

$$A\psi_m = a_m\psi_m \qquad \text{and} \qquad A\psi_n = a_n\psi_n$$

From the first equation, we get

$$(\psi_n, A\psi_m) = a_m(\psi_n, \psi_m)$$

Since A is a Hermitian operator, we have

$$(\psi_n, A\psi_m) = (A\psi_n, \psi_m) = a_n(\psi_n, \psi_m)$$

Here, we have used the fact that the eigenvalues of a Hermitian operator are real. Thus, we have

$$a_n(\psi_n, \psi_m) = a_m(\psi_n, \psi_m)$$
$$(a_n - a_m)(\psi_n, \psi_m) = 0$$

Since the eigen values a_n and a_m are different, *i.e.*, $a_n \neq a_m$, then $(\psi_n, \psi_m) = 0$. It shows that the eigenfunctions of a Hermitian operator are orthogonal to each other.

Exercise 8. If A and B are Hermitian operators, show that $(AB + BA)$ is Hermitian and $(AB - BA)$ is not.

Solution: For the operators A and B to be Hermitian, we have

$$(\psi_m, A\psi_n) = (A\psi_m, \psi_n) \qquad \text{and} \qquad (\psi_m, B\psi_n) = (B\psi_m, \psi_n)$$

Now, we have

$$\int \psi_m^*(AB + BA)\psi_n \mathrm{d}x = \int \psi_m^* AB\psi_n \mathrm{d}x + \int \psi_m^* BA\psi_n \mathrm{d}x$$

$$= \int B^* A^* \psi_m^* \psi_n \mathrm{d}x + \int A^* B^* \psi_m^* \psi_n \mathrm{d}x$$

$$= \int (B^* A^* + A^* B^*)\psi_m^* \psi_n \mathrm{d}x = \int (AB + BA)^* \psi_m^* \psi_n \mathrm{d}x$$

It shows that $(AB + BA)$ is Hermitian.

Now,

$$\int \psi_m^*(AB - BA)\psi_n \mathrm{d}x = \int \psi_m^* AB\psi_n \mathrm{d}x - \int \psi_m^* BA\psi_n \mathrm{d}x$$

$$= \int B^* A^* \psi_m^* \psi_n \mathrm{d}x - \int A^* B^* \psi_m^* \psi_n \mathrm{d}x$$

$$= \int (B^* A^* - A^* B^*)\psi_m^* \psi_n \mathrm{d}x = -\int (AB - BA)^* \psi_m^* \psi_n \mathrm{d}x$$

It shows that $(AB - BA)$ is not Hermitian.

Exercise 9. The Hamiltonian operator of a system is

$$H \equiv -\frac{\mathrm{d}^2}{\mathrm{d}x^2} + x^2$$

show that $Nx \exp(-x^2/2)$ is an eigenfunction of this operator. Find out its eigenvalue and the value of N so that the function is normalized.

Solution: We have

$$\psi = Nx \exp(-x^2/2)$$

where N is a constant. Thus,

$$H\psi = \Big(-\frac{\mathrm{d}^2}{\mathrm{d}x^2} + x^2\Big)\Big(N\, x \exp(-x^2/2)\Big)$$

$$= -N\frac{\mathrm{d}}{\mathrm{d}x}\Big[\exp(-x^2/2) - x^2 \exp(-x^2/2)\Big] + Nx^3 \exp(-x^2/2)$$

$$= Nx \exp(-x^2/2) + 2Nx \exp(-x^2/2) - Nx^3 \exp(-x^2/2) + Nx^3 \exp(-x^2/2)$$

$$= 3Nx \exp(-x^2/2) = 3\psi$$

It shows that the eigen value of H is 3. The normalization condition

$$\int_{-\infty}^{\infty} \psi^* \psi \mathrm{d}x = 1$$

gives

$$N^2 \int_{-\infty}^{\infty} x^2 \exp(-x^2) \mathrm{d}x = 1 \qquad \text{or} \qquad 2N^2 \int_{0}^{\infty} x^2 \exp(-x^2) \mathrm{d}x = 1$$

$$2N^2 \frac{\sqrt{\pi}}{4} = 1 \qquad \text{or} \qquad N = \left(\frac{2}{\sqrt{\pi}}\right)^{1/2}$$

Thus, the normalized wave function is

$$\psi = \left(\frac{2}{\sqrt{\pi}}\right)^{1/2} x \exp\left(-\frac{x^2}{2}\right)$$

4. Postulates of quantum mechanics

Investigation of a physical phenomenon in quantum mechanics is based on a number of postulates. These postulates indeed form a firm foundation of quantum mechanics and are discussed in the following.

Postulate 1: Wave function

State of a system can be described by a function Ψ. This function Ψ, called the wave function, state function or state vector, is a function of generalized coordinates $q_1, q_2, \ldots, q_n$ and time t. The function Ψ as well as its derivative in the space must be continuous, finite and single-valued over the region of the variables. All possible information about the system can be derived with the help of the wave function for the system. As such, the function Ψ does not represent any of the observable quantities or states. However, as discussed earlier, $\Psi^*\Psi$ presents the probability density. Here, Ψ^* is the complex conjugate of Ψ. Obviously, the probability density is a real quantity.

Postulate 2: Operators

Corresponding to each observable physical quantity, there is a Hermitian operator. Since an observable physical quantity is real, it can be represented by a Hermitian operator, as the eigen values of a Hermitian operator are real. Some physical quantities and the corresponding operators are as the following.

Physical quantity	Symbol in classical mechanics	Operator in quantum mechanics
Position	x	x
Position	$\vec{r}$	$\vec{r}$
Energy	E	$i\hbar\,\frac{\partial}{\partial t}$
Momentum	p_x	$-i\hbar\,\frac{\partial}{\partial x}$
Momentum	$\vec{p}$	$-i\hbar\,\nabla$
Potential energy	V	V
Time	t	t

Operators for other physical quantities can be derived in terms of these operators. Some examples of the derived operators are as the following.

(i) Kinetic energy: For a particle of mass m having momentum $\vec{p}$, an operator for the kinetic energy T is

$$T = \frac{p^2}{2m} = -\frac{\hbar^2}{2m}\nabla^2$$

(ii) Hamiltonian: For a particle of mass m with momentum $\vec{p}$, moving in a potential field $V(\vec{r})$, an operator for the Hamiltonian H is

$$H = \frac{p^2}{2m} + V(\vec{r}) = -\frac{\hbar^2}{2m}\nabla^2 + V(\vec{r})$$

(iii) Angular momentum components: For a position vector $\vec{r} = x\,\hat{e}_1 + y\,\hat{e}_2 + z\,\hat{e}_3$ and the momentum $\vec{p} = p_x\,\hat{e}_1 + p_y\,\hat{e}_2 + p_z\,\hat{e}_3$, the angular momentum $\vec{L}$ is

$$\vec{L} = \vec{r} \times \vec{p} = \begin{bmatrix} \hat{e}_1 & \hat{e}_2 & \hat{e}_3 \\ x & y & z \\ p_x & p_y & p_z \end{bmatrix}$$

For $\vec{L} = L_x\hat{e}_1 + L_y\hat{e}_2 + L_z\hat{e}_3$, the operators for the components of the angular momentum are

$$L_x = y\,p_z - z\,p_y = -i\hbar\left(y\,\frac{\partial}{\partial z} - z\,\frac{\partial}{\partial y}\right)$$

$$L_y = z\,p_x - x\,p_z = -i\hbar\left(z\,\frac{\partial}{\partial x} - x\,\frac{\partial}{\partial z}\right)$$

$$L_z = x\,p_y - y\,p_x = -i\hbar\left(x\,\frac{\partial}{\partial y} - y\,\frac{\partial}{\partial x}\right)$$

Exercise 10. Show that

(i) $[q_j, q_k] = 0$

(ii) $[q_j, p_k] = i\hbar\ \delta_{jk}$

(iii) $[p_j, p_k] = 0$

where q_1, q_2, q_3 are x, y, z and p_1, p_2, p_3 are p_x, p_y, p_z, respectively.

Solution: The relations can be verified as the following.

(i) We have

$$[q_j, q_k]\psi = (q_j q_k - q_k q_j)\psi = (q_j q_k - q_j q_k)\psi = 0$$

Therefore, $[q_j, q_k] = 0$.

(ii) We have

$$[q_j, p_k]\psi = (q_j p_k - p_k q_j)\psi = -i\hbar q_j \frac{\partial \psi}{\partial q_k} + i\hbar \frac{\partial (q_j \psi)}{\partial q_k}$$

$$= -i\hbar q_j \frac{\partial \psi}{\partial q_k} + i\hbar q_j \frac{\partial \psi}{\partial q_k} + i\hbar \psi \frac{\partial q_j}{\partial q_k} = i\hbar \delta_{jk}$$

Therefore, $[q_j, p_k] = i\hbar \delta_{jk}$.

(iii) We have

$$[p_j, p_k]\psi = (p_j p_k - p_k p_j)\psi = -\hbar^2 \frac{\partial}{\partial q_j}\left(\frac{\partial \psi}{\partial q_k}\right) + \hbar^2 \frac{\partial}{\partial q_k}\left(\frac{\partial \psi}{\partial q_j}\right)$$

$$= -\hbar^2 \frac{\partial^2 \psi}{\partial q_j \partial q_k} + \hbar^2 \frac{\partial^2 \psi}{\partial q_k \partial q_j} = 0$$

Therefore, $[p_j, p_k] = 0$.

Postulate 3: Expectation value

For a state described by a wave function Ψ, the expectation value $\langle A \rangle$ of an observable variable A is

$$\langle A \rangle = \int_{-\infty}^{\infty} \Psi^* A \Psi \ \mathrm{d}\tau$$

where A in the integral is an operator corresponding to the observable variable and the wave function is normalized. Here, the integration is taken over the complete ranges of the variables. When the wave function Ψ is not normalized, the expectation value is

$$\langle A \rangle = \int_{-\infty}^{\infty} \Psi^* A \Psi \ \mathrm{d}\tau \Big/ \int_{-\infty}^{\infty} \Psi^* \Psi \ \mathrm{d}\tau$$

Postulate 4: Eigenvalues

For an observable variable A and the wave function Ψ_i for the state i, it is possible to have a relation

$$A\Psi_i = a_i\ \Psi_i \qquad\qquad i = 1, 2, \ldots, n$$

This equation is known as an eigenvalue equation, a_i is an eigenvalue and Ψ_i the corresponding eigenfunction. The expectation value of the variable is

$$\langle A\rangle = \int_{-\infty}^{\infty} \Psi_n^* A\Psi_n\ \mathrm{d}\tau = \int_{-\infty}^{\infty} \Psi_n^* a_n \Psi_n\ \mathrm{d}\tau = a_n \int_{-\infty}^{\infty} \Psi_n^* \Psi_n\ \mathrm{d}\tau = a_n$$

The eigenvalues of an operator are the only quantities which can be measured experimentally. Hence, the eigenvalues of an observable variable is a real number, as the eigenvalues of the Hermitian operators are real. Thus, the operators associated with physical quantities must be Hermitian.

Postulate 5: Time development of a quantum mechanical system

The time development of a quantum mechanical system can be described by the time-dependent Schrödinger equation

$$i\hbar\frac{\partial}{\partial t}\Psi(\vec{r}, t) = \Big[-\frac{\hbar^2}{2m}\nabla^2 + V(\vec{r})\Big]\Psi(\vec{r}, t)$$

where the operator in the right side of the equation is independent of time t. The wave function $\Psi(\vec{r}, t)$ is a function of the space coordinates as well as the time.

4.1 Simultaneous measureability of observables

In a particular state of a given system, when two observables can be measured simultaneously, the state function is an eigenfunction of both the operators.

Theorem 3. Two operators having a common set of eigenfunctions commute with one another.

Proof: Suppose A and B are two operators having a common set of eigenfunctions Ψ_i, i = 1, 2, ..., n. Let for one of these eigenfunctions, say Ψ_j, the operators have the eigenvalues a_j and b_j, respectively. Thus, we have

$$A\Psi_j = a_j\Psi_j \qquad\qquad B\Psi_j = b_j\Psi_j \tag{3.2}$$

On multiplying second equation of (3.2) by A from left, we get

$$AB\Psi_j = A(b_j\Psi_j) = b_j(A\Psi_j) = b_ja_j\Psi_j$$

On multiplying first equation of (3.2) by B from left, we get

$$BA\Psi_j = B(a_j\Psi_j) = a_j(B\Psi_j) = a_jb_j\Psi_j$$

Since a_j and b_j are number, we have $b_ja_j = a_jb_j$. Therefore, $AB\Psi_j = BA\Psi_j$. It shows that $AB = BA$. That is, A and B commute with each other.

Theorem 4. Two commutating operators have a common set of eigenfunctions.

Proof: Suppose A and B are two commutating operators, so that $AB = BA$. Let Ψ_i, $i = 1, 2, \ldots, n$ is a set of eigenfunctions of the operator A with eigenvalues a_i, $i = 1, 2, \ldots, n$, respectively. Then we have

$$A\Psi_i = a_i\Psi_i \qquad\qquad i = 1, 2, \ldots, n$$

Operating this equation by B from left, we have

$$BA\Psi_i = Ba_i\Psi_i = a_iB\Psi_i$$

Using the relation $AB = BA$, we have

$$A(B\Psi_i) = a_i(B\Psi_i)$$

It shows that $B\Psi_i$ also is an eigenfunction of the operator A with the same eigenvalue a_i. It can only be possible when $B\Psi_i$ differs from Ψ_i by a constant multiplier, say, b_i, so that

$$B\Psi_i = b_i\Psi_i$$

This equation shows that Ψ_i is a wave function of the operator B with eigenvalue b_i. Thus, the commutating operators A and B have the same set of eigenfunctions.

5. Dirac's notation

Dirac introduced new system of notations for representing the wave functions and their complex conjugates in quantum mechanics. In this notation, the symbol $|\ \rangle$, called the ket vector, is used for representing a wave function and the symbol $\langle\ |$, called the bra vector, is used for representing the complex conjugate of the wave function. In order to distinguish between various states, a label is introduced in the ket as well as in the bra vectors. For example, the wave function Ψ_a is represented by a ket vector $|a\rangle$ and its complex conjugate Ψ_a^* is represented by the corresponding bra vector $\langle a|$. Obviously, $|a\rangle$ and $\langle a|$ are complex conjugate of each other. The words 'bra' and 'ket' are derived from the word 'bracket' by dropping the middle letter *c*.

Operation by an operator A on a ket vector produces another ket vector, as

$$A|a\rangle = |a'\rangle$$

Operation on a bra vector from right by an operator A produces another bra vector, as

$$\langle b|A = \langle b'|$$

In the bra and ket notation, inner product of two wave functions Ψ_a and Ψ_b is expressed as

$$(\Psi_a, \Psi_b) = \int \Psi_a^* \Psi_b \, d\tau = \langle a|b\rangle$$

The orthonormality relation of wave functions is expressed as

$$\langle a_i|a_j\rangle = \delta_{ij}$$

The condition for an operator A to be Hermitian is expressed as

$$\langle a|A|b\rangle = \langle b|A|a\rangle^*$$

6. Problems and questions

1. Describe eigen value and eigen function of an operator.

2. Describe a Hermitian operator. Show that eigen values of a Hermitian operator are real.

3. Show that the eigen functions of a Hermitian operator corresponding to different eigen values are orthogonal.

4. Discuss about the basic concepts of quantum mechanics.

5. Show that an operator associated with a physical quantity is Hermitian.

6. Show that various powers of the same operator commute.

7. Show that an operator commutes with its inverse.

8. Describe the basic postulates of quantum mechanics.

9. Show that the operators having a common set of eigenfunctions commute with one another.

10. Show that the commutating operators have a common set of eigenfunctions.

11. Describe the Dirac's bra and ket notations. Write down the orthonormality relation in the bra and ket notations.

12. For the operators A, B and C, show that

    (i) $[AB, C] = A[B, C] + [A, C]B$
    (ii) $[A, BC] = [A, B]C + B[A, C]$
    (iii) $\Big[[A, B], C\Big] + \Big[[B, C], A\Big] + \Big[[C, A], B\Big] = 0$

13. Show that

$$\left[x, \frac{\mathrm{d}}{\mathrm{d}x}\right] = -1 \qquad \text{and} \qquad \left[\frac{\mathrm{d}}{\mathrm{d}x}, x\right] = 1$$

14. Evaluate the following commutators:

    (i) $[x^2, p_x]$ (ii) $[x, p_x^2]$ (iii) $[x^2, p_x^2]$ (iv) $[x^n, p_x]$ (v) $[x, p_x^n]$

15. Write notes on the following

    (i) Linear vector space
    (ii) Vectors in three-dimensional space
    (iii) Vectors in n-dimensional space
    (iv) Hilbert space
    (v) Linear operators
    (vi) Basic concepts of quantum mechanics
    (vii) Eigenfunctions and eigenvalues
    (viii) Hermitian operators
    (ix) Postulates of quantum mechanics
    (x) Dirac's bra and ket notations

One-dimensional Problems

We know that solution of Schrödinger equation for a system provides information about various states of the system in the form of the eigenfunctions and eigenvalues. In the present chapter, we shall discuss solutions of one-dimensional Schrödinger equation for some steady-state systems. Time-independent, one-dimensional Schrödinger equation can be expressed as

$$-\frac{\hbar^2}{2m}\frac{\mathrm{d}^2\psi(x)}{\mathrm{d}x^2}+V(x)\psi(x)=E\psi(x)$$

Here, x is taken as the variable. We shall discuss solutions of this Schrödinger equation for some specific forms of the potential $V(x)$.

1. Square well potential with rigid walls

Let us consider a particle of mass m moving in a one-dimensional potential field (Figure 1) expressed as

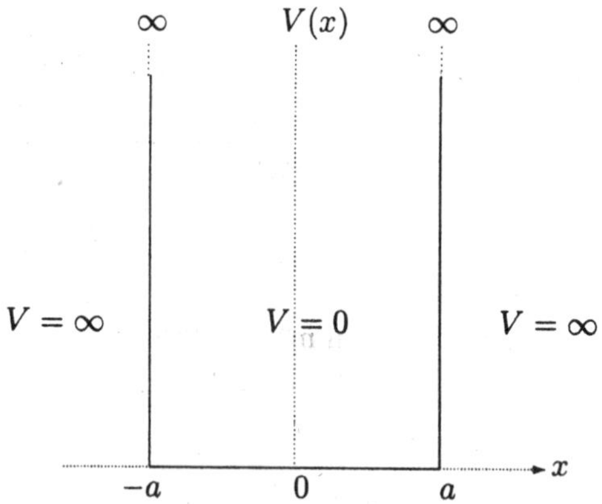

Figure 1: One-dimensional box of a width $2a$ with infinite height.

$$V(x) = \begin{cases} 0 & |x| < a \\ \infty & |x| > a \end{cases}$$

For convenience, the potential is taken symmetrical about the $x = 0$ line. Here, the potential inside the box is taken as zero whereas, for the rigid walls, it is taken as infinite outside the box. Since the potential outside the box is infinite, the probability of finding the particle outside the box is zero. Thus, the wave function outside the box can be taken as zero. The time-independent, one-dimensional Schrödinger equation for the region $|x| < a$ is

$$-\frac{\hbar^2}{2m}\frac{d^2}{dx^2}\psi(x) = E\psi(x)$$

$$\frac{d^2}{dx^2}\psi(x) + k^2\psi(x) = 0 \tag{4.1}$$

where $k^2 = 2mE/\hbar^2$. The Schrödinger equation (4.1) has a solution

$$\psi(x) = A \sin(kx) + B \cos(kx)$$

where A and B are constants. At the boundaries $x = \pm a$, the wave function $\psi(x)$ vanishes and therefore, we have

(i) at $x = a$, we have

$$A \sin(ka) + B \cos(ka) = 0 \tag{4.2}$$

(ii) at $x = -a$, we have

$$-A \sin(ka) + B \cos(ka) = 0 \tag{4.3}$$

These two equations can be satisfied simultaneously when we have separately

$$A \sin(ka) = 0 \qquad \text{and} \qquad B \cos(ka) = 0$$

The values $A = 0$ and $B = 0$ lead to a physically unacceptable situation, as the wave function $\psi(x)$ will vanish everywhere. Further, for satisfying the above two equations (4.2) and (4.3) simultaneously, both A and B cannot be non-zero at a time. Thus, the other possible cases are $A \neq 0$, $B = 0$ and $A = 0$, $B \neq 0$.

(i) For the case $A \neq 0$, and $B = 0$, we have

$$\sin(ka) = 0 \qquad ka = \frac{n\pi}{2} \qquad n = 2, 4, 6, \ldots$$

Now, we have

$$k^2 = \frac{2mE}{\hbar^2} \qquad \frac{n^2\pi^2}{4a^2} = \frac{2mE}{\hbar^2} \qquad E_n = \frac{\pi^2\hbar^2 n^2}{8ma^2}$$

where in order to distinguish between various energies, a suffix n is used with E. The corresponding wave function is

$$\psi_n(x) = A \sin\left(\frac{n\pi}{2a}x\right) \qquad n = 2, 4, 6, \ldots$$

For this wave function, we have

$$\int_{-a}^{a} \psi_n^*(x)\psi_n(x)\ \mathrm{d}x = \int_{-a}^{a} A^2 \sin^2\left(\frac{n\pi}{2a}x\right)\ \mathrm{d}x$$

$$= \frac{A^2}{2}\int_{-a}^{a}\left[1 - \cos\left(\frac{n\pi}{a}x\right)\right]\mathrm{d}x = A^2\ a$$

Thus, the normalization of this wave function gives $A^2\ a = 1$. Therefore, $A = 1/\sqrt{a}$ and the normalized wave function is

$$\psi_n(x) = \frac{1}{\sqrt{a}} \sin\left(\frac{n\pi}{2a}x\right) \qquad n = 2, 4, 6, \ldots$$

In this discussion, the case $n = 0$ is not included as its corresponds to a trivial solution $\psi(x) = 0$ or a non-harmonic solution $\psi(x) = cx + d$.

(ii) For the case $A = 0$, and $B \neq 0$, we have

$$\cos(ka) = 0 \qquad ka = \frac{n\pi}{2} \qquad n = 1, 3, 5, \ldots$$

Now, we have

$$k^2 = \frac{2mE}{\hbar^2} \qquad \frac{n^2\pi^2}{4a^2} = \frac{2mE}{\hbar^2} \qquad E_n = \frac{\pi^2\hbar^2 n^2}{8ma^2}$$

where in order to distinguish between various energies, a suffix n is used with E. The corresponding wave function is

$$\psi_n(x) = B \cos\left(\frac{n\pi}{2a}x\right) \qquad n = 1, 3, 5, \ldots$$

For this wave function, we have

$$\int_{-a}^{a} \psi_n^*(x)\psi_n(x)\ \mathrm{d}x = \int_{-a}^{a} B^2 \cos^2\left(\frac{n\pi}{2a}x\right)\ \mathrm{d}x$$

$$= \frac{B^2}{2}\int_{-a}^{a}\left[1 + \cos\left(\frac{n\pi}{a}x\right)\right]\mathrm{d}x = B^2\ a$$

Thus, the normalization of this wave function gives $B^2\,a = 1$. Therefore, $B = 1/\sqrt{a}$ and the normalized wave function is

$$\psi_n(x) = \frac{1}{\sqrt{a}}\cos\left(\frac{n\pi}{2a}x\right) \qquad n = 1, 3, 5, \ldots$$

(iii) The energy of the particle confined in the box is

$$E_n = \frac{\pi^2\hbar^2 n^2}{8ma^2} \qquad n = 1, 2, 3, 4, \ldots$$

and the corresponding wave functions are

$$\psi_n(x) = \begin{cases} \frac{1}{\sqrt{a}}\sin\left(\frac{n\pi}{2a}x\right) & n = 2, 4, 6, \ldots \\ \frac{1}{\sqrt{a}}\cos\left(\frac{n\pi}{2a}x\right) & n = 1, 3, 5, \ldots \end{cases}$$

2. Square well potential with finite walls

Let us consider a particle of mass m moving in one-dimensional potential field (Figure 2) expressed as

$$V(x) = \begin{cases} 0 & |x| < a \\ V_0 & |x| > a \end{cases}$$

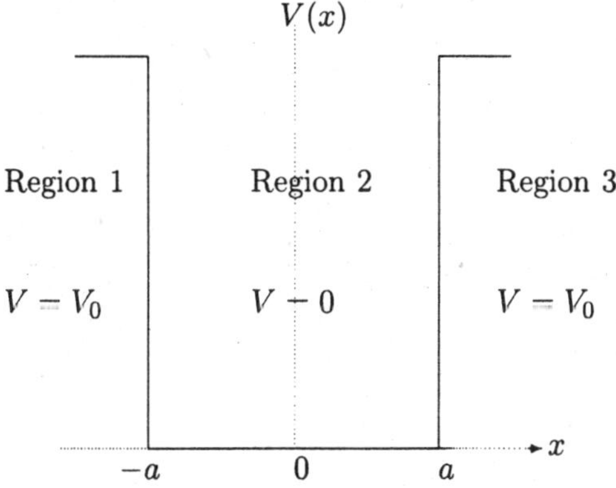

Figure 2: One-dimensional box of side $2a$ with finite height.

For convenience, the potential is taken symmetrical about the $x = 0$ line. Here, the potential inside the well is taken as zero whereas outside the well, it is taken as V_0. The time-independent, one-dimensional

Schrödinger equation for the region $|x| < a$ (Region 2) is

$$-\frac{\hbar^2}{2m}\frac{\mathrm{d}^2}{\mathrm{d}x^2}\psi(x) = E\psi(x)$$

$$\frac{\mathrm{d}^2}{\mathrm{d}x^2}\psi(x) + k^2\psi(x) = 0 \tag{4.4}$$

where $k^2 = 2mE/\hbar^2$. This equation (4.4) has a solution

$$\psi(x) = A\,\sin(kx) + B\,\cos(kx)$$

where A and B are constants.

The time-independent, one-dimensional Schrödinger equation for the region $|x| > a$ (Regions 1 and 3) is

$$-\frac{\hbar^2}{2m}\frac{\mathrm{d}^2}{\mathrm{d}x^2}\psi(x) + V_0\psi(x) = E\psi(x)$$

For the potential energy V_0, there are two possibilities: (i) $E < V_0$ and (ii) $E > V_0$. We shall consider both the cases, separately.

(i) For $E < V_0$, the Schrödinger equation can be written as

$$\frac{\mathrm{d}^2}{\mathrm{d}x^2}\psi(x) - \alpha^2\psi(x) = 0 \tag{4.5}$$

where $\alpha^2 = 2m(V_0 - E)/\hbar^2$. This equation (4.5) has a solution

$$\psi(x) = C\,\mathrm{e}^{-\alpha x} + D\,\mathrm{e}^{\alpha x} \tag{4.6}$$

where α is positive value, C and D are constants. In region 3, equation (4.6) says that for non-zero value of D, the wave function $\psi(x) \to \infty$ as $x \to \infty$, which is physically unacceptable. On the other side, in region 1, equation (4.6) says that for non-zero value of C, the wave function $\psi(x) \to \infty$ as $x \to (-\infty)$, which also is physically unacceptable. Hence, the physically acceptable solutions in the three regions are

$$\psi(x) = \begin{cases} D\,\mathrm{e}^{\alpha x} & x < -a \quad \text{Region 1} \\ \\ A\,\sin(kx) + B\,\cos(kx) & -a < x < a \quad \text{Region 2} \\ \\ C\,\mathrm{e}^{-\alpha x} & x > a \quad \text{Region 3} \end{cases}$$

The wave function in region 2 is neither symmetric nor antisymmetric about the origin $x = 0$, as it is a combination of symmetric as well as antisymmetric wave functions. Thus, at a time, we can match either the symmetric or antisymmetric wave function of region 2 to the wave functions of the regions 1 and 3. The probability of finding the particle in region 1 is $D^2 \exp(2\alpha x)$ and it decreases exponentially, as x assumes negative values. The probability of finding the particle in region 3 is $C^2 \exp(-2\alpha x)$ and it also decreases exponentially, as x assumes positive values. The rates for these exponential decays increase with the increase of the value of V_0.

The solution in region 2 can be divided into two parts: (a) symmetric function and (b) antisymmetric function.

(a) Symmetric function in region 2: Let us consider the symmetric function in region 2. It corresponds to $A = 0$. The boundary conditions, that wave function and its derivative is continuous, at $x = \pm a$ gives

$$D\,\mathrm{e}^{-\alpha a} = B\cos(ka) \qquad \text{and} \qquad D\,\alpha\,\mathrm{e}^{-\alpha a} = B\,k\,\sin(ka) \tag{4.7}$$

$$C\,\mathrm{e}^{-\alpha a} = B\cos(ka) \qquad \text{and} \qquad C\,\alpha\,\mathrm{e}^{-\alpha a} = B\,k\,\sin(ka) \tag{4.8}$$

Comparison of equations (4.7) and (4.8) obviously shows that $C = D$ and

$$ka\tan(ka) = \alpha a \tag{4.9}$$

where we have multiplied by a on both the sides.

(b) Antisymmetric function in region 2: Antisymmetric function in region 2 corresponds to $B = 0$. The boundary conditions, that wave function and its derivative is continuous, at $x = \pm a$ gives

$$D\,\mathrm{e}^{-\alpha a} = -A\sin(ka) \qquad \text{and} \qquad D\,\alpha\,\mathrm{e}^{-\alpha a} = A\,k\,\cos(ka) \tag{4.10}$$

$$C\,\mathrm{e}^{-\alpha a} = A\sin(ka) \qquad \text{and} \qquad -C\,\alpha\,\mathrm{e}^{-\alpha a} = A\,k\,\cos(ka) \tag{4.11}$$

Comparison of equations (4.10) and (4.11) obviously shows that $C = -D$ and

$$ka\cot(ka) = -\alpha a \tag{4.12}$$

where we have multiplied by a on both the sides.

The energy eigenvalues can be obtained by solving equations (4.9) and (4.12) graphically or numerically. For obtaining solutions of equations (4.9) and (4.12), let us define

$$ka = \beta \qquad \text{and} \qquad \alpha a = \gamma$$

and obtain

$$\beta^2 + \gamma^2 = k^2a^2 + \alpha^2a^2 = \frac{2mE}{\hbar^2}a^2 + \frac{2m(V_0 - E)}{\hbar^2}a^2 = \frac{2mV_0a^2}{\hbar^2}$$

This is an equation of a circle in the $\beta\gamma$-plane with radius $(2mV_0a^2/\hbar^2)^{1/2}$.

For symmetric function in region 2, equation (4.9) can be written as

$$\beta \tan\beta = \gamma \tag{4.13}$$

To get the solution, the curve (dotted curve) representing the variation of $\beta \tan\beta$ versus β is plotted along with the circles (continuous curves) of radii $(2mV_0a^2/\hbar^2)^{1/2}$ for different values of V_0a^2 (Figure 3).

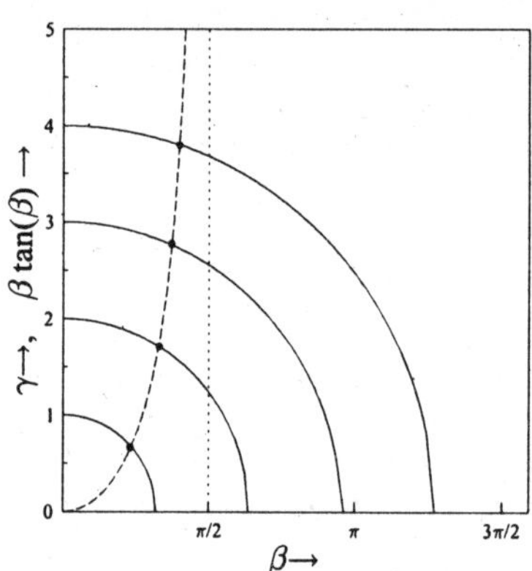

Figure 3: Variation of $\beta\tan\beta$ versus β (dotted curve) and plot of the circles (continuous curves) of radii $(2mV_0a^2/\hbar^2)^{1/2}$ for different values of V_0a^2.

Since β and γ both are positive, the interaction of two curves in the first quadrant gives the energy values for n = 1, 2, 3, ... The intersection of the circle with the curves of equation (4.13) gives β_1, β_2, ... Then

$$\beta^2 = k^2a^2 = \frac{2mEa^2}{\hbar^2} \qquad \text{or} \qquad E_n = \frac{\hbar^2\beta_n^2}{2ma^2}$$

The number of bound states of the particle depends on the height and width of the potential through V_0a^2. Since the wave function outside the well is non-zero, there is finite probability of finding the particle outside. That is, the particle leaks or tunnels into the classically forbidden region.

For the antisymmetric function in region 2, equation (4.12) can be written as

$$-\beta \cot\beta = \gamma \tag{4.14}$$

To get the solution, the curve (dotted curve) representing the variation of $-\beta \cot\beta$ versus β is plotted along with circles (continuous curves) of radii $(2mV_0a^2/\hbar^2)^{1/2}$ for different values of V_0a^2 (Figure 4).

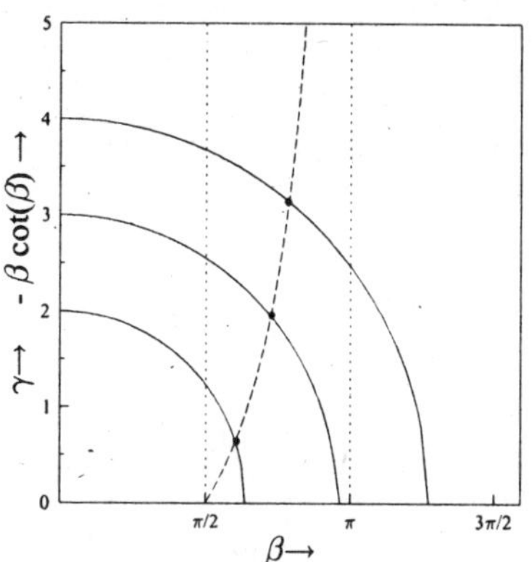

Figure 4: Variation of $(-\beta \cot\beta)$ versus β (dotted curve) and plot of the circles (continuous curves) of radii $(2mV_0a^2/\hbar^2)^{1/2}$ for different values of V_0a^2.

Since β and γ both are positive, the interaction of two curves in the first quadrant gives the energy values for n = 1, 2, 3, ... The intersection of the circles with the curve of equation (4.14) gives β_1, β_2, ... Then

$$\beta^2 = k^2a^2 = \frac{2mEa^2}{\hbar^2} \qquad \text{or} \qquad E_n = \frac{\hbar^2\beta_n^2}{2ma^2}$$

The number of bound states of the particle depends on the height and width of the potential through V_0a^2. Since the wave function outside the well is non-zero, there is finite probability of finding the particle outside also. That is, the particle leaks or tunnels into the classically forbidden region.

(ii) For $E > V_0$, the Schrödinger equation can be written as

$$\frac{d^2}{dx^2}\psi(x) + k'^2\psi(x) = 0 \tag{4.15}$$

where $k'^2 = 2m(E - V_0)/\hbar^2$. This equation (4.15) has a solution

$$\psi(x) = A' \sin(k'x) + B' \cos(k'x)$$

showing that the wave function $\psi(x)$ is sinusoidal in the regions 1 and 3 also. Thus, the probability density is distributed over all space, comprising regions 1, 2 and 3. Consequently, the particle is not bound.

3. Square potential barrier with a finite height

Let us consider a particle of mass m moving along x-direction from left strikes a potential barrier (Figure 5) expressed as

$$V(x) = \begin{cases} 0 & x < 0 \quad \text{Region 1} \\ V_0 & 0 < x < a \quad \text{Region 2} \\ 0 & x > a \quad \text{Region 3} \end{cases}$$

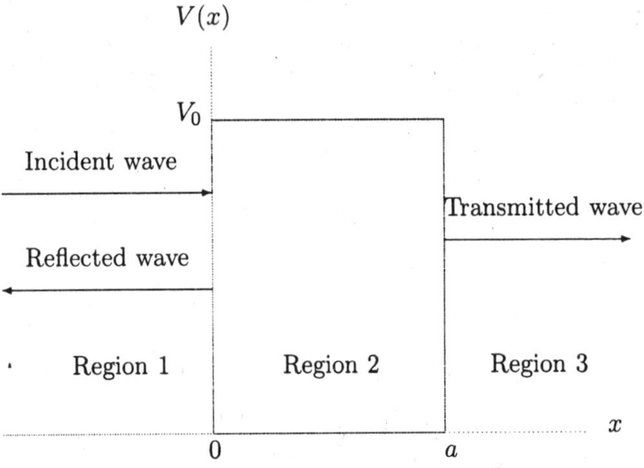

Figure 5: One-dimensional potential barrier with a constant height V_0.

Classically, when energy E of a particle is smaller than V_0, it is always reflected back whereas when E is larger than V_0, it is always transmitted through the barrier. Quantum mechanically, it can however be found that even when $E < V_0$, the particle can transmit through the barrier. This phenomenon is known as the quantum mechanical tunneling and it is possible due to the wave nature of a particle. Let us look into the situation quantum mechanically for the two cases: (i) $E < V_0$ and (ii) $E > V_0$.

Case (i): $E < V_0$

The Schrödinger equation for the particle in the three regions can be expressed as

$$\text{Region 1} \qquad \frac{d^2}{dx^2}\psi(x) + k^2\psi(x) = 0$$

Region 2 $$\frac{d^2}{dx^2}\psi(x) - \alpha^2\psi(x) = 0$$

Region 3 $$\frac{d^2}{dx^2}\psi(x) + k^2\psi(x) = 0$$

where $k^2 = 2mE/\hbar^2$ and $\alpha^2 = 2m(V_0 - E)/\hbar^2$. Solutions for these equations are

Region 1 $$\psi_1(x) = A\ e^{ikx} + B\ e^{-ikx}$$

Region 2 $$\psi_2(x) = C\ e^{\alpha x} + D\ e^{-\alpha x}$$

Region 3 $$\psi_1(x) = F\ e^{ikx} + G\ e^{-ikx}$$

where A, B, C, D, F and G are constants. The expression $\exp(ikx)$ represents a plane wave moving from left to right whereas the expression $\exp(-ikx)$ represents a plane wave moving from right to left. In region 3, we cannot expect any wave moving from right to left and therefore, the constant G is zero. In the region 1, we have waves moving from left to right (incident wave) as well as moving from right to left (reflected wave). The probability density of the incident, reflected and transmitted waves are $|A\ \exp(ikx)|^2 (\equiv |A|^2)$, $|B\ \exp(-ikx)|^2 (\equiv |B|^2)$, $|F\ \exp(ikx)|^2 (\equiv |F|^2)$, respectively. Hence, we have the reflection coefficient $R = |B|^2/|A|^2$ and the transmission coefficient $T = |F|^2/|A|^2$. The two are connected through the relation $R + T = 1$.

Considering the condition of continuity of wave function and its derivative at the boundaries $x = 0$ and $x = a$, we have

$$A + B = C + D \tag{4.16}$$

$$ikA - ikB = \alpha C - \alpha D \tag{4.17}$$

$$Ce^{\alpha a} + De^{-\alpha a} = Fe^{ika} \tag{4.18}$$

$$C\alpha e^{\alpha a} - D\alpha e^{-\alpha a} = iFke^{ika} \tag{4.19}$$

Multiplying equation(4.18) by α and adding the product to equation(4.19), we get

$$2\alpha C\ e^{\alpha a} = F(\alpha + ik)\ e^{ika}$$

$$C = \frac{F}{2\alpha}(\alpha + ik)\ e^{ika - \alpha a} \tag{4.20}$$

Multiplying equation (4.18) by α and subtracting equation (4.19) from the product, we get

$$2\alpha D\ e^{-\alpha a} = F(\alpha - ik)\ e^{ika}$$

$$D = \frac{F}{2\alpha}(\alpha - ik)\ e^{ika + \alpha a} \tag{4.21}$$

In the similar manner, from equations (4.16) and (4.17), we have

$$A = \frac{C}{2}\left(1 + \frac{\alpha}{ik}\right) + \frac{D}{2}\left(1 - \frac{\alpha}{ik}\right) \tag{4.22}$$

$$B = \frac{C}{2}\left(1 - \frac{\alpha}{ik}\right) + \frac{D}{2}\left(1 + \frac{\alpha}{ik}\right) \tag{4.23}$$

Using equations (4.20) and (4.21) in (4.22), we get

$$\begin{aligned} A &= \frac{F}{4ik\alpha}(\alpha + ik)^2 \, e^{ika-\alpha a} - \frac{F}{4ik\alpha}(\alpha - ik)^2 \, e^{ika+\alpha a} \\ &= -\frac{F}{4ik\alpha} e^{ika}\left[(\alpha^2 - k^2)(e^{\alpha a} - e^{-\alpha a}) - 2ik\alpha(e^{\alpha a} + e^{-\alpha a})\right] \\ &= -\frac{F}{2ik\alpha} e^{ika}\left[(\alpha^2 - k^2) \sinh \alpha a - 2ik \cosh \alpha a\right] \end{aligned} \tag{4.24}$$

Thus,

$$\frac{F}{A} = -\frac{2ik\alpha \exp(-ika)}{(\alpha^2 - k^2) \sinh \alpha a - 2ik \cosh \alpha a}$$

Hence, the transmission coefficient T is

$$T = \left|\frac{F}{A}\right|^2 = \frac{4k^2\alpha^2}{(\alpha^2 - k^2)^2 \sinh^2 \alpha a + 4k^2\alpha^2 \cosh^2 \alpha a}$$

Using equations (4.20) and (4.21) in (4.23), we get

$$\begin{aligned} B &= -\frac{F}{4ik\alpha}(\alpha^2 + k^2) \, e^{ika-\alpha a} + \frac{F}{4ik\alpha}(\alpha^2 + k^2) \, e^{ika+\alpha a} \\ &= \frac{F}{4ik\alpha} e^{ika}\left[-(\alpha^2 + k^2) \, e^{-\alpha a} + (\alpha^2 + k^2) \, e^{\alpha a}\right] \\ &= \frac{F}{2ik\alpha} e^{ika} \, (\alpha^2 + k^2) \sinh \alpha a \end{aligned} \tag{4.25}$$

Dividing equation (4.25) by (4.24), we have

$$\frac{B}{A} = -\frac{(\alpha^2 + k^2) \sinh \alpha a}{(\alpha^2 - k^2) \sinh \alpha a - 2ik\alpha \cosh \alpha a}$$

Hence, the reflection coefficient R is

$$R = \left|\frac{B}{A}\right|^2 = \frac{(\alpha^2 + k^2)^2 \sinh^2 \alpha a}{(\alpha^2 - k^2)^2 \sinh^2 \alpha a + 4k^2\alpha^2 \cosh^2 \alpha a}$$

Hence,

$$R+T = \frac{4\alpha^2k^2 + (\alpha^2+k^2)^2 \sinh^2 \alpha a}{(\alpha^2-k^2)^2 \sinh^2 \alpha a + 4k^2\alpha^2 \cosh^2 \alpha a}$$

$$= \frac{4\alpha^2k^2 + (\alpha^2-k^2)^2 \sinh^2 \alpha a + 4\alpha^2k^2\sinh^2 \alpha a}{(\alpha^2-k^2)^2 \sinh^2 \alpha a + 4k^2\alpha^2 \cosh^2 \alpha a}$$

$$= \frac{(\alpha^2-k^2)^2 \sinh^2 \alpha a + 4\alpha^2k^2 \cosh^2 \alpha a}{(\alpha^2-k^2)^2 \sinh^2 \alpha a + 4k^2\alpha^2 \cosh^2 \alpha a} = 1$$

For a broad and high barrier, we have $\alpha a >> 1$ and

$$\sinh \alpha a = \cosh \alpha a \approx \frac{1}{2} e^{\alpha a}$$

Thus, the transmission coefficient T is

$$T = \frac{16k^2\alpha^2}{(\alpha^2-k^2)^2 \exp(2\alpha a) + 4k^2\alpha^2 \exp(2\alpha a)}$$

$$= \frac{16k^2\alpha^2 \exp(-2\alpha a)}{(\alpha^2-k^2)^2 + 4k^2\alpha^2} = \frac{16k^2\alpha^2 \exp(-2\alpha a)}{(\alpha^2+k^2)^2}$$

Substitution of the values of k^2 and α^2 gives

$$T = \frac{16E(V_0-E)}{V_o^2} \, e^{-2\alpha a}$$

Case (ii): $E > V_0$

The Schrödinger equation for the particle in three regions can be expressed as

Region 1 $$\frac{d^2}{dx^2}\psi(x) + k^2\psi(x) = 0$$

Region 2 $$\frac{d^2}{dx^2}\psi(x) + \beta^2\psi(x) = 0$$

Region 3 $$\frac{d^2}{dx^2}\psi(x) + k^2\psi(x) = 0$$

where $k^2 = 2mE/\hbar^2$ and $\beta^2 = 2m(E-V_0)/\hbar^2$. Solutions for these equations are

Region 1 $\psi_1(x) = A\, e^{ikx} + B\, e^{-ikx}$

Region 2 $\psi_2(x) = C\, e^{i\beta x} + D\, e^{-i\beta x}$

Region 3 $\psi_1(x) = F\, e^{ikx} + G\, e^{-ikx}$

where A, B, C, D, F and G are constants. The expression $\exp(ikx)$ represents a plane wave moving from left to right whereas the expression $\exp(-ikx)$ represents a plane wave moving from right to left. In region 3, we cannot expect any wave moving from right to left and therefore, the constant G is zero. In region 1, we have waves moving from left to right (incident wave) as well as moving from right to left (reflected wave). The probability density of the incident, reflected and transmitted waves are $|A|^2$, $|B|^2$, $|F|^2$, respectively. Hence, we have the reflection coefficient $R= |B|^2/|A|^2$ and the transmission coefficient $T= |F|^2/|A|^2$. The two are connected through the relation $R+T=1$.

Considering the condition of continuity of wave function and its derivative at the boundaries $x=0$ and $x=a$, we have

$$A+B=C+D \tag{4.26}$$

$$kA-kB=\beta C-\beta D \tag{4.27}$$

$$Ce^{i\beta a}+De^{-i\beta a}=Fe^{ika} \tag{4.28}$$

$$C\beta e^{i\beta a}-D\beta e^{-i\beta a}=Fke^{ika} \tag{4.29}$$

Multiplying equation(4.28) by β and adding the product to equation(4.29), we get

$$2\beta C\ e^{i\beta a}=F(\beta+k)\ e^{ika}$$

$$C=\frac{F}{2\beta}(\beta+k)\ e^{ika-i\beta a} \tag{4.30}$$

Multiplying equation (4.28) by β and subtracting equation (4.29) from the product, we get

$$2\beta D\ e^{-i\beta a}=F(\beta-k)\ e^{ika}$$

$$D=\frac{F}{2\beta}(\beta-k)\ e^{ika+i\beta a} \tag{4.31}$$

In the similar manner, from equations (4.26) and (4.27), we have

$$A=\frac{C}{2}\left(1+\frac{\beta}{k}\right)+\frac{D}{2}\left(1-\frac{\beta}{k}\right) \tag{4.32}$$

$$B=\frac{C}{2}\left(1-\frac{\beta}{k}\right)+\frac{D}{2}\left(1+\frac{\beta}{k}\right) \tag{4.33}$$

Using equations (4.30) and (4.31) in (4.32), we get

$$\begin{aligned}A&=\frac{F}{4k\beta}(\beta+k)^2\ e^{ika-i\beta a}-\frac{F}{4k\beta}(\beta-k)^2\ e^{ika+i\beta a}\\&=-\frac{F}{4k\beta}e^{ika}\left[(\beta-k)^2\ e^{i\beta a}-(\beta+k)^2\ e^{-i\beta a}\right]\\&=-\frac{F}{2k\beta}e^{ika}\left[i(\beta^2+k^2)\ \sin\beta a-2\beta k\ \cos\beta a\right]\end{aligned} \tag{4.34}$$

Thus,

$$\frac{F}{A} = \frac{2k\beta \exp(-ika)}{2\beta k \cos \beta a - i(\beta^2 + k^2) \sin \beta a}$$

Hence, the transmission coefficient T is

$$T = \left|\frac{F}{A}\right|^2 = \frac{4k^2\beta^2}{4\beta^2 k^2 \cos^2 \beta a + (\beta^2 + k^2)^2 \sin^2 \beta a}$$

It shows that even $E > V_0$, the transmission probability is not one. Using equations (4.30) and (4.31) in (4.33), we get

$$\begin{aligned} B &= -\frac{F}{4k\beta}(\beta^2 - k^2)\, e^{ika - i\beta a} + \frac{F}{4k\beta}(\beta^2 - k^2)\, e^{ika + i\beta a} \\ &= \frac{F}{4k\beta} e^{ika}\left[(\beta^2 - k^2)\, e^{-i\beta a} - (\beta^2 - k^2)\, e^{i\beta a}\right] \\ &= \frac{F}{2k\beta} e^{ika}(\beta^2 - k^2)\, i \sin\beta a \end{aligned} \qquad (4.35)$$

Dividing equation (4.35) by (4.34), we have

$$\frac{B}{A} = -\frac{(\beta^2 - k^2)\, i \sin \beta a}{i\,(\beta^2 + k^2) \sin \beta a - 2\beta k \cos \beta a}$$

Hence, the reflection coefficient R is

$$R = \left|\frac{B}{A}\right|^2 = \frac{(\beta^2 - k^2)^2 \sin^2 \beta a}{(\beta^2 + k^2)^2 \sin^2 \beta a + 4k^2\beta^2 \cos^2 \beta a}$$

It shows that even $E > V_0$, there is a finite probability of reflection of the particle. Thus,

$$R + T = \frac{4\beta^2 k^2 + (\beta^2 - k^2)^2 \sin^2 \beta a}{(\beta^2 + k^2)^2 \sin^2 \beta a + 4k^2\beta^2 \cos^2 \beta a}$$

$$- \frac{4\beta^2 k^2 + (\beta^2 - k^2)^2 \sin^2 \beta a}{(\beta^2 - k^2)^2 \sin^2 \beta a + 4k^2\beta^2} = 1$$

4. Emission of alpha-particle from a nucleus

Inside a nucleus, an alpha-particle is held by the strong attractive short-range nuclear forces. However, when an alpha-particle is outside the nucleus, it is strongly repelled by a strong Coulomb force applied by the residual nucleus, as both of them carry positive charges. Thus, when an alpha-particles is moved from inside to the outside a nucleus, an attractive potential changes to a repulsive one. Qualitative variation of potential of alpha-particle as a function of its distance from the nucleus is shown in Figure 6.

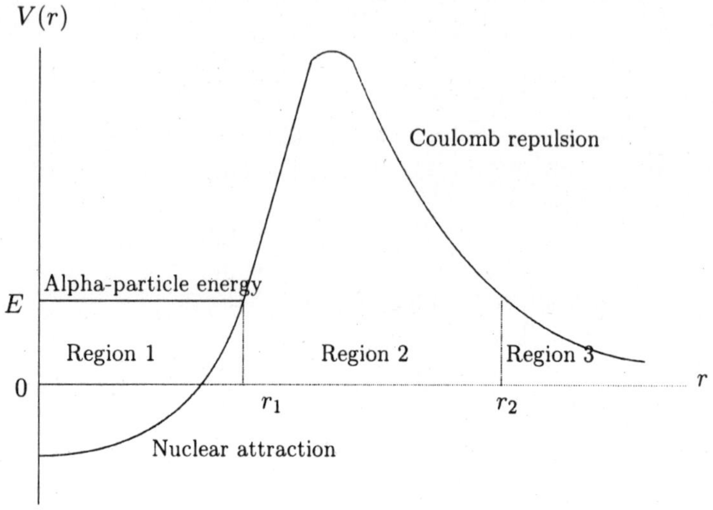

Figure 6: Variation of potential of alpha-particle as a function of its distance from the nucleus.

An alpha particle inside a nucleus moves with an energy E. Since this energy E is less than the height of the potential barrier, the classical mechanics says that the alpha particle cannot go outside the nucleus. However, as discussed in the preceding section, the quantum mechanics provides a possibility of quantum mechanical tunneling. Though the height of the potential barrier in this case is not constant, but still we can use the expression for the transmission coefficient derived in the preceding section. Thus, we have a transmission probability as

$$T = \frac{16E(V_0 - E)}{V_0^2} \exp\Big[- \sqrt{\frac{8ma^2}{\hbar^2}(V_0 - E)}\Big]$$

where V_0 is constant height of the potential barrier of width a. For $V_0 >> E$ (say, $V_0 \approx 16E$), we may have

$$T \approx \exp\Big[- \sqrt{\frac{8ma^2}{\hbar^2}(V_0 - E)}\Big]$$

Suppose charge of the residual nucleus is Ze, the repulsive potential of alpha particle at a distance r out side the nucleus is $2Ze^2/r$. The distance r_1 of the nuclear-alpha particle is approximately equal to the nuclear radius. If r_2 is the point where the alpha particle energy E is equal to the Coulomb potential, we have

$$E = \frac{2Ze^2}{r_2} \qquad \text{or} \qquad r_2 = \frac{2Ze^2}{E}$$

As the real barrier is not a square, more rigorously, we get

$$T \approx \exp\Big[- \int_{r_1}^{r_2} \sqrt{\frac{8m}{\hbar^2}\Big(\frac{2Ze^2}{r} - E\Big)}\, \mathrm{d}r \Big]$$

It is estimated that an alpha particle within a nucleus moves with a velocity of about 10^{-7} m/s. The radius of a heavy nucleus like uranium, where emission of alpha particles takes place, is about 10^{-14} m. Thus, an alpha particle strikes the wall of the barrier at the rate of about 10^{21} times per second. The probability that the alpha particle can penetrate the barrier is given by the transmission coefficient T. Hence, the probability that the alpha particle comes out of the nucleus in one second is $P \approx 10^{21}\, T$. Following the uncertainty principle, the mean life time τ is

$$\tau = \frac{1}{P} = \frac{1}{10^{21}\, T}$$

The radioactive decay constant λ is given as

$$\lambda = \frac{1}{\tau} \approx 10^{21} \exp\Big[- \sqrt{\frac{8ma^2}{\hbar^2}(V_0 - E)} \Big]$$

5. Schrödinger equation for a periodic potential

Let us now consider a Schrödinger equation where potential is a periodic function of distance. One-dimensional periodic potential is expressed as

$$V(x + n\, d) = V(x) \qquad n = 0, 1, 2, \ldots \tag{4.36}$$

where d is the period of the potential. Such periodic potential is found in a one-dimensional metal crystal consisting of a number of stationary positive ions with a separation d (Figure 7).

Let us consider a one-dimensional crystal lattice having N ions in the form of a close loop, so that the first ion is the same the $(N+1)$-th ion. The Schrödinger equation at the point x is

$$\frac{\mathrm{d}^2}{\mathrm{d}x^2}\psi(x) + \frac{2m}{\hbar^2}\Big[E - V(x)\Big]\psi(x) = 0 \tag{4.37}$$

and the Schrödinger equation at the point $(x+d)$ is

$$\frac{d^2}{dx^2}\psi(x+d) + \frac{2m}{\hbar^2}\Big[E - V(x+d)\Big]\psi(x+d) = 0 \tag{4.38}$$

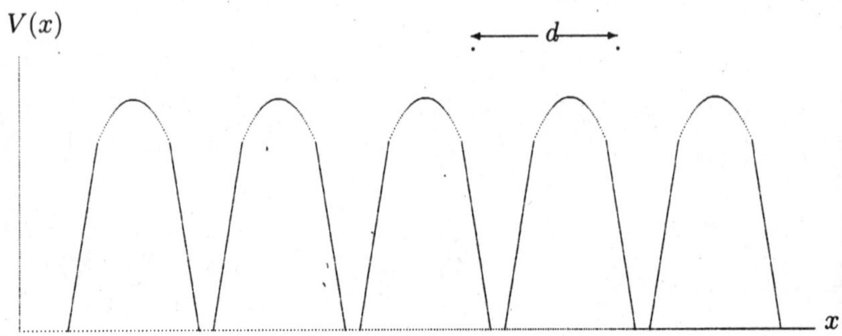

Figure 7: One-dimensional crystal lattice along with the periodic potential.

Using equation (4.36) in (4.38), we have

$$\frac{d^2}{dx^2}\psi(x+d) + \frac{2m}{\hbar^2}\Big[E - V(x)\Big]\psi(x+d) = 0 \tag{4.39}$$

Comparison of equation (4.37) and (4.39) shows that the $\psi(x)$ and $\psi(x+d)$ satisfy the same equation, and therefore they should differ by a multiplicative constant, say α. Let

$$\psi(x+d) = \alpha\ \psi(x) \tag{4.40}$$

Similarly, we have

$$\psi(x+2d) = \alpha\ \psi(x+d) = \alpha^2\ \psi(x)$$

Repetition of this process gives

$$\psi(x+Nd) = \alpha^N\ \psi(x) \tag{4.41}$$

As the lattice is assumed in the form of a ring, we have

$$\psi(x+Nd) = \psi(x)$$

Using equation (4.41) here, we have

$$\alpha^N\ \psi(x) = \psi(x)$$

so that

$$\alpha^N = 1 \qquad \text{or} \qquad \alpha^N = e^{2\pi i n} \qquad \text{for} \qquad n = 0, 1, \ldots, (N-1)$$

Therefore,

$$\alpha = e^{2\pi i n/N} \qquad \text{for} \qquad n = 0, 1, \ldots, (N-1)$$

Now, $\psi(x)$ can be expressed as

$$\psi(x) = e^{ikx}\, u(x) \tag{4.42}$$

where

$$u(x+d) = u(x) \qquad \text{and} \qquad k = \frac{2\pi n}{Nd} \tag{4.43}$$

for $n = 0, \pm 1, \pm 2, \ldots, \pm(N-1)$. A justification for equation (4.42) can easily be found by replacing x by $(x+d)$ in equation (4.42), we have

$$\begin{aligned}\psi(x+d) &= e^{ik(x+d)}\, u(x+d) = e^{ikx}\, e^{ikd}\, u(x)\\ &= e^{ikd}\psi(x) = e^{2\pi i n/N}\, \psi(x) = \alpha\, \psi(x)\end{aligned}$$

which is equation (4.40). Equation (4.42) with the condition (4.43) is known as the Bloch theorem. That is, the solution of Schrödinger equation for a periodic potential has a form of a plane wave modulated by a function having the periodicity of the lattice. Functions of the form (4.42) are sometimes referred to as the Bloch functions.

6. Kronig-Penney square-well periodic potential

In the preceding section we considered a one-dimensional metal crystal having a number of positive ions providing a periodic potential shown in Figure (7). One approximation of such periodic potential is as shown in the following Figure 8. The width of each well is a and that of each barrier is b, so that $d = a + b$. Then the periodic potential is

$$V[x + n(a+b)] = V(x) \qquad\qquad n = 0, 1, 2, \ldots$$

We have already seen that for a periodic potential, the one-electron wave functions are the Bloch functions

$$\psi(x) = e^{ikx}\, u(x)$$

where $u(x)$ has a periodicity of the lattice, *i.e.*, $u(x+d) = u(x)$. In the region $0 < x < a$, we have $V(x) = 0$ and the Schrödinger equation is

$$\frac{d^2}{dx^2}\psi_1(x) + k_1^2\psi_1(x) = 0 \tag{4.44}$$

where $k_1^2 = 2mE/\hbar^2$. In the region $-b < x < 0$, we have $V(x) = V$ and the Schrödinger equation is

$$\frac{d^2}{dx^2}\psi_2(x) - k_2^2\psi_2(x) = 0 \tag{4.45}$$

where $k_2^2 = 2m(V-E)/\hbar^2$ with $V > E$.

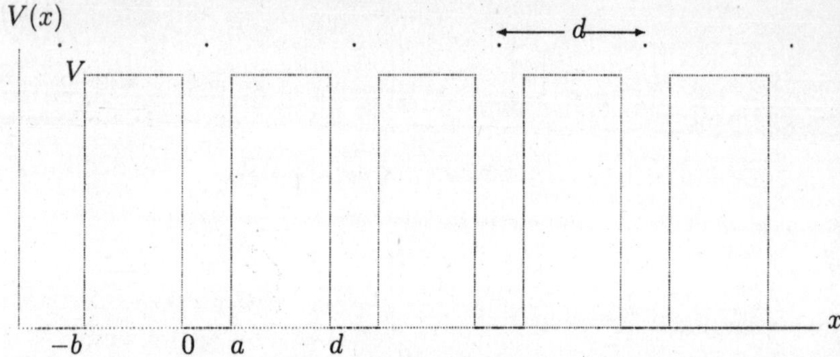

Figure 8: Kronig-Penney periodic potential.

According to the Bloch theorem, the wave functions $\psi_1(x)$ and $\psi_2(x)$ must be of the form

$$\psi_1(x) = e^{ikx}\,u_1(x) \qquad 0 < x < a \tag{4.46}$$

$$\psi_2(x) = e^{ikx}\,u_2(x) \qquad -b < x < 0 \tag{4.47}$$

Differentiating equation (4.46) with respect to x, we have

$$\frac{d}{dx}\psi_1(x) = e^{ikx}\,\frac{d}{dx}u_1(x) + ike^{ikx}\,u_1(x)$$

Again differentiating this equation with respect to x, we have

$$\frac{d^2}{dx^2}\psi_1(x) = e^{ikx}\,\frac{d^2}{dx^2}u_1(x) + 2ik\,e^{ikx}\,\frac{d}{dx}u_1(x) - k^2u_1(x)$$

Using these expressions in equation (4.44), we get

$$\frac{d^2}{dx^2}u_1(x) + 2ik\,\frac{d}{dx}u_1(x) + (k_1^2 - k^2)u_1(x) = 0 \tag{4.48}$$

Differentiating equation (4.47) with respect to x, we have

$$\frac{d}{dx}\psi_2(x) = e^{ikx}\,\frac{d}{dx}u_2(x) + ike^{ikx}\,u_2(x)$$

Again differentiating this equation with respect to x, we have

$$\frac{d^2}{dx^2}\psi_2(x) = e^{ikx}\,\frac{d^2}{dx^2}u_2(x) + 2ik\,e^{ikx}\,\frac{d}{dx}u_2(x) - k^2u_2(x)$$

Using these expressions in equation (4.45), we get

$$\frac{d^2}{dx^2}u_2(x) + 2ik\,\frac{d}{dx}u_2(x) - (k_2^2 + k^2)u_2(x) = 0 \tag{4.49}$$

For the equation (4.48), we assume a solution of the form

$$u_1(x) = e^{mx} \tag{4.50}$$

Using equation (4.50) in (4.48), we get

$$m^2\,e^{mx} + 2ikm\,e^{mx} + (k_1^2 - k^2)e^{mx} = 0$$

$$m^2 + 2ikm + (k_1^2 - k^2) = 0$$

This equation gives

$$m = i(k_1 - k) \qquad \text{and} \qquad m = -i(k_1 + k)$$

Thus, the solution of equation (4.48) is

$$u_1(x) = A\,\exp[i(k_1 - k)x] + B\,\exp[-i(k_1 + k)x]$$

For equation (4.49), we assume a solution of the form

$$u_2(x) = e^{mx} \tag{4.51}$$

Using equation (4.51) in (4.49), we get

$$m^2\,e^{mx} + 2ikm\,e^{mx} - (k_1^2 + k^2)e^{mx} = 0$$

$$m^2 + 2ikm - (k_1^2 + k^2) = 0$$

This equation gives

$$m = k_2 - ik \qquad \text{and} \qquad m = -(k_2 + ik)$$

Thus, the solution of equation (4.49) is

$$u_2(x) = C \exp[(k_2 - ik)x] + D \exp[-(k_2 + ik)x]$$

The wave functions and their derivatives must be continuous at $x = 0$. That is

$$u_1(x)\Big|_{x=0} = u_2(x)\Big|_{x=0} \qquad \text{and} \qquad \frac{du_1}{dx}\Big|_{x=0} = \frac{du_2}{dx}\Big|_{x=0}$$

These conditions give

$$A + B = C + D \tag{4.52}$$

and

$$Ai(k_1 - k) - Bi(k_1 + k) = C(k_2 - ik) - D(k_2 + ik) \tag{4.53}$$

As the potential is periodic, the value of the wave function as $x = a$ must be the same as at $x = -b$. That is

$$u_1(x)\Big|_{x=a} = u_2(x)\Big|_{x=-b} \qquad \text{and} \qquad \frac{du_1}{dx}\Big|_{x=a} = \frac{du_2}{dx}\Big|_{x=-b}$$

These conditions give

$$\begin{aligned} A \exp[i(k_1 - k)a] + B \exp[-i(k_1 + k)a] &= \\ C \exp[-(k_2 - ik)b] &+ D \exp[(k_2 + ik)b] \end{aligned} \tag{4.54}$$

and

$$\begin{aligned} Ai(k_1 - k) \exp[i(k_1 - k)a] - Bi(k_1 + k) \exp[-i(k_1 + k)a] &= \\ C(k_2 - ik) \exp[-(k_2 - ik)b] - D(k_2 + ik) &\exp[(k_2 + ik)b] \end{aligned} \tag{4.55}$$

For a nontrivial solution of equations (4.52), (4.53), (4.54) and (4.55), the determinant of the coefficients A, B, C and D should vanish.

$$\begin{vmatrix} 1 & i(k_1 - k) & \exp[i(k_1 - k)a] & i(k_1 - k) \exp[i(k_1 - k)a] \\ 1 & -i(k_1 + k) & \exp[-i(k_1 + k)a] & -i(k_1 + k) \exp[-i(k_1 + k)a] \\ 1 & (k_2 - ik) & \exp[-(k_2 - ik)b] & (k_2 - ik) \exp[-(k_2 - ik)b] \\ 1 & -(k_2 + ik) & \exp[(k_2 + ik)b] & -(k_2 + ik) \exp[(k_2 + ik)b] \end{vmatrix} = 0$$

Here, we used the fact that the rows and columns of the determinant can be interchanged. This gives the relation[1]

$$\frac{k_2^2 - k_1^2}{2k_1k_2} \sinh(k_2b)\ \sin(k_1a) + \cosh(k_2b)\ \cos(k_1a) = \cos(kd) \tag{4.56}$$

[1]Process of solution of the determinant is not given here.

Since k_1 and k_2 are functions of energy, the left side is a function of energy.

The right side of equations (4.56) can have values between $+1$ and -1 whereas there is no restriction for the value of the expression on the left side. Therefore, equation (4.56) can be satisfied for those values of E for which the value of the left side lies in between $+1$ and -1. It shows that in a crystal, electrons can have certain energy values. Further, there are energies which are forbidden. Thus, in a one-dimensional crystal, the periodicity of potential along with the condition $u(x+Nd) = u(x)$ leads to the concept of the energy bands.

7. Linear harmonic oscillator: Schrödinger equation method

In science and technology, several systems can be found in which a particle of mass m executes a motion under the force F expressed by $F = -kx$, where x is the distance of the particle from its equilibrium position and k is known as the force constant. Obviously, the particle oscillates about the equilibrium position. The equation of motion can be expressed as

$$F = -kx \qquad\qquad m\ddot{x} = -kx$$

Solution of this equation is $x = A\sin(\omega t)$, where A is the amplitude of the simple harmonic equation and $\omega(\equiv \sqrt{k/m})$ is the angular frequency of motion. The potential energy V of the system is given by

$$F = -\frac{\mathrm{d}V}{\mathrm{d}x} \qquad\qquad -\frac{\mathrm{d}V}{\mathrm{d}x} = -kx$$

On integration, we get

$$V = \frac{1}{2}\,kx^2 + C$$

Considering that $V = 0$ at $x = 0$, we get the constant of integration $C = 0$ and therefore, we have

$$V = \frac{1}{2}\,kx^2 = \frac{1}{2}\,m\omega^2x^2$$

Variation of potential energy V versus x is shown in Figure 9. For this potential, the Schrödinger equation for the linear harmonic oscillator is

$$\frac{\mathrm{d}^2\psi}{\mathrm{d}x^2} + \frac{2m}{\hbar^2}\left(E - \frac{1}{2}\,m\omega^2x^2\right)\psi \tag{4.57}$$

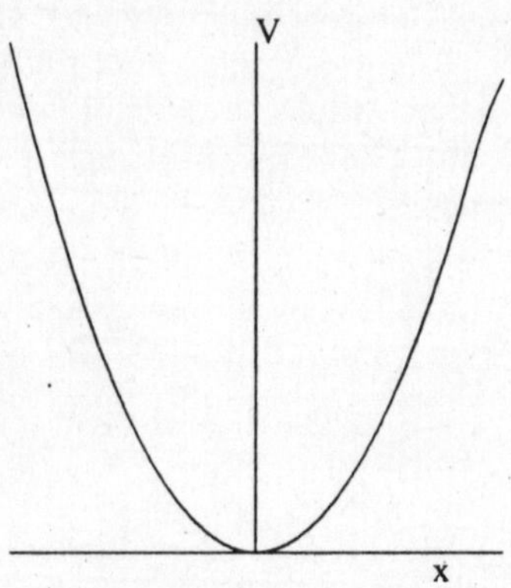

Figure 9: Variation of potential energy V versus x.

It is convenient to deal with the situation with a new variable y and a new constant λ, defined as

$$y = \left(\frac{m\omega}{\hbar}\right)^{1/2} x \qquad \text{and} \qquad \lambda = \frac{2E}{\hbar\omega}$$

After using these variable and constant in equation (4.57), we get [2]

$$\frac{d^2\psi}{dy^2} + (\lambda - y^2)\psi = 0 \tag{4.58}$$

Asymptotic solution

[2]We have

$$y = \left(\frac{m\omega}{\hbar}\right)^{1/2} x \qquad \frac{dy}{dx} = \left(\frac{m\omega}{\hbar}\right)^{1/2}$$

Thus, we have

$$\frac{d\psi}{dx} = \frac{d\psi}{dy}\frac{dy}{dx} = \left(\frac{m\omega}{\hbar}\right)^{1/2}\frac{d\psi}{dy}$$

and

$$\frac{d^2\psi}{dx^2} = \left(\frac{m\omega}{\hbar}\right)^{1/2}\frac{d^2\psi}{dy^2}\frac{dy}{dx} = \left(\frac{m\omega}{\hbar}\right)\frac{d^2\psi}{dy^2}$$

Using these relations in equation (4.57), we have

$$\left(\frac{m\omega}{\hbar}\right)\frac{d^2\psi}{dy^2} + \left(\frac{2mE}{\hbar^2} - \frac{m^2\omega^2}{\hbar^2}x^2\right)\psi = 0$$

$$\frac{d^2\psi}{dy^2} + \left(\frac{2E}{\hbar\omega} - \frac{m\omega}{\hbar}x^2\right)\psi = 0 \qquad \frac{d^2\psi}{dy^2} + (\lambda - y^2)\psi = 0$$

Let us first look for the asymptotic solution of equation (4.58) when $y \to \infty$. In the limit $y \to \infty$, we can neglect λ in comparison to y^2 and equation (4.58) becomes

$$\frac{d^2\psi}{dy^2} - y^2\psi = 0 \tag{4.59}$$

Asymptotic solutions of equation (4.59) are [3]

$$\psi = e^{\pm y^2/2} \tag{4.60}$$

Out of the two solutions, the solution $\exp(y^2/2)$ is not acceptable as it diverges as y tends to a large value.

Exact solution

Let the exact solution of equation (4.58) be written as

$$\psi = e^{-y^2/2} H(y) \tag{4.61}$$

where $H(y)$ is a function of y and $e^{-y^2/2} H(y)$ tends to zero when y tends

---

[3]Let us differentiate equation (4.60) with respect to y, we get

$$\frac{d\psi}{dy} = (\pm y)\, e^{\pm y^2/2}$$

Differentiating again with respect to y, we get

$$\frac{d^2\psi}{dy^2} = y^2\, e^{\pm y^2/2} \pm e^{\pm y^2/2} \qquad \frac{d^2\psi}{dy^2} - (y^2 \pm 1)\psi = 0$$

When y is large, we have

$$\frac{d^2\psi}{dy^2} - y^2\psi = 0$$

to infinite. Using equation (4.61) in (4.58), we get[4]

$$\frac{d^2H}{dy^2} - 2y\frac{dH}{dy} + (\lambda - 1)H = 0 \tag{4.62}$$

This is a well known Hermite equation. Let us consider a series solution of the form

$$H(y) = \sum_{n=0}^{\infty} a_n y^n \qquad a_0 \neq 0 \tag{4.63}$$

Differentiating equation (4.63) with respect to y, we have

$$\frac{dH}{dy} = \sum_{n=1}^{\infty} a_n n y^{n-1} \tag{4.64}$$

Differentiating equation (4.64) with respect to y, we have

$$\frac{d^2H}{dy^2} = \sum_{n=2}^{\infty} a_n n(n-1) y^{n-2} \tag{4.65}$$

Using equations (4.63), (4.64) and (4.65) in (4.62) we get

$$\sum_{n=2}^{\infty} a_n n(n-1)y^{n-2} - 2\sum_{n=1}^{\infty} a_n n y^n + (\lambda - 1)\sum_{n=0}^{\infty} a_n y^n = 0$$

Such equation can only be satisfied when the coefficients of various powers of y are zero separately. Equating to zero the coefficient of y^k, we have

$$(k+1)(k+2)a_{k+2} - (2k+1-\lambda)a_k = 0$$

$$a_{k+2} = \frac{2k+1-\lambda}{(k+1)(k+2)}a_k$$

---

[4]Differentiating equation (4.60) with respect to y, we get

$$\frac{d\psi}{dy} = e^{-y^2/2}\frac{dH}{dy} - ye^{-y^2/2}$$

and

$$\frac{d^2\psi}{dy^2} = e^{-y^2/2}\frac{d^2H}{dy^2} - 2ye^{-y^2/2}\frac{d\psi}{dy} + y^2e^{-y^2/2}H - e^{-y^2/2}H$$

Using this in equation (4.58), we get

$$e^{-y^2/2}\frac{d^2H}{dy^2} - 2ye^{-y^2/2}\frac{dH}{dy} + y^2e^{-y^2/2}H - e^{-y^2/2}H + (\lambda - y^2)e^{-y^2/2}H(y) = 0$$

$$\frac{d^2H}{dy^2} - 2y\frac{d\psi}{dy} + (\lambda - 1)H = 0$$

This relation allows us to calcuıate all even coefficients in terms of a_0 and all odd coefficients in terms of a_1. The series can be terminated by selecting λ so that $(2k+1-\lambda)$ vanished for $k=n$. Substituting the value of λ and $k=n$, we have

$$2n+1-\frac{2E}{\hbar\omega}=0$$

and therefore, we have

$$E_n=\left(n+\frac{1}{2}\right)\hbar\omega \qquad n=0,1,2,\ldots$$

where in order to distinguish between various values of energy E, a suffix n is used with it. In quantum theory, energy of an oscillator is

$$E_n=n\hbar\omega \qquad n=0,1,2,\ldots$$

It shows that the quantum mechanical energy of an oscillator is higher than that in the quantum theory energy by $\frac{1}{2}\hbar\omega$, which is the quantum mechanical energy of an oscillator in the lowest (ground) energy state. This energy $\frac{1}{2}\hbar\omega$, known as the zero-point energy, is a manifestation of the uncertainty principle.

Eigenfunctions

For $\lambda=2n+1$, equation (4.62) reduces to

$$\frac{\mathrm{d}^2H_n(y)}{\mathrm{d}y^2}-2y\,\frac{\mathrm{d}H_n(y)}{\mathrm{d}y}+2nH_n(y)=0 \tag{4.66}$$

Solution of this equation (4.66) is a Hermite polynomial $H_n(y)$ of degree n. The expressions for the first four Hermite polynomials are

$$H_0(y)=1$$
$$H_1(y)=2y$$
$$H_2(y)=4y^2-2$$
$$H_3(y)=8y^3-12y$$

The wave function can now be expressed as

$$\psi_n(x)=N_n\,H_n(y)\,\exp\left(-\frac{y^2}{2}\right) \qquad \text{where} \quad y^2=\frac{m\omega}{\hbar}\,x^2$$

Normalization of the wave function gives

$$\int_{-\infty}^{\infty}\psi_n^*(x)\psi_n(x)\mathrm{d}x=1$$

$$\int_{-\infty}^{\infty} |N_n|^2 \, H_n^2 \exp(-y^2)\Big(\frac{\hbar}{m\omega}\Big)^{1/2} \mathrm{d}y = 1$$

$$|N_n|^2\Big(\frac{\hbar}{m\omega}\Big)^{1/2} \int_{-\infty}^{\infty} H_n^2 \exp(-y^2)\mathrm{d}y = 1$$

$$|N_n|^2\Big(\frac{\hbar}{m\omega}\Big)^{1/2} \pi^{1/2} 2^n \, (n!) = 1$$

Thus,

$$N_n = \Big[\Big(\frac{m\omega}{\hbar\pi}\Big)^{1/2} \frac{1}{2^n(n!)}\Big]^{1/2}$$

The normalized wave function is therefore

$$\psi_n(x) = \Big[\Big(\frac{m\omega}{\hbar\pi}\Big)^{1/2} \frac{1}{2^n(n!)}\Big]^{1/2} H_n(y) \exp\Big(-\frac{m\omega x^2}{2\hbar}\Big)$$

For even values of n, the wave functions are symmetric whereas for the odd values of n, the wave functions are antisymmetric. The wave function in the ground ($n = 0$) state is

$$\psi_0(x) = \Big(\frac{m\omega}{\hbar\pi}\Big)^{1/4} \exp\Big(-\frac{m\omega x^2}{2\hbar}\Big)$$

It may also be noted that the quantum oscillator can be found outside the parabolic potential barrier as the wave function ψ does not vanish at the classical turning points. That is, the particle can penetrate the potential barrier to some extent.

8. Linear harmonic oscillator: Operator method

In the present discussion, eigenvalues and eigenfunctions for a linear harmonic oscillator would be obtained with the help of an operator method, which is based on the commutation relation

$$[x, p_x] = i\hbar$$

where x and p_x are the position and momentum operators. The Hamiltonian operator H of a linear harmonic oscillator is

$$H = \frac{p_x^2}{2m} + \frac{1}{2}\, m\omega^2 x^2$$

where $m\omega^2 x^2/2$ is the potential energy. The treatment of the problem becomes convenient when we introduce two new operators a and $a^\dagger$, defined as

$$a = \left(\frac{m\omega}{2\hbar}\right)^{1/2} x + i\left(\frac{1}{2m\hbar\omega}\right)^{1/2} p_x \tag{4.67}$$

and

$$a^\dagger = \left(\frac{m\omega}{2\hbar}\right)^{1/2} x - i\left(\frac{1}{2m\hbar\omega}\right)^{1/2} p_x \tag{4.68}$$

Here, $a^\dagger$ is Hermitian adjoint of a and vice-versa. From equations (4.67) and (4.68), we have

$$aa^\dagger = \left[\left(\frac{m\omega}{2\hbar}\right)^{1/2} x + i\left(\frac{1}{2m\hbar\omega}\right)^{1/2} p_x\right]\left[\left(\frac{m\omega}{2\hbar}\right)^{1/2} x - i\left(\frac{1}{2m\hbar\omega}\right)^{1/2} p_x\right]$$

$$= \frac{m\omega}{2\hbar}x^2 + \frac{1}{2m\hbar\omega}\, p_x^2 + \frac{i}{2\hbar}\left(p_x x - x p_x\right)$$

$$= \frac{1}{\hbar\omega}\left[\frac{1}{2}m\omega^2 x^2 + \frac{p_x^2}{2m}\right] + \frac{i}{2\hbar}(-i\,\hbar) = \frac{H}{\hbar\omega} + \frac{1}{2} \tag{4.69}$$

Similarly, we have

$$a^\dagger a = \left[\left(\frac{m\omega}{2\hbar}\right)^{1/2} x - i\left(\frac{1}{2m\hbar\omega}\right)^{1/2} p_x\right]\left[\left(\frac{m\omega}{2\hbar}\right)^{1/2} x + i\left(\frac{1}{2m\hbar\omega}\right)^{1/2} p_x\right]$$

$$= \frac{m\omega}{2\hbar}x^2 + \frac{1}{2m\hbar\omega}p_x^2 - \frac{i}{2\hbar}(p_x x - x p_x)$$

$$= \frac{1}{\hbar\omega}\left[\frac{1}{2}m\omega^2 x^2 + \frac{p_x^2}{2m}\right] - \frac{i}{2\hbar}(-i\,\hbar) = \frac{H}{\hbar\omega} - \frac{1}{2} \tag{4.70}$$

From equations (4.69) and (4.70), we get

$$aa^\dagger + a^\dagger a = \frac{2H}{\hbar\omega} \qquad\qquad H = \frac{\hbar\omega}{2}(aa^\dagger + a^\dagger a)$$

and

$$aa^\dagger - a^\dagger a = 1 \qquad\qquad [aa^\dagger] = 1$$

Therefore,

$$[a, H] = \frac{a\hbar\omega}{2}(aa^\dagger + a^\dagger a) - \frac{\hbar\omega}{2}(aa^\dagger + a^\dagger a)a$$

$$= \frac{\hbar\omega}{2}\left(aaa^\dagger + aa^\dagger a - aa^\dagger a - a^\dagger aa\right) = \frac{\hbar\omega}{2}\left(a^2 a^\dagger - a^\dagger a^2\right) = \frac{\hbar\omega}{2}\left[a^2, a^\dagger\right]$$

$$= \frac{\hbar\omega}{2}\left\{a\left[a, a^\dagger\right] + \left[a, a^\dagger\right]a\right\} = \frac{\hbar\omega}{2}\{a + a\} = \hbar\omega a$$

Thus, we have

$$aH - Ha = \hbar\omega a \tag{4.71}$$

Similarly, we have

$$\begin{aligned}
[a^\dagger, H] &= \frac{a^\dagger \hbar\omega}{2}(aa^\dagger + a^\dagger a) - \frac{\hbar\omega}{2}(aa^\dagger + a^\dagger a)a^\dagger \\
&= \frac{\hbar\omega}{2}\left(a^\dagger aa^\dagger + a^\dagger a^\dagger a - aa^\dagger a^\dagger - a^\dagger aa^\dagger\right) = \frac{\hbar\omega}{2}\left(a^{\dagger 2}a - aa^{\dagger 2}\right) \\
&= \frac{\hbar\omega}{2}\left[a^{\dagger 2}, a\right] = \frac{\hbar\omega}{2}\left\{\left[a^\dagger, a\right]a^\dagger + a^\dagger\left[a^\dagger, a\right]\right\} \\
&= \frac{\hbar\omega}{2}\left\{-a^\dagger - a^\dagger\right\} = -\hbar\omega a^\dagger
\end{aligned}$$

Hence,

$$a^\dagger H - Ha^\dagger = -\hbar\omega a^\dagger \tag{4.72}$$

To obtain the energy levels of the oscillator, we evaluate the matrix of the product $a^\dagger a$

$$\langle n|a^\dagger a|n\rangle = \left\langle n\left|\frac{H}{\hbar\omega} - \frac{1}{2}\right|n\right\rangle = \frac{E_n}{\hbar\omega} - \frac{1}{2} \tag{4.73}$$

The matrix of $a^\dagger a$ can also be expressed as

$$\begin{aligned}
\langle n|a^\dagger a|n\rangle &= \langle n|a^\dagger|m\rangle\langle m|a|n\rangle \\
&= \langle m|a|n\rangle^*\langle m|a|n\rangle = |\langle m|a|n\rangle|^2 \geq 0
\end{aligned} \tag{4.74}$$

From equations (4.73) and (4.74), we have

$$\frac{E_n}{\hbar\omega} - \frac{1}{2} \geq 0 \text{ or } E_n \geq \frac{1}{2}\hbar\omega$$

Let us consider the eigenvalue equation $H|n\rangle = E_n|n\rangle$. Multiplying this equation by a from left we get

$$\begin{aligned}
aH|n\rangle &= aE_n|n\rangle \\
aH|n\rangle &= E_n a|n\rangle
\end{aligned}$$

Using $aH = Ha + \hbar\omega a$ from equation (4.71), we have

$$\begin{aligned}
(Ha + \hbar\omega a)|n\rangle &= E_n a|n\rangle \\
Ha|n\rangle &= (E_n - \hbar\omega)a|n\rangle
\end{aligned}$$

It shows that $|n\rangle$ is an eigenvector of H with an eigenvalue E_n and $a|n\rangle$ is an eigenvector of H with an eigenvalue $E_n - \hbar\omega$, *i.e.*, the eigenvalue is lowered by $\hbar\omega$. The operator a is known as the lowering operator.

Now, we multiply the equation $H|n\rangle = E_n|n\rangle$ by $a^\dagger$ from left and obtain

$$a^\dagger H|n\rangle = a^\dagger E_n|n\rangle$$

Using $a^\dagger H = Ha^\dagger - \hbar\omega a^\dagger$ from equation (4.72), we have

$$(Ha^\dagger - \hbar\omega a^\dagger)|n\rangle = E_n a^\dagger|n\rangle$$
$$Ha^\dagger|n\rangle = (E_n + \hbar\omega)a^\dagger|n\rangle$$

It shows that $a^\dagger|n\rangle$ is an eigenvector of H with an eigenvalue $E_n + \hbar\omega$, *i.e.*, the eigenvalue is raised by $\hbar\omega$. The operator $a^\dagger$ is known as the raising operator. It shows that $a^\dagger|n\rangle$ and $a|n\rangle$ also are eigenvectors of H with the eigenvalues $(E_n + \hbar\omega)$ and $(E_n - \hbar\omega)$, respectively.

For the ground state, denoting the eigenvalue by E_0 and eigenket $|0\rangle$, we have

$$H|0\rangle = E_0|0\rangle \tag{4.75}$$

Hence, we have

$$Ha|0\rangle = (E_o - \hbar\omega)a|0\rangle$$

Since we cannot go below the ground state energy, this equation can be possible only when

$$a|0\rangle = 0 \qquad \text{or} \qquad a^\dagger a|0\rangle = 0 \tag{4.76}$$

Hence, we have

$$\left(\frac{H}{\hbar\omega} - \frac{1}{2}\right)|0\rangle = 0 \qquad \text{or} \qquad \left(\frac{E_0}{\hbar\omega} - \frac{1}{2}\right)|0\rangle = 0$$

Since $|0\rangle \neq 0$, we have

$$E_0 - \frac{1}{2}\hbar\omega = 0 \qquad \text{or} \qquad E_0 = \frac{1}{2}\hbar\omega$$

On multiplying equation (4.75) by $a^\dagger$ from left, we get

$$a^\dagger H|0\rangle = a^\dagger E_0|0\rangle$$
$$(Ha^\dagger - \hbar\omega a^\dagger)|0\rangle = E_0 a^\dagger|0\rangle$$
$$Ha^\dagger|0\rangle = (E_0 + \hbar\omega)a^\dagger|0\rangle$$

It shows that $a^\dagger|0\rangle$ is an eigenvector of H with eigenvalue $E_0 + \hbar\omega = (3/2)\hbar\omega$. Repeated multiplication by $a^\dagger$ from left increases the eigenvalue every time by $\hbar\omega$. Consequently, after n operations, the eigenvalue becomes

$$E_n = E_0 + n\hbar\omega = \left(n + \frac{1}{2}\right)\hbar\omega \tag{4.77}$$

with the eigenket $(a^\dagger)^n|0\rangle$. From equations (4.67) and (4.68), we have

$$a = \frac{1}{(2m\hbar\omega)^{\frac{1}{2}}}\Big(m\omega x + \hbar\frac{\mathrm{d}}{\mathrm{d}x}\Big), \qquad a^\dagger = \frac{1}{(2m\hbar\omega)^{\frac{1}{2}}}\Big(m\omega x - \hbar\frac{\mathrm{d}}{\mathrm{d}x}\Big)$$

Substituting the expression for a in equation (4.76), we have

$$\Big(m\omega x + \hbar\frac{\mathrm{d}}{\mathrm{d}x}\Big)\Psi_0 = 0 \qquad \text{or} \qquad \frac{d\Psi_0}{\Psi_0} = -\frac{m\omega x}{\hbar}dx$$

Integration of this equation gives

$$\Psi_0 = N_0 \exp\Big(-\frac{m\omega x^2}{2\hbar}\Big)$$

We already know

$$\Psi_1 = \frac{N_0}{(2m\hbar\omega)^{\frac{1}{2}}}\Big(m\omega x - \hbar\frac{\mathrm{d}}{\mathrm{d}x}\Big)\exp\Big(-\frac{m\omega x^2}{2\hbar}\Big)$$

Repeated operation of $a^\dagger$ from left gives

$$\Psi_n = \frac{N_0}{(2m\hbar\omega)^{\frac{n}{2}}}\Big(m\omega x - \hbar\frac{\mathrm{d}}{\mathrm{d}x}\Big)^n \exp\Big(-\frac{m\omega x^2}{2\hbar}\Big) \tag{4.78}$$

Equations (4.77) and (4.78) are the eigenvalues and eigenfunctions for the harmonic oscillator.

Exercise 1. Ground state wave function of a harmonic oscillator is

$$\psi_0(x) = \Big(\frac{m\omega}{\hbar\pi}\Big)^{1/4}\exp\Big(-\frac{m\omega x^2}{2\hbar}\Big)$$

(i) At which point is the probability density maximum? (ii) What is the value of the maximum probability density?

Solution: For the given ground state wave function, the probability density is

$$P(x) = \psi_0^*(x)\psi_0(x) = \Big(\frac{m\omega}{\hbar\pi}\Big)^{1/2}\exp\Big(-\frac{m\omega x^2}{\hbar}\Big)$$

(i) The probability density $P(x)$ will be maximum at the point where

$$\frac{\mathrm{d}P}{\mathrm{d}x} = 0$$

Substituting $P(x)$ here, we get

$$\Big(\frac{m\omega}{\hbar\pi}\Big)^{1/2}\Big(-\frac{2m\omega x}{\hbar}\Big)\exp\Big(-\frac{m\omega x^2}{\hbar}\Big) = 0$$

It gives

$$x = 0$$

Hence, the probability density is maximum at $x = 0$.

(ii) Maximum value of the probability density is

$$P(0) = \left(\frac{m\omega}{\hbar\pi}\right)^{1/2}$$

Exercise 2. Show that the zero-point energy $\hbar\omega/2$ of a linear harmonic oscillator is a manifestation of the uncertainty principle.

Solution: For a linear harmonic oscillator of mass m and angular frequency ω, the energy is

$$E = \frac{p_x^2}{2m} + \frac{1}{2}kx^2$$

where $k = m\omega^2$. Thus, a relation between the corresponding uncertainties is

$$\Delta E = \frac{(\Delta p_x)^2}{2m} + \frac{1}{2}\,k(\Delta x)^2 \tag{4.79}$$

where ΔE, Δp_x, Δx denote uncertainties in energy, momentum and position, respectively. Using the uncertainty relation

$$\Delta p_x\,\Delta x \geq \frac{\hbar}{2} \qquad \text{or} \qquad \Delta p_x \geq \frac{\hbar}{2\Delta x} \tag{4.80}$$

Using equation (4.80) in (4.79), we get

$$\Delta E \geq \frac{\hbar^2}{8m(\Delta x)^2} + \frac{1}{2}k(\Delta x)^2$$

For minimum value of ΔE with respect to Δx, we have

$$\frac{\partial \Delta E}{\partial \Delta x} = -\frac{\hbar^2}{4m(\Delta x)^3} + k\,\Delta x = 0$$

It gives

$$(\Delta x)^4 = \frac{\hbar^2}{4mk}$$

$$= \frac{\hbar^2}{4m^2\omega^2}$$

$$(\Delta x)^2 = \frac{\hbar}{2m\omega} \tag{4.81}$$

Thus, the uncertainty ΔE would be minimum when $(\Delta x)^2 = \hbar/2m\omega$. Form equation (4.80), we have

$$(\Delta p)^2 = \frac{\hbar^2}{4(\Delta x)^2} = \frac{\hbar m\omega}{2} \tag{4.82}$$

Using equations (4.81) and (4.82) in (4.79), we have

$$\Delta E = \frac{\hbar m\omega}{4m} + \frac{1}{2}\, k\, \frac{\hbar}{2m\omega} = \frac{\hbar\omega}{2}$$

It shows that the minimum uncertainty in the energy is $\hbar\omega/2$, which is the zero-point energy of the harmonic oscillator. Hence, the zero-point energy $\hbar\omega/2$ of a linear harmonic oscillator is a manifestation of the uncertainty principle.

9. Free particle

In absence of any external field, a particle moves freely. For a free particle, the potential V is zero and the Schrödinger equation is

$$-\frac{\hbar^2}{2m}\frac{\mathrm{d}^2}{\mathrm{d}x^2}\Psi(x) = E\Psi \qquad \text{or} \qquad \frac{\mathrm{d}^2\Psi}{\mathrm{d}x^2} = -k^2\Psi \tag{4.83}$$

where $k^2 = 2mE/\hbar^2$. For energy E of the particle, this equation (4.83) has two solutions

$$\Psi(x) = A\ \mathrm{e}^{ikx} \qquad \text{and} \qquad \Psi(x) = A\ \mathrm{e}^{-ikx} \tag{4.84}$$

Hence, the free particle solutions are degenerate, as the two solutions correspond to a common value. Since the energy E is real and positive, the value of k is a real number. The first solution in (4.84) corresponds to a particle moving in the positive x-direction with momentum $\hbar k$ whereas the second solution corresponds to a particle moving in the negative x-direction with momentum $-\hbar k$. The energy in both the cases is however the same. Hence, all real positive values are the allowed eigenvalues.

Box normalization

Though the particle is free to move, let us assume that the particle

is confined in a box of length L

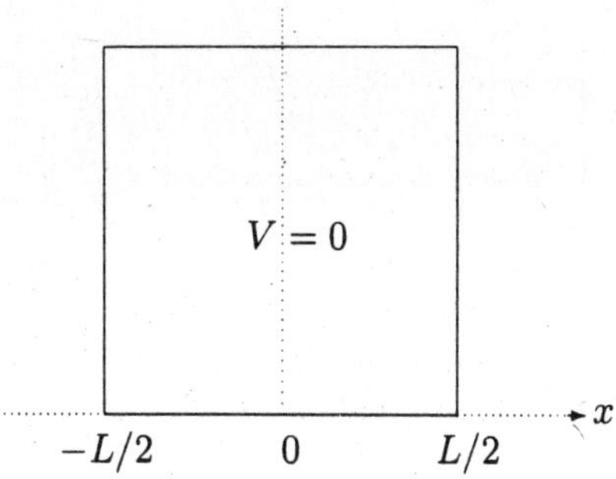

Figure 10: Particle is confined in a box of length L.

For normalization of the wave function, we have

$$\int_{-L/2}^{L/2} \Psi^*(x)\ \Psi(x)\ dx = 1 \qquad\qquad |A|^2 \int_{-L/2}^{L/2} dx = 1$$

$$|A|^2\ L = 1 \qquad\qquad A = \frac{1}{\sqrt{L}}$$

Hence, the box normalized wave functions are

$$\Psi(x) = \frac{1}{\sqrt{L}}\ e^{ikx} \qquad \text{and} \qquad \Psi(x) = \frac{1}{\sqrt{L}}\ e^{-ikx}$$

10. Problems and questions

1. Obtain energy eigenvalues and eigenfunctions of a particle trapped in a potential given by $V(x) = 0$ for $0 \leq x \leq a$ and $V(x) = \infty$ otherwise. Show that the wave functions for different energy levels are orthogonal to one another.

2. Discuss about a particle in a one-dimensional square well potential with rigid walls.

3. A particle of mass m and energy E is moving in one-dimensional potential field, of width $2a$, expressed as

$$V(x) = \begin{cases} 0 & |x| < a \\ \infty & |x| > a \end{cases}$$

   Find out eigen values and eigen functions of the particle.

4. Discuss about a particle in a one-dimensional square well potential with finite walls.

5. A particle of mass m and energy E strikes a potential barrier of height V_0 such that $E < V_0$. Derive expressions for reflection coefficient R and the transmission coefficient T. Show that $R+T = 1$.

6. A particle of mass m and energy E strikes a potential barrier of height V_0 such that $E > V_0$. Derive expressions for reflection coefficient R and the transmission coefficient T. Show that $R+T = 1$.

7. Discuss the solution of a time-independent Schrödinger equation for a periodic potential expressed by $V(x + n\,d) = V(x)$ where n can assume positive integer values and d is the period of the potential.

8. Discuss the solution of a time-independent Schrödinger equation for Kronig-Penney square-well periodic potential.

9. Obtain eigen values and eigen functions for a linear harmonic oscillator, using the time-independent Schrödinger equation.

10. Obtain eigen values and eigen functions for a linear harmonic oscillator, using the operator method.

11. For a moving particle of mass m, the Hamiltonian is

$$H = \frac{p_x^2}{2m} + \frac{1}{2}\,m\omega^2 x^2$$

Show that the operators operators a and a* defined by

$$a = \left(\frac{m\omega}{2\hbar}\right)^{1/2} x + i\left(\frac{1}{2m\hbar\omega}\right)^{1/2} p_x$$

and

$$a^{\dagger} = \left(\frac{m\omega}{2\hbar}\right)^{1/2} x - i\left(\frac{1}{2m\hbar\omega}\right)^{1/2} p_x$$

are the lowering and raising operators, respectively.

12. Show that the zero-point energy $\hbar\omega/2$ of a linear harmonic oscillator is a manifestation of the uncertainty principle.

13. Write notes on the following

(i) Quantum mechanical tunneling

(ii) Emission of alpha-particle from a nucleus

(iii) Free-particle

Three-dimensional Problems

In the preceding chapter, we discussed one-dimensional eigenvalue problems, where we solved one-dimensional Schrödinger equation for some specific forms of the potential $V(x)$. In this chapter, we shall solve three-dimensional Schrödinger equation, where we shall discuss about the rigid rotator, hydrogen atom in particular. The simplest case of the potential will be $V(r)$. Such potential, which does not depend on the angular coordinates, is known as the spherically symmetric potential. Assumption of the spherically symmetric potential helps in solving the three-dimensional Schrödinger equation.

1. Particle moving in a spherically symmetric potential

A spherically symmetric potential is that which depends only on the distance of the point form the center of the sphere. That is, the potential does not depend on the angular coordinates. In this case, the three-dimensional Schrödinger equation can be solved conveniently by making use of the spherical polar coordinates $(r,\ \theta,\ \phi)$ with the pole at the center of the sphere. In the spherical polar coordinates, the spherically symmetric potential V is a function of r only, and is independent of θ and ϕ. The time-independent Schrödinger equation in spherical polar coordinates is

$$\nabla^2\psi(r,\theta,\phi) + \frac{2m}{\hbar^2}\Big[E - V(r)\Big]\psi(r,\theta,\phi) = 0$$

In the spherical polar coordinates, this equation is expressed as

$$\frac{1}{r^2}\frac{\partial}{\partial r}\left(r^2\frac{\partial\psi}{\partial r}\right) + \frac{1}{r^2\sin\theta}\frac{\partial}{\partial\theta}\left(\sin\theta\frac{\partial\psi}{\partial\theta}\right) + \frac{1}{r^2\sin^2\theta}\frac{\partial^2\psi}{\partial\phi^2} + \frac{2m}{\hbar^2}\Big[E - V(r)\Big]\psi = 0 \quad (5.1)$$

Since the potential V depends only on r only, this equation be solved by the technique of the separation of variables. In this technique, the wave function $\psi(r,\theta,\phi)$ is expressed as a product of distinct functions, each depending on one variable, as the following.

$$\psi(r,\theta,\phi) = R(r)\ \Theta(\theta)\ \Phi(\phi) \tag{5.2}$$

Using equation (5.2) in (5.1) and multiplying by

$$\frac{r^2 \sin^2\theta}{R\ \Theta\ \Phi}$$

we get

$$\frac{\sin^2\theta}{R}\frac{\mathrm{d}}{\mathrm{d}r}\left(r^2\frac{\mathrm{d}R}{\mathrm{d}r}\right) + \frac{\sin\theta}{\Theta}\frac{\mathrm{d}}{\mathrm{d}\theta}\left(\sin\theta\frac{\mathrm{d}\Theta}{\mathrm{d}\theta}\right)$$

$$+\frac{2m}{\hbar^2}\Big[E - V(r)\Big]r^2\sin^2\theta = -\frac{1}{\Phi}\frac{\mathrm{d}^2\Phi}{\mathrm{d}\phi^2} \tag{5.3}$$

In equation (5.3), the left side depends on r and θ, whereas the right side depends only on ϕ. This can be possible only when each of the sides is equal to some constant, say m^2. Then we have

$$\frac{\mathrm{d}^2}{\mathrm{d}\phi^2}\Phi(\phi) = -m^2\Phi(\phi) \tag{5.4}$$

$$\frac{\sin^2\theta}{R}\frac{\mathrm{d}}{\mathrm{d}r}\left(r^2\frac{\mathrm{d}R}{\mathrm{d}r}\right)+\frac{\sin\theta}{\Theta}\frac{\mathrm{d}}{\mathrm{d}\theta}\left(\sin\theta\frac{\mathrm{d}\Theta}{\mathrm{d}\theta}\right)+\frac{2m}{\hbar^2}\Big[E-V(r)\Big]r^2\sin^2\theta = m^2 \tag{5.5}$$

Dividing equation (5.5) by $\sin^2\theta$ and rearranging, we get

$$\frac{1}{R}\frac{\mathrm{d}}{\mathrm{d}r}\left(r^2\frac{\mathrm{d}R}{\mathrm{d}r}\right) + \frac{2m}{\hbar^2}\Big(E - V\Big)r^2 = -\frac{1}{\Theta\sin\theta}\frac{\mathrm{d}}{\mathrm{d}\theta}\left(\sin\theta\frac{\mathrm{d}\Theta}{\mathrm{d}\theta}\right) + \frac{m^2}{\sin^2\theta} \tag{5.6}$$

In equation (5.6), the left side depends only on r, whereas the right side depends only on θ. This can be possible only when each of the side is equal to some constant, say $l(l+1)$. Then we have

$$-\frac{1}{\Theta\sin\theta}\frac{\mathrm{d}}{\mathrm{d}\theta}\left(\sin\theta\frac{\mathrm{d}\Theta}{\mathrm{d}\theta}\right) + \frac{m^2}{\sin^2\theta} = l(l+1)$$

$$\frac{1}{\sin\theta}\frac{\mathrm{d}}{\mathrm{d}\theta}\left(\sin\theta\frac{\mathrm{d}\Theta}{\mathrm{d}\theta}\right) + \Big(l(l+1) - \frac{m^2}{\sin^2\theta}\Big)\Theta = 0 \tag{5.7}$$

and

$$\frac{1}{R}\frac{\mathrm{d}}{\mathrm{d}r}\left(r^2\frac{\mathrm{d}R}{\mathrm{d}r}\right) + \frac{2m}{\hbar^2}\Big[E - V(r)\Big]r^2 = l(l+1)$$

$$\frac{1}{r^2}\frac{\mathrm{d}}{\mathrm{d}r}\left(r^2\frac{\mathrm{d}R}{\mathrm{d}r}\right) + \frac{2m}{\hbar^2}\Big[E - V(r)\Big]R - \frac{l(l+1)}{r^2}R = 0 \tag{5.8}$$

This way, the three-dimensional equation (5.1) is decomposed into three equations (5.4), (5.7) and (5.8).

Solution of the ϕ-equation

Solution of equation (5.4) is

$$\Phi(\phi) = A\ \mathrm{e}^{\pm im\phi}$$

Our basic requirement is that $\Phi(\phi)$ must be single valued, and therefore, we must have $\Phi(\phi) = \Phi(\phi + 2\pi)$. Thus, we must have

$$A\ \mathrm{e}^{\pm im\phi} = A\ \mathrm{e}^{\pm im(\phi+2\pi)} \qquad \text{or} \qquad 1 = \mathrm{e}^{\pm i2m\pi}$$

This is possible only when $m = 0, \pm 1, \pm 2, \ldots$. The quantum number m is known as the magnetic quantum number. The normalization condition gives

$$\int_0^{2\pi} \Phi^*\Phi\ \mathrm{d}\phi = 1 \qquad \text{or} \qquad |A|^2 \int_0^{2\pi} \mathrm{d}\phi = 1 \tag{5.9}$$

Thus,

$$|A|^2 2\pi = 1 \qquad \text{or} \qquad A = \frac{1}{\sqrt{2\pi}}$$

Hence, the normalized solution is

$$\Phi = \frac{1}{\sqrt{2\pi}} e^{im\phi} \qquad m = 0, \pm 1, \pm 2, \ldots$$

Solution of the θ-equation

Equation (5.7) can be solved conveniently by introducing a new variable $u = \cos\theta$. On differentiating, we get

$$\mathrm{d}u = -\sin\theta\ \mathrm{d}\theta$$

We have
$$\frac{\mathrm{d}}{\mathrm{d}\theta} = \frac{\mathrm{d}u}{\mathrm{d}\theta}\frac{\mathrm{d}}{\mathrm{d}u} = -\sin\theta\frac{\mathrm{d}}{\mathrm{d}u} = -\left(1-u^2\right)^{\frac{1}{2}}\frac{\mathrm{d}}{\mathrm{d}u}$$

On using these relations, equation (5.7) reduces to

$$\frac{\mathrm{d}}{\mathrm{d}u}\left[\left(1-u^2\right)\frac{\mathrm{d}\Theta(u)}{\mathrm{d}u}\right] + \left(l(l+) - \frac{m^2}{1-u^2}\right)\Theta(u) = 0 \tag{5.10}$$

This is an associated Legendre equation. Solution of equation (5.10) is the associated Legendre polynomial $P_l^m(x)$

$$P_l^m(x) = (1-x^2)^{m/2}\ \frac{\mathrm{d}^m}{\mathrm{d}x^m} P_l(x)$$

$$= (1-x^2)^{m/2}\ \sum_{r=0}^{M}(-1)^r \frac{(2l-2r)!}{2^l r!\ (l-r)!\ (l-2r-m)!} x^{l-2r-m}$$

where $P_l(x)$ is the Legendre polynomial of degree n and $M = (l-m)/2$ or $(l-m-1)/2$, whichever is integer. The magnetic quantum number m can assume the values

$$m = 0, \pm 1, \pm 2, \ldots, \pm l.$$

The orthonormality condition for the associated Legendre polynomial is

$$\int_{-1}^{1} P_l^m(u)\, P_k^m(u)\, du = \frac{2}{2l+1} \frac{(l+m)!}{(l-m)!}\, \delta_{lk}$$

The normalized solution is

$$\Theta(\theta) = N_{lm}\, P_l^m(u)$$

where N_{lm} is a normalization constant. The normalization condition gives

$$\int_0^{\pi} \Theta^*(\theta)\, \Theta(\theta) \sin\theta\, d\theta = 1 \qquad \text{or} \qquad \int_{-1}^{1} |N_{lm}|^2 P_l^m(z) P_l^m z)\, dz = 1$$

$$|N_{lm}|^2 \frac{2}{2l+1} \frac{(l+m)!}{(l-m)!} = 1 \qquad \text{or} \qquad N_{lm} = \sqrt{\frac{2l+1}{2} \frac{(l-m)!}{(l+m)!}}$$

The normalized solution is

$$\Theta(\theta) = \epsilon \sqrt{\frac{2l+1}{2} \frac{(l-m)!}{(l+m)!}}\, P_l^m\, (\cos\theta)$$

where ϵ is the phase factor conventionally taken as

$$\epsilon = \begin{cases} (-1)^m & \text{for } m > 0 \\ 1 & \text{for } m \leq 0 \end{cases}$$

Spherical harmonics

The solution of the angular part of the Schrödinger equation (5.1) is known as the spherical harmonics denoted by $Y_{lm}(\theta, \phi)$. Thus, the spherical harmonics satisfies the equation

$$\frac{1}{\sin\theta} \frac{\partial}{\partial\theta} \left(\sin\theta \frac{\partial Y_{lm}}{\partial\theta} \right) + \frac{1}{\sin^2\theta} \frac{\partial^2 Y_{lm}}{\partial\phi^2} + l(l+1) Y_{lm} = 0$$

and is given by

$$Y_{lm}(\theta, \phi) = \Theta(\theta)\, \Phi(\phi) = \epsilon \sqrt{\frac{2l+1}{4\pi} \frac{(l-m)!}{(l+m)!}}\, P_l^m\, (\cos\theta)\, e^{im\phi}$$

where $l = 0, 1, 2, \ldots$ $\quad m = 0, \pm 1, \pm 2, \ldots, \pm l$. Some expressions for the spherical harmonics are as the following.

$$Y_{00} = \frac{1}{2\sqrt{\pi}}$$

$$Y_{11} = -\frac{1}{2}\sqrt{\frac{3}{2\pi}} \sin\theta \; e^{i\phi}$$

$$Y_{10} = \frac{1}{2}\sqrt{\frac{3}{\pi}} \cos\theta$$

$$Y_{1,-1} = \frac{1}{2}\sqrt{\frac{3}{2\pi}} \sin\theta \; e^{-i\phi}$$

$$Y_{22} = \frac{1}{8}\sqrt{\frac{15}{2\pi}} (1 - \cos 2\theta) \; e^{i2\phi}$$

$$Y_{21} = -\frac{1}{4}\sqrt{\frac{15}{2\pi}} \sin 2\theta \; e^{i\phi}$$

$$Y_{20} = \frac{1}{8}\sqrt{\frac{5}{\pi}} (1 + 3\cos 2\theta)$$

$$Y_{2,-1} = \frac{1}{4}\sqrt{\frac{15}{2\pi}} \sin 2\theta \; e^{-i\phi}$$

$$Y_{2,-2} = \frac{1}{8}\sqrt{\frac{15}{2\pi}} (1 - \cos 2\theta) \; e^{-i2\phi}$$

$$Y_{33} = -\frac{1}{32}\sqrt{\frac{35}{\pi}} (3\sin\theta - \sin 3\theta) \;\; e^{i3\phi}$$

$$Y_{32} = \frac{1}{16}\sqrt{\frac{105}{2\pi}} (\cos\theta - \cos 3\theta) \;\; e^{i2\phi}$$

$$Y_{31} = -\frac{1}{32}\sqrt{\frac{21}{\pi}} (\sin\theta + 5\sin 3\theta) \;\; e^{i\phi}$$

$$Y_{30} = \frac{1}{16}\sqrt{\frac{7}{\pi}} (3\cos\theta + 5\cos 3\theta)$$

$$Y_{3,-1} = \frac{1}{32}\sqrt{\frac{21}{\pi}} (\sin\theta + 5\sin 3\theta) \;\; e^{-i\phi}$$

$$Y_{3,-2} = \frac{1}{16}\sqrt{\frac{105}{2\pi}}\ (\cos\theta - \cos 3\theta)\ \ e^{-i2\phi}$$

$$Y_{3,-3} = \frac{1}{32}\sqrt{\frac{35}{\pi}}\ (3\sin\theta - \sin 3\theta)\ \ e^{-i3\phi}$$

$$Y_{44} = \frac{3}{128}\sqrt{\frac{35}{2\pi}}\ (3 - 4\cos 2\theta + \cos 4\theta)\ e^{i4\phi}$$

$$Y_{43} = -\frac{3}{64}\sqrt{\frac{35}{\pi}}\ (2\sin 2\theta - \sin 4\theta)\ e^{i3\phi}$$

$$Y_{42} = \frac{3}{64}\sqrt{\frac{5}{2\pi}}\ (3 + 4\cos 2\theta - 7\cos 4\theta)\ e^{i2\phi}$$

$$Y_{41} = -\frac{3}{64}\sqrt{\frac{5}{\pi}}\ (2\sin 2\theta + 7\sin 4\theta)\ e^{i\phi}$$

$$Y_{40} = \frac{3}{128\sqrt{\pi}}\ (9 + 20\cos 2\theta + 35\cos 4\theta)$$

$$Y_{4,-1} = \frac{3}{64}\sqrt{\frac{5}{\pi}}\ (2\sin 2\theta + 7\sin 4\theta)\ e^{-i\phi}$$

$$Y_{4,-2} = \frac{3}{64}\sqrt{\frac{5}{2\pi}}\ (3 + 4\cos 2\theta - 7\cos 4\theta)\ e^{-i2\phi}$$

$$Y_{4,-3} = \frac{3}{64}\sqrt{\frac{35}{\pi}}\ (2\sin 2\theta - \sin 4\theta)\ e^{-i3\phi}$$

$$Y_{4,-4} = \frac{3}{128}\sqrt{\frac{35}{2\pi}}\ (3 - 4\cos 2\theta + \cos 4\theta)\ e^{-i4\phi}$$

$$Y_{55} = -\frac{3}{512}\sqrt{\frac{77}{\pi}}\ (10\sin\theta - 5\sin 3\theta + \sin 5\theta)\ e^{i5\phi}$$

$$Y_{54} = \frac{3}{256}\sqrt{\frac{385}{2\pi}}\ (2\cos\theta - 3\cos 3\theta + \cos 5\theta)\ e^{i4\phi}$$

$$Y_{53} = -\frac{1}{512}\sqrt{\frac{385}{\pi}}\ (6\sin\theta + 13\sin 3\theta - 9\sin 5\theta)\ e^{i3\phi}$$

$$Y_{52} = \frac{1}{128}\sqrt{\frac{1155}{2\pi}}\ (2\cos\theta + \cos 3\theta - 3\cos 5\theta)\ e^{i2\phi}$$

$$Y_{51} = -\frac{1}{256}\sqrt{\frac{165}{2\pi}}\,(2\sin\theta + 7\sin 3\theta + 21\sin 5\theta)\,e^{i\phi}$$

$$Y_{50} = \frac{1}{256}\sqrt{\frac{11}{\pi}}\,(30\cos\theta + 35\cos 3\theta + 63\cos 5\theta)$$

$$Y_{5,-1} = \frac{1}{256}\sqrt{\frac{165}{2\pi}}\,(2\sin\theta + 7\sin 3\theta + 21\sin 5\theta)\,e^{-i\phi}$$

$$Y_{5,-2} = \frac{1}{128}\sqrt{\frac{1155}{2\pi}}\,(2\cos\theta + \cos 3\theta - 3\cos 5\theta)\,e^{-i2\phi}$$

$$Y_{5,-3} = \frac{1}{512}\sqrt{\frac{385}{\pi}}\,(6\sin\theta + 13\sin 3\theta - 9\sin 5\theta)\,e^{-i3\phi}$$

$$Y_{5,-4} = \frac{3}{256}\sqrt{\frac{385}{2\pi}}\,(2\cos\theta - 3\cos 3\theta + \cos 5\theta)\,e^{-i4\phi}$$

$$Y_{5,-5} = \frac{3}{512}\sqrt{\frac{77}{\pi}}\,(10\sin\theta - 5\sin 3\theta + \sin 5\theta)\,e^{-i5\phi}$$

When $m \neq 0$, the presence of the factor $e^{im\phi}$ makes the spherical harmonics complex. However, the combinations $(Y_{lm} + Y_{l,-m})$ are real for even values of m and the combinations $(Y_{lm} - Y_{l,-m})$ are real for odd values of m.

Solution of the radial equation

Equation (5.8) can be solved conveniently by introducing a new variable $\chi(r)$ as

$$R(r) = \frac{\chi(r)}{r}$$

Differentiating this equation with respect to r, we get

$$\frac{dR}{dr} = -\frac{\chi}{r^2} + \frac{1}{r}\,\frac{d\chi}{dr}$$

$$r^2\,\frac{dR}{dr} = -\chi + r\,\frac{d\chi}{dr}$$

Differentiating this equation with respect to r, we get

$$\frac{d}{dr}\left(r^2\,\frac{dR}{dr}\right) = -\frac{d\chi}{dr} + \frac{d\chi}{dr} + r\,\frac{d^2\chi}{dr^2}$$

$$= r\,\frac{d^2\chi}{dr^2}$$

Using these relations, the radial equation reduces to

$$\frac{d^2\chi}{dr^2} + \frac{2m}{\hbar^2}\left[E - V(r) - \frac{l(l+1)\hbar^2}{2mr^2}\right]\chi = 0$$

This is Schrödinger equation of one variable r for a particle of mass m moving in a field of effective potential

$$V_{eff} = V(r) + \frac{l(l+1)\hbar^2}{2mr^2}$$

The additional potential $l(l+1)\hbar^2/(2mr^2)$ corresponds to a force $l(l+1)\hbar^2/(mr^3)$. Now, the centrifugal force can be expressed as

$$\frac{mv^2}{r} = mr\omega^2 = \frac{(mr^2\omega)^2}{mr^3} = \frac{L^2}{mr^3}$$

where L is the orbital angular momentum. Comparison of the centrifugal force with the additional force gives

$$L^2 = l(l+1)\hbar^2$$

Hence, the orbital angular momentum is $L = \sqrt{l(l+1)}\,\hbar$. In the quantum theory, the angular momentum L is $l\hbar$, where l is a non-zero positive integer. But, in the quantum mechanics, l is changed to $\sqrt{l(l+1)}$ where l can be zero also. It is one of the important contributions of the quantum mechanics.

2. System of two interacting particles

So far we have accounted for a particle moving in a given potential field. We have however a number of situations where two particles interact with each other and therefore both of them move simultaneously. Thus, the wave equation of the system has the coordinates of both the particles. This wave equation for the system of two particles can be reduced into two one-particle equations, one representing the translation motion of the center-of-mass and the other showing the motion, of a particle having reduced mass, relative to the other. The procedure of this reduction is as the following.

Reduction of a system of two particles

Let $\vec{r_1}$ and $\vec{r_2}$ be position vectors of two particles of masses m_1 and m_2, respectively (Figure 1). If $\vec{R}$ is the position vector of the center-of-mass of the system. We have

$$m_2\,(\vec{R} - \vec{r_2}) = m_1\,(\vec{r_1} - \vec{R})$$

$$\vec{R} = \frac{m_1 \vec{r_1} + m_2 \vec{r_2}}{m_1 + m_2} \tag{5.11}$$

The relative position is

$$\vec{r} = \vec{r_1} - \vec{r_2} \tag{5.12}$$

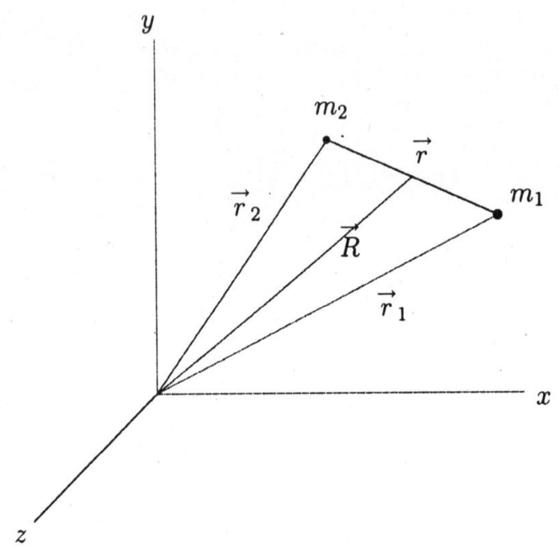

Figure 1: System of two interacting particles.

Using $\vec{r_1} = \vec{r} + \vec{r_2}$ in equation (5.11), we get

$$\vec{R} = \frac{m_1(\vec{r} + \vec{r_2}) + m_2 \vec{r_2}}{m_1 + m_2} = \frac{m_1 \vec{r} + (m_1 + m_2) \vec{r_2}}{m_1 + m_2} = \frac{m_1 \vec{r}}{m_1 + m_2} + \vec{r_2}$$

Therefore, we have

$$\vec{r_2} = \vec{R} - \frac{m_1 \vec{r}}{m_1 + m_2} \tag{5.13}$$

Using equation (5.13) in (5.12), we get

$$\vec{r_1} = \vec{r} + \vec{r_2} = \vec{r} + \vec{R} - \frac{m_1 \vec{r}}{m_1 + m_2} = \vec{R} + \frac{m_2 \vec{r}}{m_1 + m_2}$$

Therefore, we have

$$\vec{r_1} = \vec{R} + \frac{m_2 \vec{r}}{m_1 + m_2}$$

The momentums of the two particles are

$$\vec{p_1} = m_1 \dot{\vec{r_1}} = m_1 \dot{\vec{R}} + \mu \dot{\vec{r}}, \qquad \vec{p_2} = m_2 \dot{\vec{r_2}} = m_2 \dot{\vec{R}} - \mu \dot{\vec{r}} \tag{5.14}$$

where[1]

$$\mu = \frac{m_1 m_2}{m_1 + m_2} \tag{5.15}$$

is the reduced-mass of the system. Assuming that that the potential between the two particles depends only on the distance between them, the Hamiltonian of the system is

$$H = \frac{p_1^2}{2m_1} + \frac{p_2^2}{2m_2} + V(r) \tag{5.16}$$

Using equation (5.14) in (5.16), we have

$$H = \frac{m_1^2 \dot{R}^2 + \mu^2 \dot{r}^2 + 2m_1\mu \dot{\vec{R}} \cdot \dot{\vec{r}}}{2m_1} + \frac{m_2^2 \dot{R}^2 + \mu^2 \dot{r}^2 - 2m_2\mu \dot{\vec{R}} \cdot \dot{\vec{r}}}{2m_2} + V(r)$$

$$= \frac{1}{2}(m_1 + m_2)\, \dot{R}^2 + \frac{1}{2}\left(\frac{1}{m_1} + \frac{1}{m_2}\right) \mu^2 \dot{r}^2 + V(r)$$

$$= \frac{1}{2M} p_R^2 + \frac{1}{2\mu} p_r^2 + V(r)$$

where $M = m_1 + m_2$, $p_R = M \dot{\vec{R}}$ and $p_r = \mu \dot{\vec{r}}$. On replacing the dynamical quantities by the corresponding operators and writing

$$\nabla_R^2 = \frac{\partial^2}{\partial X^2} + \frac{\partial^2}{\partial Y^2} + \frac{\partial^2}{\partial Z^2} \quad \text{and} \quad \nabla_r^2 = \frac{\partial^2}{\partial x^2} + \frac{\partial^2}{\partial y^2} + \frac{\partial^2}{\partial z^2}$$

we get

$$H = -\frac{\hbar^2}{2M} \nabla_R^2 - \frac{\hbar^2}{2\mu} \nabla_r^2 + V(r)$$

The time-independent Schrödinger equation has the form

$$\left[-\frac{\hbar^2}{2M} \nabla_R^2 - \frac{\hbar^2}{2\mu} \nabla_r^2 + V(r)\right] \Psi_T(\vec{R}, \vec{r}) = E_T\, \Psi_T(\vec{R}, \vec{r}) \tag{5.17}$$

---

[1] Let us take $m_1 < m_2$, and rearrange equation (5.15) to get

$$\mu = \frac{m_1}{1 + \frac{m_1}{m_2}}$$

Since m_1/m_2 is positive, we have $\mu < m_1$. As m_1 is taken smaller than m_2, it shows that the reduced-mass μ is less than both m_1 and m_2.

Equation (5.17) can be decomposed into two equations with the help of the separation of variables by writing

$$\Psi_T(\vec{R}, \vec{r}) = \chi(\vec{R})\ \psi(\vec{r}) \tag{5.18}$$

Using equation (5.18) in (5.17), we get

$$\left[-\frac{\hbar^2}{2M}\nabla_R^2 - \frac{\hbar^2}{2\mu}\nabla_r^2 + V(r)\right] \chi(\vec{R})\ \psi(\vec{r}) = E_T\ \chi(\vec{R})\ \psi(\vec{r})$$

$$-\frac{\hbar^2}{2M}\psi(\vec{r})\ \nabla_R^2\chi(\vec{R}) - \frac{\hbar^2}{2\mu}\chi(\vec{R})\ \nabla_r^2\psi(\vec{r})$$

$$+ V(r)\chi(\vec{R})\ \psi(\vec{r}) = E_T\ \chi(\vec{R})\ \psi(\vec{r}) \tag{5.19}$$

dividing this equation by $\chi(\vec{R})\ \psi(\vec{r})$ and rearranging, we get

$$-\frac{\hbar^2}{2M\chi(\vec{R})}\ \nabla_R^2\chi(\vec{R}) = \frac{\hbar^2}{2\mu\psi(\vec{r})}\ \nabla_r^2\psi(\vec{r}) + E_T - V(r)$$

The left side of this equation depends on $\vec{R}$ whereas the right side depends on $\vec{r}$. It can only be possible when both sides are equal to some constant, say E_t. Thus, we have

$$-\frac{\hbar^2}{2M\chi(\vec{R})}\ \nabla_R^2\chi(\vec{R}) = E_t$$

$$-\frac{\hbar^2}{2M}\ \nabla_R^2\chi(\vec{R}) = E_t\chi(\vec{R}) \tag{5.20}$$

and

$$\frac{\hbar^2}{2\mu\psi(\vec{r})}\ \nabla_r^2\psi(\vec{r}) + E_T - V(r) = E_t$$

$$-\frac{\hbar^2}{2\mu}\nabla_r^2\psi(\vec{r}) + V(r)\psi(\vec{r}) = (E_T - E_t)\psi(\vec{r}) \tag{5.21}$$

Equation (5.20) is the same as a free-particle equation of mass M. This equation describes the translational motion of the system in space. Equation (5.21) is the same as the Schrödinger equation of a particle of mass μ moving in a potential $V(r)$. This equation represents the relative motion of the two particles, where the energy of the relative motion is $E = E_T - E_t$. Thus, the Schrödinger equation for relative motion is

$$-\frac{\hbar^2}{2\mu}\nabla^2\psi(\vec{r}) + V(r)\psi(\vec{r}) = E\psi(\vec{r})$$

3. Rigid rotator

A rigid rotator consists of two masses m_1 and m_2 separated by a fixed distance r. Let this system rotates about an axis passing through its center-of-mass and perpendicular to the plane containing the masses. For free motion of the system, we have $V(r) = 0$ and as the distance r is fixed, the Schrödinger equation of motion in spherical polar coordinates is

$$\frac{1}{r^2 \sin\theta}\frac{\partial}{\partial\theta}\left(\sin\theta\frac{\partial\psi}{\partial\theta}\right) + \frac{1}{r^2\ \sin^2\theta}\frac{\partial^2\psi}{\partial\phi^2} + \frac{2\mu E}{\hbar^2}\psi = 0$$

$$\frac{1}{\sin\theta}\frac{\partial}{\partial\theta}\left(\sin\theta\frac{\partial\psi}{\partial\theta}\right) + \frac{1}{\sin^2\theta}\frac{\partial^2\psi}{\partial\phi^2} + \frac{2IE}{\hbar^2}\psi = 0 \qquad (5.22)$$

Here, $I = \mu r^2$ is the moment of inertia of the system. If we take

$$\frac{2IE}{\hbar^2} = l(l+1) \qquad (5.23)$$

the solution of equation (5.22) is the spherical harmonics $Y_{lm}(\theta, \phi)$, as we discussed earlier. Equation (5.23) gives the energy eigenvalues of the rigid rotator as

$$E_l = \frac{l(l+1)\hbar^2}{2I} \qquad \text{where} \qquad l = 0, 1, 2, \ldots$$

Since the energy does not depend on the magnetic quantum number m, the energy levels are degenerate and each one has $(2l+1)$ fold degeneracy.

Exercise 1. A rigid rotator is constrained to rotate about a fixed axis. Obtain its normalized eigen functions and eigenvalues.

Solution: Since the rotator is rotating about a fixed axis, its motion lies in a plane. Let us take it as the xy-plane so that θ is 90°. Further, as the rotator is rigid, the coordinate r is constant. Hence, the Schrödinger equation has only one variable ϕ and is

$$\frac{1}{r^2}\frac{\partial^2\psi}{\partial\phi^2} + \frac{2\mu E}{\hbar^2}\psi = 0$$

or

$$\frac{\partial^2\psi}{\partial\phi^2} = -\frac{2IE}{\hbar^2}\psi = -m^2\psi$$

where $I = \mu r^2$ is the moment of inertia of the rotator and $m^2 = 2IE/\hbar^2$. The solution of this equation is

$$\psi = A\ e^{im\phi}, \qquad m = 0, \pm 1, \pm 2, \ldots$$

The energy eigen values are

$$E_m = \frac{\hbar^2 m^2}{2I} \qquad m = 0, \pm 1, \pm 2, \ldots$$

The normalization condition gives

$$\int_0^{2\pi} \psi^* \psi \, d\phi = 1 \qquad \text{or} \qquad |A|^2 \int_0^{2\pi} d\phi = 1$$

Thus,

$$|A|^2 2\pi = 1 \qquad \text{or} \qquad A = \frac{1}{\sqrt{2\pi}}$$

Hence, the normalized eigen functions are

$$\Phi = \frac{1}{\sqrt{2\pi}} e^{im\phi} \qquad m = 0, \pm 1, \pm 2, \ldots$$

4. Hydrogen atom

Hydrogen atom is the only atom for which exact solution of Schrödinger equation is possible. In a similar manner, we can solve Schrödinger equation exactly for hydrogen-like atoms. Thus, here we shall discuss about hydrogen-like atom which consists of a nucleus of charge Ze and an electron of charge $-e$ which is revolving around the nucleus at a distance r. For the hydrogen atom, the atomic number Z is one. The potential energy for hydrogen-like atom is

$$V(r) = -\frac{Ze^2}{r}$$

The time-independent Schrödinger equation for relative motion is

$$\left(-\frac{\hbar^2}{2\mu} \nabla^2 - \frac{Ze^2}{r} \right) \psi(\vec{r}) = E \psi(\vec{r})$$

Since the potential is spherically symmetric, as it is independent of θ as well as ϕ, the Schrödinger equation can be solved by the technique of the separation of variables

$$\psi(r, \theta, \phi) = R(r) \, \Theta(\theta) \, \Phi(\phi)$$

The θ as well as ϕ, dependent equations can be solved in the same manner as discussed earlier. Hence the solution of the angular part is the spherical harmonics $Y_{lm}(\theta, \phi)$. The radial part is

$$\frac{1}{r^2} \frac{\mathrm{d}}{\mathrm{d}r} \left(r^2 \frac{\mathrm{d}R}{\mathrm{d}r} \right) + \frac{2\mu}{\hbar^2} \left[E - \frac{l(l+1)\hbar^2}{2\mu r^2} + \frac{Ze^2}{r} \right] R = 0$$

For convenience, let us introduce

$$\alpha = \sqrt{\frac{-8\mu E}{\hbar^2}} \qquad \lambda = \frac{Ze^2}{\hbar}\sqrt{\frac{\mu}{-2E}}$$

For the bound states, the energy E is negative and therefore both α and λ are real, and we get

$$\frac{d^2R}{dr^2} + \frac{2}{r}\frac{dR}{dr} + \left[-\frac{\alpha^2}{4} - \frac{l(l+1)}{r^2} + \frac{\lambda\alpha}{r}\right]R = 0 \tag{5.24}$$

For $\rho = \alpha r$, we have [2]

$$\frac{d^2R}{d\rho^2} + \frac{2}{\rho}\frac{dR}{d\rho} + \left[-\frac{1}{4} - \frac{l(l+1)}{\rho^2} + \frac{\lambda}{\rho}\right]R = 0 \tag{5.25}$$

Asymptotic solution

Let us first investigate asymptotic solution of this equation (5.25). When $\rho \to \infty$, equation (5.25) has the asymptotic form

$$\frac{d^2R}{d\rho^2} - \frac{1}{4}R = 0 \tag{5.26}$$

This equation has two solutions, $R = e^{\rho/2}$ and $R = e^{-\rho/2}$. Since $e^{\rho/2}$ diverges, as ρ tends to a large value, the first solution is not acceptable. Hence, the asymptotic solution of equation (5.25) is $R = e^{-\rho/2}$. Let the exact solution of equation (5.25) be

$$R(\rho) = e^{-\rho/2}\, F(\rho)$$

---

[2]On differentiating $\rho = \alpha r$ with respect to r, we have

$$\frac{d\rho}{dr} = \alpha$$

On differentiating with respect to r, we get

$$\frac{dR}{dr} = \frac{dR}{d\rho}\frac{d\rho}{dr} = \alpha\frac{dR}{d\rho}$$

Differentiating further with respect to r, we get

$$\frac{d^2R}{dr^2} = \alpha\frac{d^2R}{d\rho^2}\frac{d\rho}{dr} = \alpha^2\frac{d^2R}{d\rho^2}$$

Using the above expressions in equation (5.24), we get

$$\alpha^2\frac{d^2R}{d\rho^2} + \frac{2}{\rho}\alpha^2\frac{dR}{d\rho} + \left[-\frac{\alpha^2}{4} - \frac{l(l+1)\alpha^2}{\rho^2} + \frac{\lambda\alpha^2}{\rho}\right]R = 0$$

After cancellation of the term α^2, we get equation (5.25).

Therefore,

$$\frac{dR}{d\rho} = e^{-\rho/2}\frac{dF}{d\rho} - \frac{1}{2}e^{-\rho/2}F(\rho)$$

$$\frac{d^2R}{d\rho^2} = e^{-\rho/2}\frac{d^2F}{d\rho^2} - e^{-\rho/2}\frac{dF}{d\rho} + \frac{1}{4}e^{-\rho/2}F(\rho)$$

Using these relations in equation (5.25), we get

$$\rho^2\frac{d^2F}{d\rho^2} + \rho(2-\rho)\frac{dF}{d\rho} + [\rho\lambda - l(l+1) - \rho]\,F(\rho) = 0 \tag{5.27}$$

When $\rho = 0$, we get

$$l(l+1)F(0) = 0 \qquad \text{or} \qquad F(0) = 0, \quad \text{when} \quad l \neq 0$$

Therefore, we try a series solution $F(\rho)$ having no constant term, so that $F(\rho)$ is zero when $\rho = 0$:

$$F(\rho) = \sum_{k=0}^{\infty} a_k\rho^{c+k}$$

Therefore,

$$\frac{dF}{d\rho} = \sum_{k=0}^{\infty} a_k(c+k)\,\rho^{c+k-1}$$

$$\frac{d^2F}{d\rho^2} = \sum_{k=0}^{\infty} a_k(c+k)(c+k-1)\,\rho^{c+k-2}$$

Using these relations in equation (5.27), we get

$$\sum_{k=0}^{\infty} a_k(c+k)\,\rho^{c+k-1}\,\rho^{c+k} + (2-\rho)\sum_{k=0}^{\infty} a_k(c+k)\,\rho^{c+k}$$
$$+\Big[\rho\lambda - l(l+1) - \rho\Big]\sum_{k=0}^{\infty} a_k\rho^{c+k} = 0$$

$$\sum_k a_k(\lambda - 1 - c - k)\rho^{c+k+1}$$
$$+\sum_k a_k(c^2 + 2ck + k^2 + c + k - l^2 - l)\rho^{c+k} = 0 \tag{5.28}$$

Equation (5.28) is a polynomial equation and it can be possible when coefficient of each power of ρ is equal to zero. Equating the coefficient of ρ^c equal to zero, we get

$$a_0(c^2 + c - l^2 - l) = 0$$

Since $a_0 \neq 0$, we have

$$c^2 + c - l^2 - l = 0$$
$$(c - l)(c + l + 1) = 0$$

It gives

$$c = l \qquad \text{or} \qquad c = -(l+1)$$

For $c = -(l+1)$, we have

$$F(\rho) = \sum_{k=0}^{\infty} a_k\ \rho^{-(l+1)+k} = \frac{1}{\rho^{(l+1)}} \sum_{k=0}^{\infty} a_k\ \rho^k$$

It shows that F diverges with the increase of the value of ρ and therefore this solution is not acceptable.

For $c = l$, we have

$$F(\rho) = \sum_{k=0}^{\infty} a_k\ \rho^{l+k} = \rho^l \sum_{k=0}^{\infty} a_k\ \rho^k \tag{5.29}$$

and equation (5.28) becomes

$$\sum^{k} a_k(\lambda - 1 - l - k)\ \rho^{l+k+1} + \sum^{k} a_k\ k(k + 2l + 1)\ \rho^{l+k} = 0$$

Setting the coefficient of ρ^{l+k+1} equal to zero, we have

$$a_k(\lambda - 1 - l - k) + a_{k+1}(k+1)(k + 2l + 2) = 0$$

Thus, we get a recursion relation

$$a_{k+1} = \frac{l + k + 1 - \lambda}{(k+1)(k+2l+2)}\ a_k \tag{5.30}$$

With the help of this recursion relation the coefficients a_1, a_2, ... can be obtained in terms of a_0. For large values of k, we have

$$\frac{a_{k+1}}{a_k} = \frac{1}{k}$$

We know the expansion

$$e^{\rho} = \sum_{k=0}^{\infty} \frac{1}{k!}\ \rho^k = \sum_{k=0}^{\infty} a_k \rho^k$$

Equation (5.29) can be expressed as

$$F(\rho) = \rho^l \sum_{k=0}^{\infty} a_k \, \rho^k = \rho^l \, e^{\rho}$$

where

$$\frac{a_{k+1}}{a_k} = \frac{k!}{(k+1)!} = \frac{1}{k+1} \underset{k \to \infty}{\longrightarrow} \frac{1}{k}$$

It shows that when $k \to \infty$, the series for $F(\rho)$ behaves like $\rho^l e^{\rho}$ and therefore we have

$$R(\rho) = e^{-\rho/2} \rho^l e^{\rho} = \rho^l \, e^{\rho/2}$$

Since it gives an infinite series, the expression for $R(\rho)$ is not acceptable. The series must terminate after a certain value of k, say n'. For this $a_{n'+1}$ must be zero. From equation (5.30), we have

$$l + n' + 1 - \lambda = 0 \qquad n' = 0, 1, 2, \ldots$$

Energy eigenvalues

Introducing a new quantum number n as

$$n = l + n' + 1 = \lambda = \frac{Ze^2}{\hbar} \sqrt{\frac{\mu}{-2E}}$$

we have

$$E_n = -\frac{\mu Z^2 e^4}{2\hbar^2 n^2} \tag{5.31}$$

Since n' and l are integers including zero, we have

$$n = 1, 2, \ldots$$

As $n \geq (l+1)$, the highest value of l is $(n-1)$ and we have

$$l = 0, 1, 2, \ldots, (n-1)$$

This new quantum number n, known as the 'principal quantum number', determines the energy. For hydrogen atom, we have $Z = 1$ and the reduced mass μ is the mass of the electron m. This expression (5.31) for energy E_n is the same as obtained by Bohr for hydrogen atom on the basis of quantum ideas.

Since the energy E_n neither depends on l nor m, the energy levels are degenerate. Since for a given l, the magnetic quantum number m

can assume the values from $-l$ to l, the degeneracy due to m is $(2l+1)$ fold. Now, for a given value of n, the azimuthal quantum number l can assume the values from 0 to $(n-1)$. Thus, the combined degeneracy due to both m and l is

$$\sum_{l=0}^{n-1}(2l+1)$$

$$= 2\sum_{l=0}^{n-1} l + \sum_{l=0}^{n-1} 1 = 2\Big[\frac{(n-1)n}{2}\Big] + n = n^2$$

It shows an n^2 fold degeneracy of energy levels in hydrogen atom.[3]

Radial wave functions

The above restriction in energy reduces the series for $F(\rho)$ into a polynomial (5.29)

$$F(\rho) = \rho^l \sum_{k=0}^{n'} a_k\, \rho^k$$

Expressing

$$F(\rho) = \rho^l L(\rho)$$

we have

$$\frac{\mathrm{d}F}{\mathrm{d}\rho} = \rho^l\,\frac{\mathrm{d}L}{\mathrm{d}\rho} + l\,\rho^{l-1}\,L$$

$$\frac{\mathrm{d}^2F}{\mathrm{d}\rho^2} = \rho^l\,\frac{\mathrm{d}^2L}{\mathrm{d}\rho^2} + 2l\,\rho^{l-1}\,\frac{\mathrm{d}L}{\mathrm{d}\rho} + l(l-1)\,\rho^{l-2}\,L$$

Using these relations in equation (5.27), we get

$$\rho^{l+2}\,\frac{\mathrm{d}^2L}{\mathrm{d}\rho^2} + 2l\,\rho^{l+1}\,\frac{\mathrm{d}L}{\mathrm{d}\rho} + l(l+1)\rho^l\,L + (2-\rho)\rho^{l+1}\,\frac{\mathrm{d}L}{\mathrm{d}\rho}$$

$$+l(2-\rho)\rho^l\,L + [\rho\lambda - l(l+1) - \rho]\,\rho^l\,L = 0$$

On rearranging these equation, putting $\lambda = n$ and canceling a factor ρ^{l+1} throughout, we get

$$\rho\frac{\mathrm{d}^2L(\rho)}{\mathrm{d}\rho^2} + (2l+2-\rho)\frac{\mathrm{d}L(\rho)}{\mathrm{d}\rho} + (n-l-1)\,L(\rho) = 0 \qquad (5.32)$$

[3]When we account for the electron spin, each level further becomes two fold degenerate, as spin quantum number s is 1/2. After accounting for the electron spin, the degeneracy of the energy levels in the hydrogen atom becomes $2n^2$.

Let us consider the associated Laguerre polynomial of order p and degree q, denoted by $L_q^p(\rho)$, satisfying equation

$$\rho\frac{\mathrm{d}^2L_q^p}{\mathrm{d}\rho^2}+(p+1-\rho)\frac{\mathrm{d}L_q^p}{\mathrm{d}\rho}+q\ L_q^p(\rho)=0 \tag{5.33}$$

Comparison of equations (5.32) and (5.33) gives $p=(2l+1)$ and $q=(n-l-1)$. Hence,

$$R_{nl}(r)=N\ \mathrm{e}^{-\rho/2}\rho^l\ L_{n-l-1}^{2l+1}(\rho)$$

The normalization gives

$$\int_0^\infty R_{nl}^2(r)r^2\ \mathrm{d}r=1$$

$$\frac{n^3\hbar^6}{8\mu^3Z^3\mathrm{e}^6}|N^2|\int_0^\infty \mathrm{e}^{-\rho}\rho^{2l}\left(L_{n-l-1}^{2l+1}\right)^2\ \rho^2\ \mathrm{d}\rho=1 \tag{5.34}$$

Using the orthonormalization properties of the associated Laguerre polynomials, we get [4]

$$\frac{n^3\hbar^6}{8\mu^3Z^3\mathrm{e}^6}|N^2|\frac{2n\ (n+l)!}{(n-l-1)!}=1$$

$$|N|=\pm\left\{\left(\frac{2Z\mu\mathrm{e}^2}{n\hbar^2}\right)^3\frac{(n-l-1)!}{2n(n+1)!}\right\}^{\frac{1}{2}}$$

[4] Orthonormality criterion for associated Laguerre polynomials is

$$\int_0^\infty \mathrm{e}^{-x}\ x^k\ L_n^k(x)\ L_m^k(x)\ \mathrm{d}x=\frac{(m+k)!}{m!}\ \delta_{mn}$$

Now,

$$x\ L_m^k(x)=(2m+k+1)\ L_m^k(x)-(m+k)\ L_{m-1}^k(x)-(m+1)\ L_{m+1}^k(x)$$

Thus,

$$\int_0^\infty \mathrm{e}^{-x}\ x^{k+1}\ L_n^k(x)\ L_m^k(x)\ \mathrm{d}x=\int_0^\infty \mathrm{e}^{-x}\ x^k\Big[(2m+k+1)\ L_m^k(x)$$

$$-(m+k)\ L_{m-1}^k(x)-(m+1)\ L_{m+1}^k(x)\Big]L_m^k(x)\ \mathrm{d}x$$

$$=(2m+k+1)\int_0^\infty \mathrm{e}^{-x}\ x^k\ L_m^k(x)\ L_m^k(x)\ \mathrm{d}x=(2m+k+1)\frac{(m+k)!}{m!}$$

Hence, the normalized radial wave function is

$$R_{nl}(r) = -\left\{ \left(\frac{2\mu e^2 Z}{n\hbar^2} \right)^3 \frac{(n-l-1)!}{2n(n+1)!} \right\}^{\frac{1}{2}} e^{-\rho/2} \rho^l L_{n-l-1}^{2l+1}(\rho)$$

The negative sign is chosen to make R_{10} positive. The presence of the factor r^l makes the wave function zero at $r = 0$ except for the s-state.

Wave functions of hydrogen-like atom

The complete wave function for hydrogen-like atom is

$$\psi_{nlm}(r, \theta, \phi) = R_{nl}(r) Y_{lm}(\theta, \phi)$$

where $n = 1, 2, \ldots$ $\quad l = 0, 1, 2, \ldots$ $\quad m = 0, \pm 1, \pm 2, \ldots, \pm l$.
Some explicit forms of the wave functions are

$$\psi_{100} = \frac{1}{\sqrt{\pi}} \left(\frac{Z}{a_0} \right)^{3/2} e^{-Zr/a_0}$$

$$\psi_{200} = \frac{1}{\sqrt{\pi}} \left(\frac{Z}{2a_0} \right)^{3/2} \left(1 - \frac{Zr}{2a_0} \right) e^{-Zr/2a_0}$$

$$\psi_{211} = \frac{1}{8\sqrt{\pi}} \left(\frac{Z}{a_0} \right)^{5/2} r \, e^{-Zr/2a_0} (\sin\theta) e^{i\phi}$$

$$\psi_{210} = \frac{1}{\sqrt{\pi}} \left(\frac{Z}{2a_0} \right)^{5/2} r \, e^{-Zr/2a_0} \cos\theta$$

$$\psi_{21,-1} = \frac{1}{8\sqrt{\pi}} \left(\frac{Z}{a_0} \right)^{5/2} r \, e^{-Zr/2a_0} (\sin\theta) e^{-i\phi}$$

Exercise 2. Show that the eigen functions of hydrogen atom

$$\psi_{200} = \frac{1}{\sqrt{\pi}} \left(\frac{Z}{2a_0} \right)^{3/2} \left(1 - \frac{Zr}{2a_0} \right) e^{-Zr/2a_0}$$

$$\psi_{211} = \frac{1}{8\sqrt{\pi}} \left(\frac{Z}{a_0} \right)^{5/2} r \, e^{-Zr/2a_0} (\sin\theta) e^{i\phi}$$

$$\psi_{210} = \frac{1}{\sqrt{\pi}} \left(\frac{Z}{2a_0} \right)^{5/2} r \, e^{-Zr/2a_0} \cos\theta$$

$$\psi_{21,-1} = \frac{1}{8\sqrt{\pi}} \left(\frac{Z}{a_0} \right)^{5/2} r \, e^{-Zr/2a_0} (\sin\theta) e^{-i\phi}$$

are orthogonal to each other.

Solution: To check the orthogonality of two eigen functions ψ_i and ψ_j, we calculate the integral

$$\int_0^\infty \int_0^\pi \int_0^{2\pi} \psi_i^* \, \psi_j \, r^2 \sin\theta \, dr \, d\theta \, d\phi$$

For any pair of the eigen functions, the ϕ-dependence part has an integral of the form

$$\int_0^{2\pi} e^{im\phi} d\phi$$

where m is a non-zero integer. The value of this integral is always zero as the following.

$$\int_0^{2\pi} e^{im\phi} d\phi = \left[\frac{e^{im\phi}}{im}\right]_0^{2\pi} = \frac{e^{i2m\pi} - 1}{im} = 0$$

It shows that the given eigen functions are orthogonal to each other.

Exercise 3. The normalized eigen function for the ground state of hydrogen atom is

$$\psi_{100} = \frac{1}{\sqrt{\pi}} \left(\frac{1}{a_0}\right)^{3/2} e^{-Zr/a_0}$$

Calculate the expectation value of the radius vector r of the electron in the ground state.

Solution: The expectation value $\langle r \rangle$ is

$$\langle r \rangle = \int_0^\infty \int_0^\pi \int_0^{2\pi} \psi_{100}^* \, r \, \psi_{100} \, r^2 \sin\theta \, dr \, d\theta \, d\phi$$

Since the wave function ψ_{100} does not depend on θ as well as ϕ, the integration over ϕ gives 2π and the integration over θ gives 2. Using the expression for the ground state eigen function, we have

$$\langle r \rangle = \frac{1}{\pi a_0^3} 4\pi \int_0^\infty r^3 \, e^{-2Zr/a_0} \, dr = \frac{4}{a_0^3} \frac{3!}{(2/a_0)^4} = \frac{3}{2} a_0$$

Hence, the expectation value of the radius vector r of electron in the ground state of hydrogen atom is $3a_0/2$.

5. Free particle

In the preceding chapter, we discussed a free-particle in one-dimensional space. Here, we shall consider a free particle in a three-dimensional space. For a free particle, potential energy is zero, and the Schrödinger equation is

$$-\frac{\hbar^2}{2m}\nabla^2\psi(\vec{r}) = E\psi(\vec{r}) \tag{5.35}$$

Plane wave solution

For a plane wave solutions, in the Cartesian coordinates, we express the wave function as

$$\psi(\vec{r}) = X(x)\, Y(y)\, Z(z) \tag{5.36}$$

where X, Y, Z are functions of x, y, z, respectively. Using equation (5.36) in (5.35), we can write

$$\frac{\mathrm{d}^2X(x)}{\mathrm{d}x^2} + k_x^2 X(x) = 0, \qquad k_x^2 = \frac{2mE_x}{\hbar^2} \tag{5.37}$$

$$\frac{\mathrm{d}^2Y(y)}{\mathrm{d}y^2} + k_y^2 Y(y) = 0, \qquad k_y^2 = \frac{2mE_y}{\hbar^2}$$

$$\frac{\mathrm{d}^2Z(z)}{\mathrm{d}z^2} + k_z^2 Z(z) = 0, \qquad k_z^2 = \frac{2mE_z}{\hbar^2}$$

$$\tag{5.38}$$

where

$$E = E_x + E_y + E_z = \frac{\hbar^2}{2m}(k_x^2 + k_y^2 + k_z^2) = \frac{\hbar^2k^2}{2m}$$

Solution of equation (5.37) is

$$X(x) = C_1\ \mathrm{e}^{ik_xx}$$

and $X(x)$ will be finite for all real values of k_x. Similarly, we have

$$Y(y) = C_2\ \mathrm{e}^{ik_yy} \qquad Z(z) = C_3\ \mathrm{e}^{ik_zz}$$

On combining these three solutions, we have

$$\psi(\vec{r}) = X(x)\, Y(y)\, Z(z) = A\ \mathrm{e}^{ik_xx}\ \mathrm{e}^{ik_yy}\ \mathrm{e}^{ik_zz} = A\ \mathrm{e}^{i\vec{k}\cdot\vec{r}}$$

where A is the normalization constant. Since the plane waves extend up to the infinite distances, the eigen functions for a plane wave are not normalizable. Under such situation, we consider box normalization by restricting the domain of $\psi(\vec{r})$ as taken to be finite cube of side L centered at the origin. The normalization constant for each coordinate is $1/\sqrt{L}$. Thus, the box normalized plane wave function is

$$\psi(\vec{r}) = \frac{1}{\sqrt{L^3}}\, e^{i\vec{k}.\vec{r}}$$

Spherical wave solution

In the spherical polar coordinates (r, θ, ϕ), the Schrödinger equation for a free particle is

$$\frac{1}{r^2}\frac{\partial}{\partial r}\left(r^2\frac{\partial\psi}{\partial r}\right) + \frac{1}{r^2\,\sin\theta}\frac{\partial}{\partial\theta}\left(\sin\theta\frac{\partial\psi}{\partial\theta}\right) + \frac{1}{r^2\,\sin^2\theta}\frac{\partial^2\psi}{\partial\phi^2} + k^2\psi = 0$$

where $k^2 = 2mE/\hbar^2$. This Schrödinger equation can be solved by the technique of the separation of variables

$$\psi(r,\theta,\phi) = R(r)\;\Theta(\theta)\;\Phi(\phi)$$

The θ as well as ϕ, dependent equations can be solved in the same manner as discussed earlier. Hence, the solution of the angular part is the spherical harmonics $Y_{lm}(\theta,\phi)$.

$$\frac{1}{\sin\theta}\frac{\partial}{\partial\theta}\left(\sin\theta\frac{\partial Y_{lm}}{\partial\theta}\right) + \frac{1}{\sin^2\theta}\frac{\partial^2 Y_{lm}}{\partial\phi^2} + l(l+1)Y_{lm} = 0$$

The radial part is

$$\frac{1}{r^2}\frac{d}{dr}\left(r^2\frac{dR_l}{dr}\right) + \left[\frac{2mE}{\hbar^2} - \frac{l(l+1)}{r^2}\right]R_l(r) = 0$$

$$\frac{d^2R_l}{dr^2} + \frac{2}{r}\frac{dR_l}{dr} + \left[k^2 - \frac{l(l+1)}{r^2}\right]R_l(r) = 0 \qquad (5.39)$$

where $k^2 = 2mE/\hbar^2$. Introducing a new variable $\rho = kr$, equation (5.39) reduces to[5]

$$\frac{d^2R_l(\rho)}{d\rho^2} + \frac{2}{\rho}\frac{dR_l}{d\rho} + \left[1 - \frac{l(l+1)}{\rho^2}\right]R_l(\rho) = 0 \qquad (5.40)$$

---

[5]For $\rho = kr$, we have

$$\frac{d\rho}{dr} = k$$

Hence,

$$\frac{dR_l}{dr} = \frac{dR_l}{d\rho}\frac{d\rho}{dr} = k\,\frac{dR_l}{d\rho}$$

By writing

$$R(\rho) = Z(\rho)\ \rho^{-1/2}$$

we have

$$\frac{dR}{d\rho} = \frac{dZ}{d\rho}\ \rho^{-1/2} - \frac{1}{2}\ \rho^{-3/2}\ Z$$

and

$$\frac{d^2R}{d\rho^2} = \frac{d^2Z}{d\rho^2}\ \rho^{-1/2} - \rho^{-3/2}\ \frac{dZ}{d\rho} + \frac{3}{4}\ \rho^{-5/2}\ Z$$

Using these expressions in equation (5.40), we get

$$\frac{d^2Z(\rho)}{d\rho^2} + \frac{1}{\rho}\frac{dZ(\rho)}{d\rho} + \left\{1 - \left[\frac{l + (1/2)}{\rho^2}\right]^2\right\} Z(\rho) = 0$$

This is a Bessel equation and its general solution is

$$Z(\rho) = A' J_{l+(1/2)}(\rho) + B'\ J_{-[l+(1/2)]}(\rho)$$

Thus,

$$R(\rho) = \frac{A'}{\rho^{1/2}} J_{l+(1/2)}(\rho) + \frac{B'}{\rho^{1/2}} J_{-[l+(1/2)]}(\rho)$$

$$= A\ j_l(\rho) + B\ \eta_l(\rho)$$

where A and B are constants, $j_l(\rho)$ and $\eta_l(\rho)$ are the spherical Bessel functions and the spherical Neumann functions, respectively. These functions are defined as

$$j_l(\rho) = \left(\frac{\pi}{2\rho}\right)^{1/2}\ J_{(l+1/2)}(\rho)$$

and

$$\eta_l(\rho) = (-1)^{l+1}\left(\frac{\pi}{2\rho}\right)^{1/2}\ J_{-(l+1/2)}(\rho)$$

Differentiating again, we get

$$\frac{d^2R_l}{dr^2} = k\ \frac{d^2R_l}{d\rho^2}\ \frac{d\rho}{dr} = k^2\ \frac{d^2R_l}{d\rho^2}$$

Using these expressions in equation (5.39), we get

$$k^2\ \frac{d^2R_l}{d\rho^2} + \frac{2k^2}{\rho}\frac{dR_l}{d\rho} + \left[k^2 - \frac{l(l+1)k^2}{\rho^2}\right] R_l(r) = 0$$

On canceling a factor k^2 throughout, we get equation (5.40).

Explicit expressions for some spherical Bessel functions and spherical Neumann functions are

$$j_0(\rho) = \frac{\sin\rho}{\rho}$$
$$j_1(\rho) = \frac{\sin\rho}{\rho^2} - \frac{\cos\rho}{\rho}$$
$$j_2(\rho) = \left(\frac{3}{\rho^3} - \frac{1}{\rho}\right)\sin\rho - \frac{3}{\rho^2}\cos\rho$$

and

$$\eta_0(\rho) = -\frac{\cos\rho}{\rho}$$
$$\eta_1(\rho) = -\frac{\cos\rho}{\rho^2} - \frac{\sin\rho}{\rho}$$
$$\eta_2(\rho) = -\left(\frac{3}{\rho^3} - \frac{1}{\rho}\right)\cos\rho - \frac{3}{\rho^2}\sin\rho$$

For small and large values of ρ, the functions j_l and η_l vary as

$$j_l(\rho) \underset{\rho\to 0}{\longrightarrow} \frac{\rho^l}{1.3.5\ldots(2l+1)}; \qquad j_l(\rho) \underset{\rho\to\infty}{\longrightarrow} \frac{1}{\rho}\sin\left(\rho - \frac{l\pi}{2}\right)$$

$$\eta_l(\rho) \underset{\rho\to 0}{\longrightarrow} \frac{1.3.5\ldots(2l-1)}{\rho^l+1}; \qquad \eta_l(\rho) \underset{\rho\to\infty}{\longrightarrow} -\frac{1}{\rho}\cos\left(\rho - \frac{l\pi}{2}\right)$$

The solution of the free particle equation for a definite energy $E = \hbar^2k^2/2m$ and a definite angular momentum $\sqrt{l(l+1)}\,\hbar$ can be expressed as

$$\psi_{k,l}(r,\theta,\phi) = [A_l\ j_l(kr) + B_l\ \eta_l(kr)]Y_{lm}(\theta,\phi)$$

A general solution for a definite energy is

$$\psi_k(r,\theta,\phi) = \sum_{l=0}^{\infty}[A_l\cdot j_l(kr) + B_l\ \eta_l(kr)]Y_{lm}(\theta,\phi)$$

Since $j_l(kr)$ becomes infinite at the origin, the requirement that ψ must be finite everywhere makes the constant $B_l = 0$. Thus, the wave function for a free particle is

$$\psi_k(r,\theta,\phi) = \sum_{l=0}^{\infty} A_l\ j_l(kr)\ Y_{lm}(\theta,\phi)$$

6. Three-dimensional spherically symmetric square-well potential

Let us consider a three-dimensional spherically symmetric square-well potential (Figure 2) expressed as

$$V(r) = \begin{cases} -V_0 & 0 < r < a \\ 0 & r > a \end{cases}$$

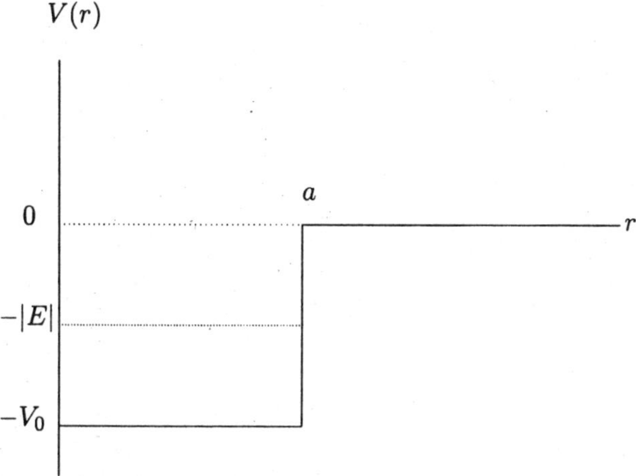

Figure 2: Three-dimensional spherically symmetric square-well potential.

Total energy of a particle in the potential well is negative and its magnitude less than the depth of the well. Since the potential is spherically symmetric, as it is independent of θ as well as ϕ, the Schrödinger equation can be solved by the technique of the separation of variables

$$\psi(r,\theta,\phi) = R(r)\ \Theta(\theta)\ \Phi(\phi)$$

The θ as well as ϕ, dependent equations can be solved in the same manner as discussed earlier. Hence the solution of the angular part is the spherical harmonics $Y_{lm}(\theta,\phi)$. The radial part is

$$\frac{1}{r^2}\frac{\mathrm{d}}{\mathrm{d}r}\left(r^2\frac{\mathrm{d}R_l}{\mathrm{d}r}\right) + \left[\frac{2\mu(V_o - |E|)}{\hbar^2} - \frac{l(l+1)}{r^2}\right] R_l = 0, \qquad 0 < r < a \quad (5.41)$$

and

$$\frac{1}{r^2}\frac{\mathrm{d}}{\mathrm{d}r}\left(r^2\frac{\mathrm{d}R_l}{\mathrm{d}r}\right) + \left[\frac{-2\mu|E|}{\hbar^2} - \frac{l(l+1)}{r^2}\right] R_l = 0, \qquad r > a \qquad (5.42)$$

As discussed in the preceding section, consideration of a finite value for l would involve the use of the spherical Bessel and spherical Neumann functions. For a simple treatment, let us consider the case of the s-state ($l = 0$) which is sufficient for understanding the ground state of most of the systems of our interest.

For solving equations (5.41) and (5.42), let us use

$$R_l = \frac{u}{r} \qquad k_1^2 = \frac{2\mu|E|}{\hbar^2} \qquad k_2^2 = \frac{2\mu(V_o - |E|)}{\hbar^2} \tag{5.43}$$

so the equations (5.41) and (5.42) for the s-state are[6]

$$\frac{d^2u}{dr^2} + k_2^2\ u = 0 \qquad 0 < r < a \tag{5.44}$$

and

$$\frac{d^2u}{dr^2} - k_1^2\ u = 0 \qquad r > a \tag{5.45}$$

Solutions of equations (5.44) and (5.45) are, respectively

$$u(r) = A\ \sin(k_2 r) + B\ \cos(k_2 r)$$

and

$$u(r) = C\ e^{-k_1 r} + D\ e^{k_1 r}$$

where A, B, C, D are constants. Since $u(r)$ must tend to zero as r tends to zero and therefore, the constant B has to be zero. Further, as r tends to infinite, $u(r)$ must tend to zero. This says that the constant D should be zero. Hence, the final solutions are

$$u(r) = A\ \sin(k_2 r) \qquad 0 < r < a$$

and

$$u(r) = C\ e^{-k_1 r} \qquad r > a$$

Applying the condition of continuity at $x = a$ for $u(r)$ and its derivative, we get

$$A\ \sin(k_2 a) = C\ e^{-k_1 a} \tag{5.46}$$

---

[6]For $R_l - u/r$, we have

$$\frac{dR_l}{dr} = \frac{1}{r}\frac{du}{dr} - \frac{u}{r^2} \qquad r^2\frac{dR_l}{dr} = r\frac{du}{dr} - u$$

$$\frac{d}{dr}\left(r^2\frac{dR_l}{dr}\right) = r\frac{d^2u}{dr^2} + \frac{du}{dr} - \frac{du}{dr} \qquad \frac{1}{r^2}\frac{d}{dr}\left(r^2\frac{dR_l}{dr}\right) = \frac{1}{r}\frac{d^2u}{dr^2}$$

Further, for the s state, we put $l = 0$.

and

$$A\ k_2\ \cos(k_2 a) = -C\ k_1\ e^{-k_1 a} \tag{5.47}$$

Dividing equation (5.47) by (5.46), we get

$$k_2\ \cot(k_2 a) = -k_1 \tag{5.48}$$

As was done in the case of one-dimensional square-well potential, this equation (5.48) can be solved graphically and can be found that no solution exists unless $V_0 a^2 > (\pi^2\hbar^2/8\mu)$ and there will be a bound state when

$$\frac{\pi^2\hbar^2}{8\mu} < V_0 a^2 \leq \frac{9\pi^2\hbar^2}{8\mu} \tag{5.49}$$

7. Deutron

An application of three-dimensional spherically symmetric square-well potential is the ground state of the deutron nucleus. Nucleus of deutron is the smallest one where nucleons of two kinds (proton and neutron) are held together. There is one proton and one neutron. Study of this two particle system helps us in understanding the nature of nuclear force, nuclear size and other parameters about the nucleus. The binding energy of deutron $|E|$ is 2.226 MeV and thus k_1 can be calculated with the help of the equation (5.43).

Equation (5.49) can be used to calculate the parameters of the square-well potential. Equation (5.48) gives

$$\cot(k_2 a) = -\frac{k_1}{k_2} = \left(\frac{|E|}{V_o - |E|}\right)^{1/2}$$

Since the energy $|E|$ is small as compared to the potential depth V_o, we get an approximate range-depth relation and can set

$$\frac{|E|}{V_o - |E|} \approx 0$$

and therefore,

$$\cot(k_2 a) = 0 \qquad \text{or} \qquad k_2 a = \frac{\pi}{2}$$

Hence,

$$\frac{2\mu V_0}{\hbar^2}\ a^2 \approx \frac{\pi^2}{4} \qquad \text{or} \qquad V_0\ a^2 = \frac{\pi^2\hbar^2}{8\mu} \tag{5.50}$$

where $m = m_p = m_n$, the mass of a nucleon. The reduced mass μ is $m/2$. The range of the nuclear potential is approximately 1.4×10^{-15} m. Therefore,

$$V_0 = \frac{\pi^2\hbar^2}{4ma^2} = \frac{(6.626 \times 10^{-34})^2}{16 \times (1.673 \times 10^{-27}) \times (1.4 \times 10^{-15})^2}$$

$$= 8.37 \times 10^{-12}\text{J} = \frac{8.37 \times 10^{-12}}{1.602 \times 10^{-19}} = 52.2 \text{ MeV}$$

It justifies the assumption $V_o >> |E|$. Further information can be obtained from the wave function in the region $r > a$. The form of the wave function $\exp(-k_1 r)$ suggests that $1/k_1$ can be taken as a measure of the spatial extent of the deutron. Hence, the probability of finding the deutron is maximum in the region $0 < r < (1/k_1)$. Now,

$$\frac{1}{k_1} = \left(\frac{\hbar^2}{m|E|}\right)^{1/2} \tag{5.51}$$

From equations (5.50) and (5.51), we have

$$\frac{1/k_1}{a} = \frac{2}{\pi}\sqrt{\frac{V_0}{|E|}} = \frac{2}{\pi}\sqrt{\frac{52.2}{2.226}} \approx 3$$

It shows that in the range of the wave function of the deutron, the $(1/k_1)$ is considerably greater than the range a of the potential. Hence, the deutron is loosely bound structure with the nucleons spending considerable time outside the range of the potential.

8. Problems and questions

1. Show that for a spherically symmetric potential, Schrödinger equation can be decomposed into three equations, each of them having one variable.

2. Solve the Schrödinger equation for a particle moving in a spherically symmetric potential.

3. Derive energies and wave functions for motion of electron in a hydrogen atom.

4. What are spherical harmonics? Are they mutually orthogonal?

5. Derive an expression for a spherical harmonics.

6. Show that the equation of motion of a system of two interacting particles can be resolved into two one-particle equations, one representing the translational motion of the center-of-mass and the other the relative motion of the two particles.

7. Discuss motion of a rigid rotator. Sow that the energy levels of the rotator are degenerate.

8. Neglecting the electron spin degeneracy, show that the energy levels in a hydrogen atom have n^2 fold degeneracy.

9. Solve the Schrödinger equation for a particle moving in a three-dimensional spherically symmetric square-well potential.

10. Write notes on the following

    (i) Plane wave solution for a free particle

    (ii) Spherical wave solution for a free particle

    (iii) Rigid rotator

    (iv) Free particle

    (v) Deutron

Conservation Laws

Treatment of a system in science and technology becomes quite simple when its symmetries are known. Knowledge of such symmetries of the system helps in getting information about certain constants of motion of the system, which help in deriving information about the system. In the present chapter, we shall discuss about the connection between some important symmetry operations and the associated conservation laws of a system.

1. Symmetry transformations

When a unitary transformation U is very close to a unit operation I, it is said to be infinitesimal. We can then write

$$U = I + i\epsilon G$$

where G is a Hermitian operator and ϵ is a real small parameter. Let us consider

$$UU^\dagger = (I + i\epsilon G)(I - i\epsilon G^\dagger) \approx I + i\epsilon(G - G^\dagger)$$

where the term containing ϵ^2 is neglected. For U to be unitary, we have $UU^\dagger = I$, showing that

$$G - G^\dagger = 0 \qquad \text{or} \qquad G = G^\dagger$$

The operator G is known as the generator of the infinitesimal unitary transformation.

Suppose in an infinitesimal unitary transformation, a wave function Ψ is transformed to Ψ', so that

$$\Psi' = (I + i\epsilon G)\Psi$$

Let the change in Ψ under infinitesimal unitary transformation is expressed as $\delta\Psi$, so that $\Psi' = \Psi + \delta\Psi$. Thus, we have

$$\Psi + \delta\Psi = (I + i\epsilon G)\Psi$$
$$\delta\Psi = i\epsilon G\Psi$$

For a dynamical operator, the transformed operator A' is

$$\begin{aligned} A' = UAU^{\dagger} &= (I + i\epsilon G)A(I - i\epsilon G) \\ &\approx A + i\epsilon(GA - AG) = A + i\epsilon[G, A] \end{aligned} \tag{6.1}$$

where the terms containing ϵ^2 is neglected. A dynamical variable A is said to be invariant under the unitary transformation when A' is itself A. That is,

$$A' = UAU^{\dagger} = A \tag{6.2}$$

From equations (6.1) and (6.2), we get

$$[G, A] = 0$$

Thus, a dynamical variable A is said to be invariant when it commutes with the generator G of the infinitesimal unitary transformation.

A case of interest is where a dynamical variable is Hamiltonian H of the system. Then the condition for conservation of H becomes

$$[G, H] = 0$$

Thus, when the generator G of the infinitesimal unitary transformation commutes with the Hamiltonian H of a system, the Hamiltonian is constant of motion. The transformation that leaves the Hamiltonian invariant is known as a symmetry transformation. Hence, the existence of a symmetry transformation implies the conservation of a dynamical variable of the system.

The symmetries of a system may be associated with geometrical operations, such as displacement, rotation, and inversion in space and time. Such symmetries are known as the geometrical symmetries. Out of them, the symmetries associated with displacement in space as well as in time and the rotation in space come under the category of continuous symmetries whereas those associated with space inversion and time reversal come under the category of discrete symmetries. In the following sections, we shall discuss about nature and properties of some operators and the conservation laws associated with these space-time symmetries.

2. Conservation of momentum

(i) Translation in space

To account for the momentum of a particle, let us consider a particle moving, say, along x-axis, as shown in Figure 1. Our aim here is to find out the effect of linear motion on the wave function of a particle. For that, let us consider two inertial reference frames S and S′ where S′ is moving relative to S along the x-axis. Let x and x' be x-coordinates of a point P, placed on the x-axis, measured in S and S′, respectively, and at this moment the distance between O and O′ is ρ. Suppose ψ and ψ' are the wave functions of the particle in S and S′. We assume that the physical properties of an isolated system do not change during linear motion by an arbitrary displacement ρ. Thus, for the system at P, the wave functions in these two reference frames must be the same. Thus, we have

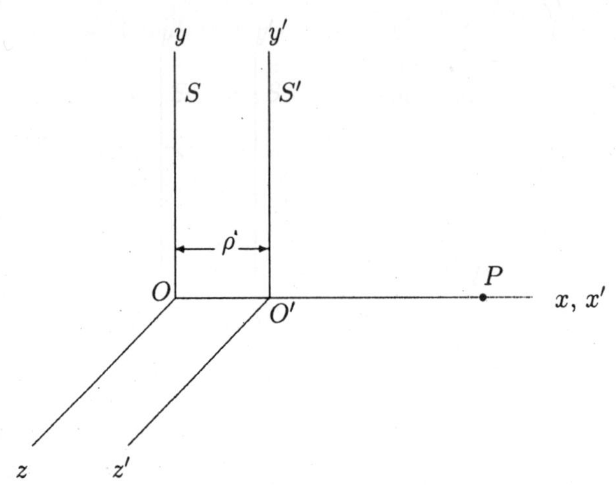

Figure 1: Reference frame S′ is shifted through a distance ρ relative to S along the direction of x-axis.

$$\psi'(x') = \psi(x) \qquad \text{and} \qquad x' = x - \rho$$

Therefore,

$$\psi'(x-\rho) = \psi(x) \qquad \text{or} \qquad \psi'(x) = \psi(x+\rho) \tag{6.3}$$

When ρ is infinitesimal, right side of equation (6.3) can be expanded to get

$$\psi'(x) = \psi(x) + \rho\,\frac{\partial\psi(x)}{\partial x} + \ldots \;= (1 + i\,\rho\,G_x)\psi(x)$$

where G_x is defined as

$$G_x \equiv -i\,\frac{\partial}{\partial x} = \frac{p_x}{\hbar} \tag{6.4}$$

As ψ and ψ' are wave functions of the same particle in S and S′, respectively, they are expressing the same physical system in terms of two different bases in the Hilbert space. The wave function ψ is transformed to ψ' by the action of the operator $i\rho G_x$ on ψ. Hence, the G_x behaves as the generator of infinitesimal translation in the x-direction. Equation (6.4) shows that the momentum operator behaves as the generator of infinitesimal translation in space. We have

$$\psi'(x) = \left(1 + \frac{i\rho p_x}{\hbar}\right)\psi(x) \tag{6.5}$$

Translation through some finite distance ρ' can be considered as a succession of N infinitesimal translations, so that $\rho' = N\rho$ and equation (6.5) gives

$$\begin{aligned}
\psi'(x) &= \lim_{N\to\infty}\left(1 + \frac{i\rho p_x}{\hbar}\right)^N \psi(x) \\
&= \lim_{N\to\infty}\Big[1 + N\left(\frac{i\rho p_x}{\hbar}\right) + \frac{N(N-1)}{2!}\left(\frac{i\rho p_x}{\hbar}\right)^2 \\
&\qquad + \frac{N(N-1)(N-2)}{3!}\left(\frac{i\rho p_x}{\hbar}\right)^3 + \ldots\Big] \\
&= 1 + N\left(\frac{i\rho p_x}{\hbar}\right) + \frac{N^2}{2!}\left(\frac{i\rho p_x}{\hbar}\right)^2 + \frac{N^3}{3!}\left(\frac{i\rho p_x}{\hbar}\right)^3 + \ldots \\
&= \sum_{r=0}^{\infty}\frac{1}{r!}\left(\frac{iN\rho p_x}{\hbar}\right)^r \psi(x) \\
&= \exp\left(-\frac{iN\rho p_x}{\hbar}\right)\psi(x) = \exp\left(-\frac{i\rho' p_x}{\hbar}\right)\psi(x)
\end{aligned}$$

This relation can be generalized to three-dimensional space as

$$\psi'(\vec{r}) = \exp\left(\frac{i\,\vec{\rho}\,'\cdot\vec{p}}{\hbar}\right)\psi(\vec{r})$$

where $\vec{p}$ is the momentum of the particle. For a system of n particles with momenta $\vec{p}_1, \vec{p}_2, \vec{p}_3, \ldots, \vec{p}_n$, total momentum $\vec{p}$ is $\vec{p} = \vec{p}_1 + \vec{p}_2 + \vec{p}_3 + \ldots + \vec{p}_n$ and $\vec{\rho}$ is the position vector of O$'$ relative to O.

(ii) Conservation of momentum

From generalization of equations (6.5) in three-dimensional space, the infinitesimal unitary translation operator is

$$U = I + \left(\frac{i\,\vec{\rho}\,'\cdot\vec{p}}{\hbar}\right)$$

From this we have

$$H' = UHU^{\dagger} = \left(I + \frac{i\,\vec{\rho}'\,.\,\vec{p}}{\hbar}\right) H \left(I - \frac{i\,\vec{\rho}'\,.\,\vec{p}}{\hbar}\right)$$

$$\approx H + \frac{i\,\vec{\rho}'}{\hbar}.\left(\vec{p}\,H - H\,\vec{p}\right) = H + \frac{i\,\vec{\rho}'}{\hbar}.\left[\vec{p}\;H\right]$$

Thus, invariance of the Hamiltonian under translation in space requires that the momentum operator $\vec{p}$ must commute with the Hamiltonian H. It shows that the linear momentum $\vec{p}$ of a system is conserved when $\vec{p}$ commutes with H. We know that $\vec{p}$ commutes with H when the H does not contain a potential energy term which generally depend on the position coordinates of the system.

3. Conservation of energy

(i) Translation in time

To account for the energy of a particle, let us consider a particle placed at a point x is observed at two different times t and $t + \tau$. Our aim here is to find out the effect of time translation on the wave function of the particle. Let $\Psi(x, t)$ and $\Psi(x, t + \tau)$ be the wave functions of the particle at two different times. We assume that the physical properties of an isolated system do not change during an arbitrary time translation τ. Thus, we have

$$\Psi'(x, t) = \Psi(x, t + \tau) \tag{6.6}$$

When τ is infinitesimal, right side of equation (6.6) can be expanded to get

$$\Psi'(x, t) = \Psi(x, t) + \tau \frac{\partial \Psi(x, t)}{\partial t} + \ldots$$

$$= \Psi(x, t) + \frac{i\tau}{\hbar}\left(-i\hbar\frac{\partial}{\partial t}\right)\Psi(x, t)$$

$$= \left[1 + i\tau\left(-\frac{H}{\hbar}\right)\right]\Psi(x, t)$$

Thus, $(-H/\hbar)$ may be taken as the generator of infinitesimal translation in time.

(ii) Conservation of energy

For the translation in time, unitary operator may be expressed as

$$U = I - \left(\frac{i\tau H}{\hbar}\right)$$

where H is the Hamiltonian which is independent of time. From this we have

$$H' = UHU^{\dagger} = \left(I - \frac{i\tau H}{\hbar}\right)H\left(I + \frac{i\tau H}{\hbar}\right) = H$$

Thus, we have

$$UH = HU \qquad \text{or} \qquad [U, H] = 0$$

This shows that U commutes with H. The independence of H means that the total energy of the system is conserved. Hence, total energy of a system is conserved when the system is invariant under translation in time. When the Hamiltonian H depends on time, total energy of the system will not be conserved.

4. Conservation of angular momentum

(i) Rotation in space

In order to account for the angular momentum of a particle, let us consider two inertial reference frames S and S′ having a common origin at O (Figure 2).

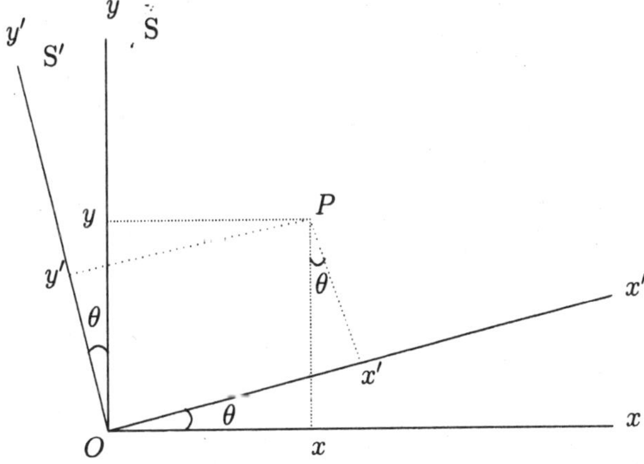

Figure 2: Reference frame S′ is rotated through an angle θ relative to S.

Suppose, the reference frame S′ is rotating in the anti-clockwise direction about the z-axis and in time t, the S′ is rotated through an angle θ relative to S. At this time, let the coordinates of the point P be (x, y, z) and (x', y', z') in S and S′, respectively. From Figure 2, we have

$$x' = x\cos\theta + y\sin\theta, \qquad y' = -x\sin\theta + y\cos\theta, \qquad z' = z$$

When the rotation angle θ is small, we have $\cos\theta \approx 1$ and $\sin\theta \approx \theta$ and thus, we get

$$x' = x + y\,\theta, \qquad y' = -x\,\theta + y, \qquad z' = z \tag{6.7}$$

The wave function at the point P does not depend on the coordinate system. Thus, the wave functions in the reference frames S and S$'$ are related as the following.

$$\psi'(x', y', z') = \psi(x, y, z)$$

Substituting x', y', z' from equation(6.7), we get

$$\psi'(x + y\theta,\ y - x\theta,\ z') = \psi(x, y, z) \tag{6.8}$$

Treatment of the problem becomes convenient, when we replace x by $(x - y\theta)$ and y by $(y + x\theta)$ on both sides of equation (6.8). Thus, we have

$$\psi'(x, y, z) = \psi(x - y\theta, y + x\theta, z) \tag{6.9}$$

Taking Taylor series expansion of the right side of (6.9), we have

$$\begin{aligned}\psi'(x, y, z) &= \psi(x, y, z) - y\theta\,\frac{\partial\psi}{\partial x} + x\theta\,\frac{\partial\psi}{\partial y} + \ldots\\ &= \psi(x, y, z) + \theta\Big(-y\,\frac{\partial\psi}{\partial x} + x\,\frac{\partial\psi}{\partial y}\Big)\\ &= \psi(x, y, z) + \frac{i\theta}{\hbar}\Big[x\Big(-i\hbar\frac{\partial}{\partial y}\Big) - y\Big(-i\hbar\frac{\partial}{\partial x}\Big)\Big]\psi(x, y, z)\\ &= \psi(x, y, z) + \frac{i\theta}{\hbar}\,L_z\psi(x, y, z)\\ &= \Big(1 + \frac{i\theta}{\hbar}L_z\Big)\psi(x, y, z)\end{aligned} \tag{6.10}$$

where

$$L_z = -i\hbar\Big(x\frac{\partial}{\partial y} - y\frac{\partial}{\partial x}\Big)$$

is an operator associated with z-component of angular momentum. From equation (6.10), $L_z/\hbar$ is the generator of infinitesimal rotation about the z-axis. On generalization of equation (6.10), for rotation about any arbitrary direction, we have

$$\psi'(x, y, z) = \Big(1 + \frac{i\theta}{\hbar}\hat{n}\,.\,\vec{L}\,\Big)\psi(x, y, z)$$

where $\hat{n}$ is the unit vector along the arbitrary direction and $\vec{L}$ the angular momentum. For a system of n particles with angular momenta $\vec{l}_1$, $\vec{l}_2$, $\vec{l}_3, \ldots, \vec{l}_n$, total momentum $\vec{L}$ is obtained as $\vec{L} = \vec{l}_1 + \vec{l}_2 + \vec{l}_3 + \ldots + \vec{l}_n$.

Rotation through a finite angle θ' can be considered as a succession of N infinitesimal rotations, so that $\theta' = N\theta$ and equation (6.10) gives

$$\begin{aligned}
\psi' &= \lim_{N\to\infty}\Big(1+\frac{i\theta\hat{n}\,.\,\vec{L}}{\hbar}\Big)^N\psi \\
&= \lim_{N\to\infty}\Big[1+N\Big(\frac{i\theta\hat{n}\,.\,\vec{L}}{\hbar}\Big)+\frac{N(N-1)}{2!}\Big(\frac{i\theta\hat{n}\,.\,\vec{L}}{\hbar}\Big)^2 \\
&\qquad\qquad +\frac{N(N-1)(N-2)}{3!}\Big(\frac{i\theta\hat{n}\,.\,\vec{L}}{\hbar}\Big)^3\ldots\Big]\psi \\
&= \lim_{N\to\infty} 1+N\Big(\frac{i\theta\hat{n}\,.\,\vec{L}}{\hbar}\Big)+\frac{N^2}{2!}\Big(\frac{i\theta\hat{n}\,.\,\vec{L}}{\hbar}\Big)^2+\frac{N^3}{3!}\Big(\frac{i\theta\hat{n}\,.\,\vec{L}}{\hbar}\Big)^3+\ldots\Big]\psi \\
&= \lim_{N\to\infty}\sum_{r=0}^{\infty}\frac{1}{r!}\Big(\frac{iN\theta\hat{n}.\,\vec{L}}{\hbar}\Big)^r\psi = \lim_{N\to\infty}\exp\Big(\frac{iN\theta\hat{n}\,.\,\vec{L}}{\hbar}\Big)\psi \\
&= \exp\Big(\frac{i\theta'\hat{n}\,.\,\vec{L}}{\hbar}\Big)\psi = \Big[1+\frac{i\theta'\hat{n}\,.\,\vec{L}}{\hbar}\Big]\psi
\end{aligned}$$

(ii) Conservation of angular momentum

The infinitesimal unitary rotation operator is

$$U = I + \Big(\frac{i\theta'}{\hbar}\hat{n}\,.\,\vec{L}\Big)$$

From this we have

$$\begin{aligned}
H' &= U\;H\;U^{\dagger} = \Big(I+\frac{i\theta'}{\hbar}\hat{n}\,.\,\vec{L}\Big)H\Big(I-\frac{i\theta'}{\hbar}\hat{n}\,.\,\vec{L}\Big) \\
&= H+\frac{i\theta'}{\hbar}\hat{n}\,.(\vec{L}\,H - H\,\vec{L}) = H+\frac{i\theta'}{\hbar}\hat{n}\,.[\vec{L},H]
\end{aligned}$$

Thus, the invariance of Hamiltonian under rotation in space requires that the angular momentum operator $\vec{L}$ must commute with the Hamiltonian H. In shows that the angular momentum $\vec{L}$ of a system is conserved when $\vec{L}$ commutes with the Hamiltonian H.

5. Space inversion: Parity conservation

Variation of linear momentum, energy or angular momentum are continuous processes. Conservation laws about these continuous processes have been discussed in the preceding sections. Now we consider the phenomenon of reflection about the origin. This operation, generally called the 'space inversion' or the 'parity operation', is a discrete process and has a unitary operator, known as the'parity operator', denoted by P. Application of the parity operator on a single particle wave function Ψ is defined as

$$P\Psi(\vec{r}) = \Psi(-\vec{r})$$

When the parity operator is applied on a multi-particle wave function Ψ, we have

$$P\Psi(\vec{r_1}, \vec{r_2}, \ldots, \vec{r_n}) = \Psi(-\vec{r_1}, -\vec{r_2}, \ldots, -\vec{r_n})$$

Thus, we have

$$P^2\Psi(\vec{r}) = \Psi(\vec{r}) \tag{6.11}$$

Equation (6.11) shows that Ψ is an eigen equation of the operator P^2 with eigenvalue 1. Thus, the eigenvalues of P are -1 or $+1$. That is,

$$P\Psi(\vec{r}) = \pm\Psi(\vec{r})$$

Hence, the eigen function either changes the sign (odd parity) or does not change the sign (even parity) under inversion operation. The eigen functions having even and odd parities are denoted by $\Psi_+(\vec{r})$ or $\Psi_-(\vec{r})$, respectively. Since these wave functions have different eigenvalues $+1$ and -1, respectively, they must be orthogonal to each other.

$$\left\langle \Psi_+(\vec{r}) \mid \Psi_-(\vec{r}) \right\rangle = 0$$

It can easily be proved that the parity operator P is Hermitian and unitary. Thus, we have

$$P = P^\dagger \qquad \text{and} \qquad PP^\dagger = P^\dagger P = 1$$

Now, if the parity operator P leaves the Hamiltonian H invariant, *i.e.*, $PHP^\dagger = H$, then we say that the system has space inversion symmetry and the parity operator commutes with H: In other words, the parity is conserved when P commutes with H.

The parity of an atomic wave function is calculated by its angular factor and is given by $(-1)^l$. For a many electrons atom, the parity is given by $(-1)^{\sum l}$. The concept of parity is used for classification of electronic states in molecules.

6. Time reversal

Under this transformation, the time is reversed. This transformation also is discrete. Let after the time reversal, denoted by T, the wave function $\Psi(\vec{r}, t)$ changes to a new wave function $\Psi'(\vec{r}, t')$, where $t' = -t$. Then we have

$$\Psi'(\vec{r}, t') = T\Psi(\vec{r}, t)$$

Let A be an time-independent operator associated with an observable and A' is its transform. That is,

$$A' = TAT^{-1}$$

Consider the relation

$$A\Psi = \phi$$

Then

$$\begin{aligned}\phi' &= T\phi = TA\Psi = TAT^{-1}T\Psi \\ &= (TAT^{-1})T\Psi = (TAT^{-1})\Psi' = A'\Psi'\end{aligned}$$

Effect of the time reversal on observables

In conformity with the time-reversal invariance in classical mechanics, a requirement in quantum mechanics is that the sign of the position operator $\vec{r}$ must not change whereas that of the operators for the linear momentum and angular momentum must change. That is,

$$\begin{aligned}\vec{r}' &= T\,\vec{r}\,T^{-1} = \vec{r} \\ \vec{p}' &= T\,\vec{p}\,T^{-1} = -\,\vec{p} \\ \vec{L}' &= T\,\vec{L}\,T^{-1} = -\,\vec{L}\end{aligned}$$

7. Problems and questions

1. Show that for a dynamical variable A to be invariant it must commute with the generator G of the infinitesimal unitary transformation.

2. Show that linear momentum $\vec{p}$ of a system is conserved when $\vec{p}$ commutes with the Hamiltonian H.

3. When Hamiltonian H of a system does not depend on time, show that the total energy of the system is conserved.

4. Show that angular momentum $\vec{L}$ of a system is conserved when $\vec{L}$ commutes with the Hamiltonian H.

5. Write notes on the following

   (i) Symmetry transformations

   (ii) Conservation of linear momentum

   (iii) Conservation of energy

   (iv) Conservation of angular momentum

   (v) Parity conservation

   (vi) Time reversal

Angular Momentum

In classical mechanics as well as in quantum mechanics, angular momentum of a physical system is an interesting property. In quantum mechanics, we deal with operators, corresponding to the angular momentum and its components. In this chapter, we shall discuss about the properties of the operators of angular momentum and its components. Then we shall calculate their eigenvalues and eigen functions. During discussion, we shall also come across the half-integral quantum numbers. Finally, we shall discuss about addition of angular momenta.

1. Angular momentum operators

In classical mechanics, angular momentum $\vec{L}$ of a particle is given by the vector product of position vector $\vec{r}$ and linear momentum $\vec{p}$ of the particle:

$$\begin{aligned}\vec{L} &= \vec{r} \times \vec{p} \\ &= (y\,p_z - z\,p_y)\hat{i} + (z\,p_x - x\,p_z)\hat{j} + (x\,p_y - y\,p_x)\hat{k}\end{aligned}$$

where $\vec{r} = x\hat{i} + y\hat{j} + z\hat{k}$ and $\vec{p} = p_x\hat{i} + p_y\hat{j} + p_z\hat{k}$. Replacing the coordinates and momentum components by their respective operators, the operators for the components of angular momentum in the Cartesian coordinates can be expressed as

$$L_x = -i\hbar\left(y\frac{\partial}{\partial z} - z\frac{\partial}{\partial y}\right)$$

$$L_y = -i\hbar\left(z\frac{\partial}{\partial x} - x\frac{\partial}{\partial z}\right)$$

$$L_z = -i\hbar\left(x\frac{\partial}{\partial y} - y\frac{\partial}{\partial x}\right)$$

Here, we have $\vec{L} = L_x\hat{i} + L_y\hat{j} + L_z\hat{k}$. When we use spherical polar coordinates (r, θ, ϕ), the expressions for L_x, L_y, L_z and square of angular momentum L^2 are

$$L_x = i\hbar \left(\sin\phi \frac{\partial}{\partial\theta} + \cot\theta \, \cos\phi \frac{\partial}{\partial\phi} \right)$$

$$L_y = i\hbar \left(-\cos\phi \frac{\partial}{\partial\theta} + \cot\theta \, \sin\phi \frac{\partial}{\partial\phi} \right)$$

$$L_z = -i\hbar \frac{\partial}{\partial\phi}$$

and

$$L^2 = -\hbar^2 \left[\frac{1}{\sin\theta} \frac{\partial}{\partial\theta} \left(\sin\theta \frac{\partial}{\partial\theta} \right) + \frac{1}{\sin^2\theta} \frac{\partial^2}{\partial\phi^2} \right]$$

Notice that here we have θ and ϕ and not r. For the detailed study of angular momentum, let us first derive commutation relations between these operators.

2. Commutation relations for angular momentum

When we use Cartesian coordinates, the commutation relations between the components of angular momentum $\vec{L}$ can be obtained rather conveniently. We have

$$\begin{aligned}[L_x, \ L_y] &= [(y\ p_z - z\ p_y), \ (z\ p_x - x\ p_z)] \\ &= [y\ p_z, \ z\ p_x] - [y\ p_z, \ x\ p_z] - [z\ p_y, \ z\ p_x] + [z\ p_y, \ x\ p_z] \quad (7.1)\end{aligned}$$

Let us evaluate the terms on the right side of equation (7.1) as the following.[1]

$$\begin{aligned}[y\ p_z, \ z\ p_x] &= y\ p_z\ z\ p_x - z\ p_x\ y\ p_z = p_z\ y\ z\ p_x - z\ p_x\ p_z\ y \\ &= p_z\ z\ y\ p_x - z\ p_z\ p_x\ y = p_z\ z\ y\ p_x - z\ p_z\ y\ p_x \\ &= (p_z\ z - z\ p_z) y\ p_x = -i\hbar y p_x\end{aligned}$$

[1]We shall make use of the following commutation relations, derived earlier, between the position coordinates q's and the components of linear momentum p's:

$[q_j, q_k] = 0$	that is	$q_j q_k = q_k q_j$
$[p_j, p_k] = 0$	that is	$p_j p_k = p_k p_j$
$[q_j, p_k] = i\hbar\delta_{jk}$	that is	$q_j p_k - p_k q_j = i\hbar\delta_{jk}$

$$\begin{aligned}[y\,p_z,\ x\,p_z] &= y\,p_z\,x\,p_z - x\,p_z\,y\,p_z = y\,x\,p_z\,p_z - x\,y\,p_z\,p_z \\ &= x\,y\,p_z\,p_z - x\,y\,p_z\,p_z = 0\end{aligned}$$

$$\begin{aligned}[z\,p_y,\ z\,p_x] &= z\,p_y\,z\,p_x - z\,p_x\,z\,p_y = z\,z\,p_y\,p_x - z\,z\,p_x\,p_y \\ &= z\,z\,p_x\,p_y - z\,z\,p_x\,p_y = 0\end{aligned}$$

$$\begin{aligned}[z\,p_y,\ x\,p_z] &= z\,p_y\,x\,p_z - x\,p_z\,z\,p_y = z\,p_y\,p_z\,x - p_z\,x\,z\,p_y \\ &= z\,p_z\,p_y\,x - p_z\,z\,x\,p_y = z\,p_z\,x\,p_y - p_z\,z\,x\,p_y \\ &= (z\,p_z - p_z\,z)x\,p_y = i\hbar x p_y\end{aligned}$$

Thus, we have

$$[L_x,\ L_y] = -i\hbar y p_x + i\hbar x p_y = i\hbar(xp_y - yp_x) = i\hbar L_z$$

Similarly, we can obtain

$$[L_y, L_z] = i\hbar L_x \qquad\qquad [L_z, L_x] = i\hbar L_y$$

Hence, the components of angular momentum do not commute with one another and therefore cannot be measured simultaneously.

We have

$$\begin{aligned}[L_x,\ L_x] &= [(y\,p_z - z\,p_y),\ (y\,p_z - z\,p_y)] \\ &= [y\,p_z,\ y\,p_z] - [y\,p_z,\ z\,p_y] - [z\,p_y,\ y\,p_z] + [z\,p_y,\ z\,p_y] \qquad (7.2)\end{aligned}$$

Let us evaluate the terms on the right side of equation (7.2) as the following.

$$[y\,p_z,\ y\,p_z] = y\,p_z\,y\,p_z - y\,p_z\,y\,p_z = 0$$

$$\begin{aligned}[y\,p_z,\ z\,p_y] &= y\,p_z\,z\,p_y - z\,p_y\,y\,p_z = p_z\,y\,z\,p_y - p_y\,z\,y\,p_z \\ &= p_z\,y\,z\,p_y - p_y\,y\,z\,p_z\end{aligned}$$

$$\begin{aligned}[z\,p_y,\ y\,p_z] &= z\,p_y\,y\,p_z - y\,p_z\,z\,p_y = p_y\,z\,y\,p_z - p_z\,y\,z\,p_y \\ &= p_y\,y\,z\,p_z - p_z\,y\,z\,p_y\end{aligned}$$

$$[z\,p_y,\ z\,p_y] = z\,p_y\,z\,p_y - z\,p_y\,z\,p_y = 0$$

Thus, we have

$$[L_x,\ L_x] = -p_z\,y\,z\,p_y + p_y\,y\,z\,p_z - p_y\,y\,z\,p_z + p_z\,y\,z\,p_y = 0$$

Similarly, we can obtain

$$[L_y,\ L_y] = 0 \qquad\qquad [L_z,\ L_z] = 0$$

Let us consider the following relations of the angular momentum operator $\vec{L}$

$$\begin{aligned}\vec{L} \times \vec{L} &= (L_y\, L_z - L_z\, L_y)\hat{i} + (L_z L_x - L_x L_z)\hat{j} + (L_x L_y - L_y L_x)\hat{j} \\ &= [L_y, L_z]\hat{i} + [L_z, L_z]\hat{j} + [L_y, L_y]\hat{k} = i\,\hbar\, L_x\hat{i} + i\hbar L_y\hat{j} + i\hbar L_z\hat{k} \\ &= i\hbar(L_x\hat{i} + L_y\hat{j} + L_z\hat{k} = i\hbar\,\vec{L}\end{aligned}$$

We know that cross product of a vector with itself is zero. Here, we have to consider $\vec{L}$ as an operator and not as a vector.

Let us evaluate commutator of L^2 with the components of $\vec{L}$. Here, we shall use the relation [2] expressed as $[ab, c] = a[b, c] + [a, c]b$.

$$\begin{aligned}[L^2,\, L_x] &= [L_x^2,\, L_x] + [L_y^2,\, L_x] + [L_z^2,\, L_x] \\ &= 0 + L_y[L_y,\, L_x] + [L_y,\, L_x]\, L_y + L_z[L_z,\, L_x] + [L_z,\, L_x]\, L_z \\ &= L_y\, (-i\hbar L_z) + (-i\hbar L_z)L_y + i\hbar L_z L_y + i\hbar L_y L_z = 0\end{aligned}$$

Similarly, we can obtain

$$[L^2,\, L_y] = 0 \qquad\qquad [L^2,\, L_z] = 0$$

It shows that the square of angular momentum commutes with the components of angular momentum. Since the components among themselves do not commutate, we cannot measure L^2, L_x, L_y and L_z simultaneously. Hence, the total angular momentum can be measured simultaneously with any one of the components. It is well accepted option that L^2 and L_z are measured simultaneously. This choice of z-component is completely arbitrary. However, acceptable to all over the world. Thus, they have common wave functions.

2.1 Lowering and raising operators

Now, we define two operators L_+ and L_- as

$$L_+ = L_x + iL_y, \qquad\qquad L_- = L_x - iL_y$$

The operators L_+ and L_- are known as the raising operator and lowering operator, respectively. Let us consider the following commutator

---

[2] We have

$$\begin{aligned}[ab, c] &= abc - cab = abc - acb + acb - cab \\ &= a(bc - cb) + (ac - ca)b = a[b, c] + [a, c]b\end{aligned}$$

relations.[3]

$$[L^2, L_+] = [L^2, L_x + iL_y] = [L^2, L_x] + i[L^2, L_y] = 0$$

$$[L^2, L_-] = [L^2, L_x - iL_y] = [L^2, L_x] - i[L^2, L_y] = 0$$

$$\begin{aligned}[L_z, L_+] = [L_z, L_x + iL_y] &= [L_z, L_x] + i[L_z, L_y] \\ &= i\hbar L_y + i(-i\hbar L_x) = \hbar(L_x + iL_y) = \hbar L_+\end{aligned}$$

$$\begin{aligned}[L_z, L_-] = [L_z, L_x - iL_y] &= [L_z, L_x] - i[L_z, L_y] \\ &= i\hbar L_y - i(-i\hbar L_x) = -\hbar(L_x - iL_y) = -\hbar L_-\end{aligned}$$

$$[L_x, L_+] = [L_x, L_x + iL_y] = [L_x, L_x] + i[L_x, L_y] = ii\hbar L_z = -\hbar L_z$$

$$[L_x, L_-] = [L_x, L_x - iL_y] = [L_x, L_x] - i[L_x, L_y] = -ii\hbar L_z = \hbar L_z$$

$$[L_y, L_+] = [L_y, L_x + iL_y] = [L_y, L_x] + i[L_y, L_y] = -i\hbar L_z = -i\hbar\ L_z$$

$$[L_y, L_-] = [L_y, L_x - iL_y] = [L_y, L_x] - i[L_y, L_y] = -i\hbar L_z = -i\hbar L_z$$

$$\begin{aligned}[L_+, L_-] &= [L_x + iL_y, L_x - iL_y] \\ &= [L_x, L_x] + i[L_y, L_x] - i[L_x, L_y] + [L_y, L_y] \\ &= i(-i\hbar L_z) - i(i\hbar L_z) = 2\hbar L_z\end{aligned}$$

We have two more relations

$$\begin{aligned}L_+L_- &= (L_x + iL_y)(L_x - iL_y) = L_x^2 + L_y^2 - i(L_xL_y - L_yL_x) \\ &= L^2 - L_z^2 + \hbar L_z \\ L_-L_+ &= (L_x - iL_y)(L_x + iL_y) = L_x^2 + L_y^2 + i(L_xL_y - L_yL_x) \\ &= L^2 - L_z^2 - \hbar L_z\end{aligned}$$

---

[3]We have

$$\begin{aligned}[a+b, c] &= (a+b)c - c(a+b) = ac + bc - ca - cb \\ &= (ac - ca) + (bc - cb) = [a, c] + [b, c]\end{aligned}$$

$$\begin{aligned}[a+b, c+d] &= (a+b)(c+d) - (c+d)(a+b) \\ &= (ac + ad + bc + bd) - (ca + cb + da + db) \\ &= (ac - ca) + (ad - da) + (bc - cb) + (bd - db) \\ &= [a, c] + [a, d] + [b, c] + [b, d]\end{aligned}$$

All these commutation relations form a strong foundation of the theory of angular momentum.

Exercise 1. Suppose the components of angular momentum operator obey the rule $[J_x, J_y] = -i\hbar L_z$ and similar commutation relations, calculate the commutators $[J^2, J_x]$ and $[J^2, J_+]$. What would be the role of J_+ and J_- in this new scenario.

Solution: We have

$$\begin{aligned}[J^2, J_x] &= [J_x^2, J_x] + [J_y^2, J_x] + [J_z^2, J_x] \\ &= 0 + J_y[J_y, J_x] + [J_y, J_x]J_y + J_z[J_z, J_x] + [J_z, J_x]J_z. \\ &= i\hbar J_y J_z + i\hbar J_z J_y - i\hbar J_z J_y - i\hbar J_y J_z = 0\end{aligned}$$

Similarly, we can have $[J^2, J_y] = 0$. Therefore,

$$[J^2, J_+] = [J^2, J_x + iJ_y] = [J^2, J_x] + [J^2, J_y] = 0$$

Let us now calculate $[J_z, J_+]$ and $[J_z, J_-]$.

$$[J_z, J_+] = [J_z, J_x] + i[J_z, J_y] = -i\hbar J_y - \hbar J_x = -\hbar J_+$$

and

$$[J_z, J_-] = [J_z, J_x] - i[J_z, J_y] = -i\hbar J_y + \hbar J_x = \hbar J_+$$

In this new scenario, J_+ would be a lowering operator and J_- would be a raising operator.

Exercise 2. Components of the angular momenta $\vec{J}_1$ and $\vec{J}_1$ satisfy the commutation rules. Show that the components of the sum $\vec{J} = \vec{J}_1 + \vec{J}_2$ also satisfy the commutation rules. Whether the difference $\vec{J}_1 - \vec{J}_2$ is an angular momentum.

Solution: For the components of $\vec{J}$, we have

$$\begin{aligned}[J_x, J_y] &= [J_{1x} + J_{2x}, J_{1y} + J_{2y}] \\ &= [J_{1x}, J_{1y}] + [J_{1x}, J_{2y}] + [J_{2x}, J_{1y}] + [J_{2x}, J_{2y}] \\ &= i\hbar J_{1z} + 0 + 0 + i\hbar J_{2z} = i\hbar J_z\end{aligned}$$

Similarly, we can find out $[J_y, J_z] = J_x$ and $[J_z, J_x] = J_y$. It shows that the components of $\vec{J}$ satisfy the commutation rules.

Expressing $\vec{J}' = \vec{J}_1 - \vec{J}_2$, we have

$$\begin{aligned}[J_x', J_y'] &= [J_{1x} - J_{2x}, J_{1y} - J_{2y}] \\ &= [J_{1x}, J_{1y}] - [J_{1x}, J_{2y}] - [J_{2x}, J_{1y}] + [J_{2x}, J_{2y}] \\ &= i\hbar J_{1z} - 0 - 0 + i\hbar J_{2z} = i\hbar(J_{1z} + J_{2z})\end{aligned}$$

which is not the operator J'_z. Since the commutation rules are not satisfied by $\vec{J}_1 - \vec{J}_2$, it is not angular momentum.

3. Eigenvalues and Eigenfunctions of L^2 and L_z

The eigenvalue equation of L^2 in spherical polar coordinates can be expressed as

$$-\hbar^2 \left[\frac{1}{\sin\theta}\frac{\partial}{\partial\theta}\left(\sin\theta\frac{\partial}{\partial\theta}\right) + \frac{1}{\sin^2\theta}\frac{\partial^2}{\partial\phi^2}\right] Y = \lambda\hbar^2 Y$$

where $\lambda\hbar^2$ is eigenvalue of L^2 and Y the corresponding eigenfunction. On rearranging this equation, we get

$$\left[\frac{1}{\sin\theta}\frac{\partial}{\partial\theta}\left(\sin\theta\frac{\partial}{\partial\theta}\right) + \frac{1}{\sin^2\theta}\frac{\partial^2}{\partial\phi^2}\right] Y + \lambda Y = 0$$

which is the same as the angular part of the Schrödinger equation of a system moving in a spherically symmetric potential $V(r)$. When we take $\lambda = l(l+1)$, this equation has a solution

$$Y_{lm}(\theta, \phi) = \Theta(\theta)\ \Phi(\phi) = \epsilon \sqrt{\frac{2l+1}{4\pi}\frac{(l-m)!}{(l+m)!}}\ P_l^m\ (\cos\theta)\ e^{im\phi}$$

where $l = 0, 1, 2, \ldots$ $m = 0, \pm1, \pm2, \ldots, \pm l$ and ϵ is the phase factor conventionally taken as

$$\epsilon = \begin{cases} (-1)^m & \text{for } m > 0 \\ 1 & \text{for } m \leq 0 \end{cases}$$

Now, we have

$$L_z Y_{lm}(\theta, \phi) = -i\hbar\frac{\partial}{\partial\phi} Y_{lm}(\theta, \phi) = m\hbar Y_{lm}(\theta, \phi)$$

Hence, the spherical harmonics $Y_{lm}(\theta, \phi)$ are eigenfunctions of L^2 as well as of L_z with eigen values $l(l+1)\hbar^2$ and $m\hbar$, respectively.

4. General angular momentum

A number of experimental observations, such as spectra of alkali atoms, anomalous Zeeman effect, Stern-Gerlach experiment, etc., required an idea of inclusion the concept that an electron in an atom possesses an additional intrinsic angular momentum where half-integer quantum number was essentially needed. This angular momentum, called the spin angular momentum, has 1/2 quantum number. Therefore, an idea of general angular momentum J where integer as well half-integer quantum numbers could be incorporated was introduced. The general angular momentum J as well as its Cartesian components J_x, J_y and J_z were considered to follow the commutation relations as discussed for the angular momentum L and its components L_x, L_y and L_z

5. Eigenvalues of J^2 and J_z

As in the case the angular momentum L, the general angular momentum J commutes with its Cartesian components J_x, J_y and J_z. However, the components, among themselves are noncommutating. Further, as earlier, J^2 and J_z have simultaneous eigenfunctions (eigenkets). Denoting the simultaneous eigenkets by $|\lambda m\rangle$, the eigenvalue equations for J^2 and J_z can be expressed as

$$J^2|\lambda m\rangle = \lambda|\lambda m\rangle \qquad\qquad J_z|\lambda m\rangle = m|\lambda m\rangle \tag{7.3}$$

First of equation (7.3) can be written as

$$\begin{gathered}(J_x^2 + J_y^2 + J_z^2)|\lambda m\rangle = \lambda|\lambda m\rangle \\ (J_x^2 + J_y^2)|\lambda m\rangle + m^2|\lambda m\rangle = \lambda|\lambda m\rangle\end{gathered}$$

Multiplying this equation from left by $\langle\lambda m|$, we get

$$\langle\lambda m|(J_x^2 + J_y^2)|\lambda m\rangle + m^2\langle\lambda m|\lambda m\rangle = \lambda\langle\lambda m|\lambda m\rangle$$

$$\langle\lambda m|J_x^2|\lambda m\rangle + \langle\lambda m|J_y^2|\lambda m\rangle = (\lambda - m^2) \tag{7.4}$$

As the J_x and J_y are Hermitian, their eigenvalues are real and therefore, the left side of equation (7.4) must be positive. Thus,

$$(\lambda - m^2) \geq 0 \qquad \text{or} \qquad \lambda \geq m^2 \tag{7.5}$$

Operating first equation (7.3) from left by J_+, we have

$$J_+J^2|\lambda m\rangle = \lambda J_+|\lambda m\rangle$$

Since J^2 commutes with J_+, we have

$$J^2J_+|\lambda m\rangle = \lambda J_+|\lambda m\rangle$$

This shows that $|\lambda m\rangle$ and $J_+|\lambda m\rangle$ are eigenkets of J^2 with the same eigenvalue λ. Now, operating second of equation (7.3) from left by J_+, we have

$$J_+J_z|\lambda m\rangle = mJ_+|\lambda m\rangle$$

Using $[J_z, J_+] = \hbar J_+$, *i.e.*, $J_+J_z = J_zJ_+ - \hbar J_+$ in this equation, we have

$$\begin{aligned}(J_zJ_+ - \hbar J_+)|\lambda m\rangle &= mJ_+|\lambda m\rangle \\ J_zJ_+|\lambda m\rangle &= (m+\hbar)J_+|\lambda m\rangle\end{aligned}$$

This shows that $J_+|\lambda m\rangle$ is eigenket of J_z with the eigenvalue $(m + \hbar)$ where $J_+|\lambda m\rangle$ is eigenket of J^2 with the same eigenvalue λ. As the operation by J_+ on $|\lambda m\rangle$ produces a new eigenket with the same eigenvalue λ of J^2, but eigenvalue of J_z increases by $\hbar$, it is called a raising operator.

Repeated operation by J_+ increases the eigenvalue of J_z by $\hbar$ each time. For satisfaction of equation (7.5), this process of increment must stop at some point. Let the maximum permissible value of the eigenvalue of J_z be μ with eigenket $|\lambda\mu\rangle$. Then we have

$$J^2|\lambda\mu\rangle = \lambda|\lambda\mu\rangle \tag{7.6}$$

$$J_z|\lambda\mu\rangle = \mu|\lambda\mu\rangle \tag{7.7}$$

Multiplying this equation from left by J_+, we get

$$J_+J_z|\lambda\mu\rangle = \mu J_+|\lambda\mu\rangle$$

Using $J_+J_z = J_zJ_+ - \hbar J_+$ in this equation, we get

$$\begin{aligned}(J_zJ_+ - \hbar J_+)|\lambda\mu\rangle &= \mu\, J_+|\lambda\mu\rangle \\ J_zJ_+|\lambda\mu\rangle &= (\mu+\hbar)J_+|\lambda\mu\rangle\end{aligned}$$

Eigenvalue $(\mu+\hbar)$ of J_z is not possible, as it is larger than the maximum permissible value of μ. Therefore, this equation can only be satisfied when

$$J_+|\lambda\mu\rangle = 0$$

Multiplying this equation from left by J_-, we get

$$J_-J_+|\lambda\mu\rangle = 0$$

Using $J_-\ J_+ = J^2 - J_z^2 - \hbar\ J_z$ in this equation, we get

$$\begin{aligned}(J^2 - J_z^2 - \hbar\ J_z)|\lambda\mu\rangle &= 0 \\ (\lambda - \mu^2 - \hbar\mu)|\lambda\mu\rangle &= 0\end{aligned}$$

Since $|\lambda\mu\rangle \neq 0$, equation (7.7), we have

$$\lambda - \mu^2 - \hbar\,\mu = 0$$
$$\lambda = \mu(\mu + \hbar)$$

Multiplying equation (7.7) from left by J_-, we get

$$J_- J_z |\lambda\mu\rangle = \mu J_- |\lambda\mu\rangle$$

Using $J_- J_z = J_z J_- + \hbar J_-$ in this equation, we get

$$(J_z J_- + \hbar J_-)|\lambda\mu\rangle = \mu\ J_- |\lambda\mu\rangle$$
$$J_z J_- |\lambda\mu\rangle = (\mu - \hbar) J_- |\lambda\mu\rangle$$

Multiplying equation (7.6) from left by J_-, we get

$$J_- J^2 |\lambda\mu\rangle = \lambda J_- |\lambda\mu\rangle$$

Since J^2 commutes with J_-, we have

$$J^2 J_- |\lambda\mu\rangle = \lambda J_- |\lambda\mu\rangle$$

Comparison of this equation with equation (7.7) shows that $J_-\ |\lambda\mu\rangle$ is an eigenket of J_z with eigen value $(\mu - \hbar)$ and of J^2 with the same eigenvalue λ. Since operation by J_- generates a state with the same eigen value of J^2, but of J_z with a value lower by $\hbar$, the operator J_- is known as the lowering operator. Repeated operation by J_- decreases the eigenvalue of J_z by $\hbar$ each time. For satisfaction of equation (7.5), this process of decrement must stop at some point. Let the minimum permissible value of the eigenvalue of J_z be $(\mu - n\hbar)$ such that

$$J_z J_-^n |\lambda\mu\rangle = (\mu - n\hbar) J_-^n |\lambda\mu\rangle$$

Multiplying this equation from left by J_-, we get

$$J_- J_z J_-^n |\lambda\mu\rangle = (\mu - n\hbar) J_-^{n+1} |\lambda\mu\rangle$$

Using $J_- J_z = J_z J_- + \hbar J_-$ in this equation, we get

$$(J_z J_- + \hbar J_-) J_-^n |\lambda\mu\rangle = (\mu - n\hbar) J_-^{n+1} |\lambda\mu\rangle$$
$$J_z J_-^{n+1} |\lambda\mu\rangle = (\mu - n\hbar - \hbar) J_-^{n+1} |\lambda\mu\rangle$$

Eigenvalue $(\mu - n\hbar - \hbar)$ of J_z is not possible, as it is smaller than the minimum permissible value of μ. Therefore, this equation can only be satisfied when

$$J_-^{n+1}|\lambda\mu\rangle = 0$$
$$J_-J_-^n|\lambda\mu\rangle = 0$$

Multiplying this equation from left by J_+, we get

$$J_+J_-J_-^n|\lambda\mu\rangle = 0$$

Using $J_+J_- = J^2 - J_z^2 + \hbar J_z$ in this equation, we get

$$(J^2 - J_z^2 + \hbar J_z)J_-^n|\lambda\mu\rangle = 0$$
$$\left[\lambda - (\mu - n\hbar)^2 + \hbar(\mu - n\hbar)\right] J_-^n|\lambda\mu\rangle = 0$$

Since $J_-^n|\lambda\mu\rangle \neq 0$, we have

$$\lambda - (\mu - n\hbar)^2 + \hbar(\mu - n\hbar) = 0$$

Using $\lambda = \mu(\mu + \hbar)$ in this equation, we get

$$\mu(\mu + \hbar) - (\mu - n\hbar)^2 + \hbar(\mu - n\hbar) = 0$$
$$(n + 1)(2\mu - n\,\hbar) = 0$$

Since $(n + 1) \neq 0$, we have

$$2\mu - n\hbar = 0 \qquad \text{or} \qquad \mu = \frac{n}{2}\,\hbar$$

The number of steps n is always a positive integer, including zero. Writing j for $n/2$, we have $\mu = j\hbar$ and $(\mu - n\hbar) = (j\hbar - 2j\hbar) = -j\hbar$. Thus, the largest and smallest eigenvalues of J_z are $j\hbar$ and $-j\hbar$. Hence, for a given value of j, the eigenvalues of J_z are $m\hbar$, where m varies from $-j$ to j in the steps of unity.[4] Now, for the simultaneous eigenvector $|jm\rangle$ for J^2 and J_z, we have

$$J^2|jm\rangle = j(j + 1)\hbar^2|jm\rangle \tag{7.8}$$
$$J_z|jm\rangle = m\hbar|jm\rangle \tag{7.9}$$

where $j = 0, 1/2, 1, 3/2, 2, \ldots$ and $m = -j, -j + 1, \ldots, j$.

6. Angular momentum matrices

The eigenkets $|jm\rangle$ form a complete orthonormal set and they can be used as a basis for the matrix representation of angular momentum. In the matrix representation a function F of angular momentum is expressed by a matrix with the elements $\langle j'm'|F|jm\rangle$. The rows of an matrix will be labeled by the values of j' and m' and the columns by the values of j and m.

[4]Note that this m and that in equation (7.3) are different.

6.1 Matrices for J^2 and J_z

Multiplying equation (7.8) from left by $\langle j'm'|$, we get

$$\langle j'm'|\ J^2|jm\rangle = j(j+1)\hbar^2\langle j'm'|jm\rangle = j(j+1)\hbar^2\delta_{jj'}\delta mm'$$

For $j \neq j'$, the matrix for J^2, *i.e.*, $\langle j'm'|J^2|jm\rangle$, is a null matrix. For $j = j' = 0$, the matrix for J^2 is

$$\hbar^2 \begin{bmatrix} 0 \end{bmatrix}$$

For $j = j' = 1/2$, the matrix for J^2 is

$$3\hbar^2/4 \begin{bmatrix} 1 & 0 \\ 0 & 1 \end{bmatrix}$$

For $j = j' = 1$, the matrix for J^2 is

$$2\hbar^2 \begin{bmatrix} 1 & 0 & 0 \\ 0 & 1 & 0 \\ 0 & 0 & 1 \end{bmatrix}$$

For other values of $j = j'$, the matrices for J^2 can be obtained in a similar manner.

Multiplying equation (7.9) from left by $\langle j'm'|$, we get

$$\langle j'm'|J_z|jm\rangle = m\hbar\langle j'm'|jm\rangle = m\hbar\delta_{jj'}\delta mm'$$

For $j \neq j'$, the matrix for J_z, *i.e.*, $\langle j'm'|J_z|jm\rangle$, is a null matrix. For $j = j' = 0$, the matrix for J_z is

$$\hbar \begin{bmatrix} 0 \end{bmatrix}$$

For $j = j' = 1/2$, the matrix for J_z is

$$\hbar/2 \begin{bmatrix} 1 & 0 \\ 0 & 1 \end{bmatrix}$$

For $j = j' = 1$, the matrix for J_z is

$$\hbar \begin{bmatrix} 1 & 0 & 0 \\ 0 & 0 & 0 \\ 0 & 0 & 1 \end{bmatrix}$$

For other values of $j = j'$, the matrices for J_z can be obtained in a similar manner.

6.2 Matrices for J_+, J_-, J_x and J_y

Multiplying equation (7.9) from left by J_+, we get

$$J_+ \, J_z|jm\rangle = m\hbar J_+|jm\rangle$$

Using $J_+J_z = J_zJ_+ - \hbar J_+$ in this equation, we get

$$(J_zJ_+ - \hbar J_+)|jm\rangle = m\hbar J_+|jm\rangle$$

$$J_zJ_+|jm\rangle = (m+1)\hbar J_+|jm\rangle \tag{7.10}$$

This shows that $J_+|jm\rangle$ is an eigenket of J_z with eigenvalue $(m+1)\hbar$. This eigen value $(m+1)\hbar$ of J_z can also be expressed as

$$J_z|j, m+1\rangle = (m+1)\hbar|j, m+1\rangle \tag{7.11}$$

As the eigenvalues of J_z in equations (7.10) and (7.11) are equal, it shows that the two eigenkets must be proportional to each other. Thus, we have

$$J_+|jm\rangle = a_m|j, m+1\rangle \tag{7.12}$$

Multiplying equation (7.9) from left by J_-, we get

$$J_-J_z|jm\rangle = m\hbar J_-|jm\rangle$$

Using $J_-J_z = J_zJ_- + \hbar J_-$ in this equation, we get

$$(J_zJ_- + \hbar J_-)|jm\rangle = m\hbar J_-|jm\rangle$$

$$J_zJ_-|jm\rangle = (m-1)\hbar J_-|jm\rangle \tag{7.13}$$

This shows that $J_-|jm\rangle$ is an eigenket of J_z with eigenvalue $(m-1)\hbar$. This eigen value $(m-1)\hbar$ of J_z can also be expressed as

$$J_z|j, m-1\rangle = (m-1)\hbar|j, m-1\rangle \tag{7.14}$$

As the eigenvalues of J_z in equations (7.13) and (7.14) are equal, it shows that the two eigenkets must be proportional to each other. Thus,we have

$$J_-|jm\rangle = b_m|j, m-1\rangle$$

Replacing m by $m+1$ here, we get

$$J_-|jm+1\rangle = b_{m+1}|j, m\rangle \tag{7.15}$$

Multiplying equation (7.12) from left by $\langle j, m+1|$, we get

$$\langle j, m+1|J_+|jm\rangle = a_m\langle j, m+1|j, m+1\rangle = a_m$$

Taking complex conjugate of this equation, we have

$$a_m^* = \langle j, m|J_-|jm+1\rangle \tag{7.16}$$

Multiplying equation (7.15) from left by $\langle j, m|$, we get

$$\langle j, m|J_-|jm+1\rangle = b_{m+1}\langle j, m|j, m\rangle = b_{m+1} \tag{7.17}$$

Comparison of equations (7.16) and (7.17) shows that

$$a_m^* = b_{m+1}$$

Now, multiplying equation (7.12) from left by J_-, we get

$$J_-J_+|jm\rangle = a_m J_-|j, m+1\rangle$$

Using $J_-J_+ = J^2 - J_z^2 - \hbar J_z$ and equation (7.15) in this equation, we get

$$(J^2 - J_z^2 - \hbar\ J_z)|jm\rangle = a_m\ b_{m+1}|j, m\rangle$$

$$[j(j+1) - m^2 - m]\hbar^2|jm\rangle = |a_m|^2|j, m\rangle$$

Thus, we have

$$a_m = [j(j+1) - m^2 - m]^{1/2}\hbar$$

Using this value in equation (7.12), we have

$$J_+|jm\rangle = [j(j+1) - m^2 - m]^{1/2}\hbar|j, m+1\rangle \tag{7.18}$$

Multiplying equation (7.18) from left by $\langle j'm'|$, we get

$$\begin{aligned}\langle j'm'|J_+|jm\rangle &= [j(j+1) - m^2 - m]^{1/2}\hbar\langle j'm'|j, m+1\rangle \\ &= [j(j+1) - m^2 - m]^{1/2}\hbar\ \delta_{jj'}\delta_{m',m+1}\end{aligned} \tag{7.19}$$

Using $b_{m+1} = a_m^* = [j(j+1) - m^2 - m]^{1/2}\hbar$ in equation (7.15), we have

$$J_-|jm+1\rangle = [j(j+1) - m(m+1)]^{1/2}\hbar|j, m\rangle$$

Replacing m by $m - 1$, we get

$$J_-|jm\rangle = [j(j+1) - m(m-1)]^{1/2}\hbar|j, m-1\rangle \tag{7.20}$$

Multiplying equation (7.20) from left by $\langle j'm'|$, we get

$$\begin{aligned}\langle j'm'|J_-|jm\rangle &= [j(j+1) - m(m-1)]^{1/2}\hbar\langle j'm'|j, m-1\rangle \\ &= [j(j+1) - m(m-1)]^{1/2}\hbar\ \delta_{jj'}\delta_{m',m-1}\end{aligned} \tag{7.21}$$

Equations (7.19) and (7.21) can be used to get the matrix elements for J_+ and J_-. From these matrices, the elements for J_x and J_y can be obtain as

$$J_x = \frac{1}{2}(J_+ + J_-) \qquad J_y = \frac{1}{2i}(J_+ - J_-)$$

For $j = 0$, we have

$$J_+ = 0 \qquad J_- = 0 \qquad J_x = 0 \qquad J_y = 0$$

For $j = 1/2$, we have

$$J_+ = \hbar \begin{bmatrix} 0 & 1 \\ 0 & 0 \end{bmatrix} \qquad J_- = \hbar \begin{bmatrix} 0 & 0 \\ 1 & 0 \end{bmatrix}$$

$$J_x = \frac{\hbar}{2} \begin{bmatrix} 0 & 1 \\ 1 & 0 \end{bmatrix} \qquad J_y = \frac{\hbar}{2} \begin{bmatrix} 0 & -i \\ i & 0 \end{bmatrix}$$

For $j = 1$, we have

$$J_+ = \sqrt{2}\hbar \begin{bmatrix} 0 & 1 & 0 \\ 0 & 0 & 1 \\ 0 & 0 & 0 \end{bmatrix} \qquad J_- = \sqrt{2}\hbar \begin{bmatrix} 0 & 0 & 0 \\ 1 & 0 & 0 \\ 0 & 1 & 0 \end{bmatrix}$$

$$J_x = \frac{\hbar}{\sqrt{2}} \begin{bmatrix} 0 & 1 & 0 \\ 1 & 0 & 1 \\ 0 & 1 & 0 \end{bmatrix} \qquad J_y = \frac{\hbar}{\sqrt{2}} \begin{bmatrix} 0 & -i & 0 \\ i & 0 & -i \\ 0 & i & 0 \end{bmatrix}$$

Exercise 3. The matrix elements of J^2, J_z, J_+, J_- are expressed as

$$\langle j'm'|J^2|jm\rangle = j(j+1)\hbar^2\delta_{jj'}\delta mm'$$

$$\langle j'm'|J_z|jm\rangle = m\hbar\delta_{jj'}\delta mm'$$

$$\langle j'm'|J_+|jm\rangle = [j(j+1) - m^2 - m]^{1/2}\hbar\ \delta_{jj'}\delta_{m',m+1}$$

$$\langle j'm'|J_-|jm\rangle = [j(j+1) - m(m-1)]^{1/2}\hbar\delta_{jj'}\delta_{m',m-1}$$

derive matrices for J^2, J_z, J_x and J_y for $j = 3/2$.

Solution: For $j = 3/2$, the allowed value of m are 3/2, 1/2, −1/2, −3/2. For these values of j and m, the matrices for J^2 and J_z are

$$J^2 = 15\frac{\hbar^2}{4} \begin{bmatrix} 1 & 0 & 0 & 0 \\ 0 & 1 & 0 & 0 \\ 0 & 0 & 1 & 0 \\ 0 & 0 & 0 & 1 \end{bmatrix} \qquad J_z = \frac{\hbar}{2} \begin{bmatrix} 3 & 0 & 0 & 0 \\ 0 & 1 & 0 & 0 \\ 0 & 0 & -1 & 0 \\ 0 & 0 & 0 & -3 \end{bmatrix}$$

and the matrices for J_+ and J_- are

$$J_+ = \hbar \begin{bmatrix} 0 & \sqrt{3} & 0 & 0 \\ 0 & 0 & 2 & 0 \\ 0 & 0 & 0 & \sqrt{3} \\ 0 & 0 & 0 & 0 \end{bmatrix} \qquad J_- = \hbar \begin{bmatrix} 0 & 0 & 0 & 0 \\ \sqrt{3} & 0 & 0 & 0 \\ 0 & 2 & 0 & 0 \\ 0 & 0 & \sqrt{3} & 0 \end{bmatrix}$$

Using the relations

$$J_x = \frac{1}{2}(J_+ + J_-) \qquad J_y = \frac{1}{2i}(J_+ - J_-)$$

we get

$$J_x = \frac{\hbar}{2} \begin{bmatrix} 0 & \sqrt{3} & 0 & 0 \\ \sqrt{3} & 0 & 2 & 0 \\ 0 & 2 & 0 & \sqrt{3} \\ 0 & 0 & \sqrt{3} & 0 \end{bmatrix} \quad J_y = \frac{\hbar}{2i} \begin{bmatrix} 0 & \sqrt{3} & 0 & 0 \\ -\sqrt{3} & 0 & 2 & 0 \\ 0 & -2 & 0 & \sqrt{3} \\ 0 & 0 & -\sqrt{3} & 0 \end{bmatrix}$$

7. Spin angular momentum

For accounting the multiplicity of atomic states, Uhlenbeck and Goudsmitt proposed in 1925 that an electron in an atom possesses an intrinsic angular momentum in addition to the orbital angular momentum. This intrinsic angular momentum $\vec{S}$ is known as the spin angular momentum and its projection on the z-axis can have the values $S_z = m_s\hbar$, with $m_s = \pm 1/2$. They also suggested that the spin angular momentum produces an intrinsic magnetic moment $\vec{\mu}_s$

$$\vec{\mu}_s = -\frac{e}{m}\vec{S}$$

With the assumption that all the stable and unstable particles do have spin angular momentum $\vec{S}$, its Cartesian components S_x, S_y, S_z are considered to follow the general commutation rules

$$[S_x, S_y] = i\hbar S_z \qquad [S_y, S_z] = i\hbar S_x \qquad [S_z, S_x] = i\hbar S_y$$

and S^2 and S_z are to have eigenvalues $s(s+1)\hbar^2$ and $m_s\hbar$, respectively.

$$S^2|sm_s\rangle = s(s+1)\hbar^2|sm_s\rangle \qquad \text{and} \qquad S_z|sm_s\rangle = m_s\hbar|sm_s\rangle$$

The spin magnetic quantum number m_s can assume the value from $-s$ to s in the steps of unity.

7.1 One-half spin systems

Most of the stable elementary particles, like electron, proton, neutron, etc. have (1/2) spin. Such particles are known as the fermions and they follow Fermi-Dirac statistics and obey the Pauli exclusion principle. The matrices representing S_x, S_y and S_z are the same as for J_x, J_y and J_z corresponding to $j = 1/2$. Thus, we have

$$S_x = \frac{\hbar}{2}\begin{bmatrix} 0 & 1 \\ 1 & 0 \end{bmatrix} \qquad S_y = \frac{\hbar}{2}\begin{bmatrix} 0 & -i \\ i & 0 \end{bmatrix} \qquad S_z = \frac{\hbar}{2}\begin{bmatrix} 1 & 0 \\ 0 & -1 \end{bmatrix}$$

As a common practice, we work with a matrix σ which is related to the matrix S as

$$S = \frac{\hbar}{2}\sigma$$

The components σ_x, σ_y, σ_z of σ are known as the Pauli's spin matrices, given by

$$\sigma_x = \begin{bmatrix} 0 & 1 \\ 1 & 0 \end{bmatrix} \qquad \sigma_y = \begin{bmatrix} 0 & -i \\ i & 0 \end{bmatrix} \qquad \sigma_z = \begin{bmatrix} 1 & 0 \\ 0 & -1 \end{bmatrix}$$

Pauli was the first to introduce two-component state vectors for explaining certain observed features of atomic spectra. Since the eigenvalues of S_x, S_y, S_z are $\pm\hbar/2$, therefore the eigenvalues of σ_x, σ_y, σ_z are ± 1. These matrices satisfy the relations

$$\sigma_x^2 = \sigma_y^2 = \sigma_z^2 = I$$
$$\sigma_x\sigma_y = i\sigma_z \qquad \sigma_y\sigma_z = i\sigma_x \qquad \sigma_z\sigma_x = i\sigma_y$$
$$\sigma_x\sigma_y + \sigma_y\sigma_x = \sigma_y\sigma_z + \sigma_z\sigma_y = \sigma_z\sigma_x + \sigma_x\sigma_z = 0$$

Exercise 4. For the Pauli's spin matrices σ_x, σ_y, σ_z, show that

$$\sigma_x^2 = \sigma_y^2 = \sigma_z^2 = I$$
$$\sigma_x\sigma_y = i\sigma_z \qquad \sigma_y\sigma_z = i\sigma_x \qquad \sigma_z\sigma_x = i\sigma_y$$
$$\sigma_x\sigma_y + \sigma_y\sigma_x = \sigma_y\sigma_z + \sigma_z\sigma_y = \sigma_z\sigma_x + \sigma_x\sigma_z = 0$$

Solution: Pauli's spin matrices σ_x, σ_y, σ_z are given by

$$\sigma_x = \begin{bmatrix} 0 & 1 \\ 1 & 0 \end{bmatrix} \qquad \sigma_y = \begin{bmatrix} 0 & -i \\ i & 0 \end{bmatrix} \qquad \sigma_z = \begin{bmatrix} 1 & 0 \\ 0 & -1 \end{bmatrix}$$

Thus, we have

$$\sigma_x^2 = \begin{bmatrix} 0 & 1 \\ 1 & 0 \end{bmatrix} \begin{bmatrix} 0 & 1 \\ 1 & 0 \end{bmatrix} = \begin{bmatrix} 1 & 0 \\ 0 & 1 \end{bmatrix}$$

$$\sigma_y^2 = \begin{bmatrix} 0 & -i \\ i & 0 \end{bmatrix} \begin{bmatrix} 0 & -i \\ i & 0 \end{bmatrix} = \begin{bmatrix} 1 & 0 \\ 0 & 1 \end{bmatrix}$$

$$\sigma_z^2 = \begin{bmatrix} 1 & 0 \\ 0 & -1 \end{bmatrix} \begin{bmatrix} 1 & 0 \\ 0 & -1 \end{bmatrix} = \begin{bmatrix} 1 & 0 \\ 0 & 1 \end{bmatrix}$$

It gives $\sigma_x^2 = \sigma_y^2 = \sigma_z^2 = I$. Now, we have

$$\sigma_x \sigma_y = \begin{bmatrix} 0 & 1 \\ 1 & 0 \end{bmatrix} \begin{bmatrix} 0 & -i \\ i & 0 \end{bmatrix} = \begin{bmatrix} i & 0 \\ 0 & -i \end{bmatrix} = i\,\sigma_z$$

$$\sigma_y \sigma_z = \begin{bmatrix} 0 & -i \\ i & 0 \end{bmatrix} \begin{bmatrix} 1 & 0 \\ 0 & -1 \end{bmatrix} = \begin{bmatrix} 0 & i \\ i & 0 \end{bmatrix} = i\,\sigma_x$$

$$\sigma_z \sigma_x = \begin{bmatrix} 1 & 0 \\ 0 & -1 \end{bmatrix} \begin{bmatrix} 0 & 1 \\ 1 & 0 \end{bmatrix} = \begin{bmatrix} 0 & 1 \\ -1 & 0 \end{bmatrix} = i\,\sigma_y$$

Further, we have

$$\sigma_y \sigma_x = \begin{bmatrix} 0 & -i \\ i & 0 \end{bmatrix} \begin{bmatrix} 0 & 1 \\ 1 & 0 \end{bmatrix} = \begin{bmatrix} -i & 0 \\ 0 & i \end{bmatrix} = -i\,\sigma_z$$

$$\sigma_z \sigma_y = \begin{bmatrix} 1 & 0 \\ 0 & -1 \end{bmatrix} \begin{bmatrix} 0 & -i \\ i & 0 \end{bmatrix} = \begin{bmatrix} 0 & -i \\ -i & 0 \end{bmatrix} = -i\,\sigma_x$$

$$\sigma_x \sigma_z = \begin{bmatrix} 0 & 1 \\ 1 & 0 \end{bmatrix} \begin{bmatrix} 1 & 0 \\ 0 & -1 \end{bmatrix} = \begin{bmatrix} 0 & -1 \\ 1 & 0 \end{bmatrix} = -i\,\sigma_y$$

Therefore, we have

$$\sigma_x \sigma_y + \sigma_y \sigma_x = \sigma_y \sigma_z + \sigma_z \sigma_y = \sigma_z \sigma_x + \sigma_x \sigma_z = 0$$

Exercise 5. For the Pauli's spin matrices σ_x, σ_y, σ_z, show that

$$[\sigma_x, \sigma_y] = 2i\sigma_z \qquad \text{and} \qquad \sigma_x\, \sigma_y\, \sigma_z = i$$

Solution: We have

$$[\sigma_x, \sigma_y] = \sigma_x, \sigma_y - \sigma_y, \sigma_x$$

Now,

$$\sigma_x\sigma_y = \begin{bmatrix} 0 & 1 \\ 1 & 0 \end{bmatrix}\begin{bmatrix} 0 & -i \\ i & 0 \end{bmatrix} = \begin{bmatrix} i & 0 \\ 0 & -i \end{bmatrix} = i\,\sigma_z$$

and

$$\sigma_y\sigma_x = \begin{bmatrix} 0 & -i \\ i & 0 \end{bmatrix}\begin{bmatrix} 0 & 1 \\ 1 & 0 \end{bmatrix} = \begin{bmatrix} -i & 0 \\ 0 & i \end{bmatrix} = -i\,\sigma_z$$

Therefore, we have

$$[\sigma_x, \sigma_y] = i\,\sigma_z + i\,\sigma_z = 2i\,\sigma_z$$

Now,

$$\sigma_x\sigma_y\sigma_y = \begin{bmatrix} 0 & 1 \\ 1 & 0 \end{bmatrix}\begin{bmatrix} 0 & -i \\ i & 0 \end{bmatrix}\begin{bmatrix} 1 & 0 \\ 0 & -1 \end{bmatrix}$$

$$= \begin{bmatrix} i & 0 \\ 0 & -i \end{bmatrix}\begin{bmatrix} 1 & 0 \\ 0 & -1 \end{bmatrix} = i\begin{bmatrix} 1 & 0 \\ 0 & -1 \end{bmatrix}$$

Therefore, we have

$$\sigma_x\sigma_y\sigma_z = i$$

8. Eigenvectors for spin matrices

After inclusion of spin, an electron now has four degrees of freedom, the three position coordinates $(x,\ y,\ z)$ and another observable pertaining to spin, which is generally taken the z-component S_z of the spin S. The wave function for an electron can be written as $\psi(\vec{r}, S_z)$ or $\psi(\vec{r}, m_s)$. For one-half spin systems, the coordinate m_s can take the values $1/2$ and $-1/2$, defined as the spin-up and spin-down or vice-versa. For a weak interaction between the spin and space parts, the wave function of an electron can be expressed as a product of two wave functions

$$\psi(\vec{r}, m_s) = \phi(\vec{r})\chi(m_s)$$

where $\phi(\vec{r})$ is a part depending on the space coordinates and $\chi(m_s)$ is the part depending on the spin coordinates.

For the eigenvalues $\pm\hbar/2$ of S_x, S_y, S_z, the eigenvectors can be easily obtained by writing the eigenvalue equation. The eigenvalue equation for S_x with eigenvalue $\hbar/2$ is

$$\frac{\hbar}{2}\begin{bmatrix} 0 & 1 \\ 1 & 0 \end{bmatrix}\begin{bmatrix} a_1 \\ a_2 \end{bmatrix} = \frac{\hbar}{2}\begin{bmatrix} a_1 \\ a_2 \end{bmatrix}$$

$$\begin{bmatrix} a_2 \\ a_1 \end{bmatrix} = \begin{bmatrix} a_1 \\ a_2 \end{bmatrix}$$

giving $a_1 = a_2$. Now, the normalization relation $a_1^2 + a_2^2 = 1$, gives $a_1 = a_2 = 1/\sqrt{2}$. Thus, the eigenvector for S_x with eigenvalue $\hbar/2$ is

$$\frac{1}{\sqrt{2}}\begin{bmatrix} 1 \\ 1 \end{bmatrix}$$

The eigenvalue equation for S_x with eigenvalue $-\hbar/2$ is

$$\frac{\hbar}{2}\begin{bmatrix} 0 & 1 \\ 1 & 0 \end{bmatrix}\begin{bmatrix} a_1 \\ a_2 \end{bmatrix} = -\frac{\hbar}{2}\begin{bmatrix} a_1 \\ a_2 \end{bmatrix}$$

$$\begin{bmatrix} a_2 \\ a_1 \end{bmatrix} = -\begin{bmatrix} a_1 \\ a_2 \end{bmatrix}$$

giving $a_1 = -a_2$. Now, the normalization relation $a_1^2 + a_2^2 = 1$, gives $a_1 = 1/\sqrt{2}$ and $a_2 = -1/\sqrt{2}$. Thus, the eigenvector for S_x with eigenvalue $\hbar/2$ is

$$\frac{1}{\sqrt{2}}\begin{bmatrix} 1 \\ -1 \end{bmatrix}$$

The eigenvalue equation for S_y with eigenvalue $\hbar/2$ is

$$\frac{\hbar}{2}\begin{bmatrix} 0 & -i \\ i & 0 \end{bmatrix}\begin{bmatrix} a_1 \\ a_2 \end{bmatrix} = \frac{\hbar}{2}\begin{bmatrix} a_1 \\ a_2 \end{bmatrix}$$

$$\begin{bmatrix} -i\,a_2 \\ i\,a_1 \end{bmatrix} = \begin{bmatrix} a_1 \\ a_2 \end{bmatrix}$$

giving $a_1 = -ia_2$. Now, the normalization relation $a_1^2 + a_2^2 = 1$, gives $a_1 = 1/\sqrt{2}$ and $a_2 = i/\sqrt{2}$. Thus, the eigenvector for S_y with eigenvalue $\hbar/2$ is

$$\frac{1}{\sqrt{2}}\begin{bmatrix} 1 \\ i \end{bmatrix}$$

The eigenvalue equation for S_y with eigenvalue $-\hbar/2$ is

$$\frac{\hbar}{2}\begin{bmatrix} 0 & -i \\ i & 0 \end{bmatrix}\begin{bmatrix} a_1 \\ a_2 \end{bmatrix} = -\frac{\hbar}{2}\begin{bmatrix} a_1 \\ a_2 \end{bmatrix}$$

$$\begin{bmatrix} -i\,a_2 \\ i\,a_1 \end{bmatrix} = -\begin{bmatrix} a_1 \\ a_2 \end{bmatrix}$$

giving $a_1 = ia_2$. Now, the normalization relation $a_1^2 + a_2^2 = 1$, gives $a_1 = -i/\sqrt{2}$ and $a_2 = 1/\sqrt{2}$. Thus, the eigenvector for S_x with eigen value $\hbar/2$ is

$$\frac{1}{\sqrt{2}}\begin{bmatrix} 1 \\ -i \end{bmatrix}$$

The eigenvalue equation for S_z with eigenvalue $\hbar/2$ is

$$\frac{\hbar}{2}\begin{bmatrix} 1 & 0 \\ 0 & -1 \end{bmatrix}\begin{bmatrix} a_1 \\ a_2 \end{bmatrix} = \frac{\hbar}{2}\begin{bmatrix} a_1 \\ a_2 \end{bmatrix}$$

$$\begin{bmatrix} a_1 \\ -a_2 \end{bmatrix} = \begin{bmatrix} a_1 \\ a_2 \end{bmatrix}$$

giving $a_2 = 0$. Now, the normalization relation $a_1^2 + a_2^2 = 1$, gives $a_1 = 1$ and $a_2 = 0$. Thus, the eigenvector for S_x with eigenvalue $\hbar/2$ is (denoted by α)

$$\alpha = \begin{bmatrix} 1 \\ 0 \end{bmatrix}$$

The eigenvalue equation for S_z with eigenvalue $-\hbar/2$ is

$$\frac{\hbar}{2}\begin{bmatrix} 1 & 0 \\ 0 & -1 \end{bmatrix}\begin{bmatrix} a_1 \\ a_2 \end{bmatrix} = -\frac{\hbar}{2}\begin{bmatrix} a_1 \\ a_2 \end{bmatrix}$$

$$\begin{bmatrix} -a_1 \\ a_2 \end{bmatrix} = \begin{bmatrix} a_1 \\ a_2 \end{bmatrix}$$

giving $a_1 = 0$. Now, the normalization relation $a_1^2 + a_2^2 = 1$, gives $a_2 = 1$ and $a_1 = 0$. Thus, the eigenvector for S_x with eigenvalue $\hbar/2$ is (denoted by β)

$$\beta = \begin{bmatrix} 0 \\ 1 \end{bmatrix}$$

9. Addition of angular momenta

In an atom, there are several electrons and their angular momenta generally combine. Similarly, in a nucleus, angular momenta of nucleons (protons and neutrons) combine. Thus, we need to study the addition of angular momenta. Addition of angular momenta plays very significant role in the study of atomic spectra, structure of nuclei, etc. Here, we shall discuss a general problem of addition of two angular momenta.

Let us consider two noninteracting systems having angular momenta $\vec{J}_1$ and $\vec{J}_2$ and eigenkets $|j_1m_1\rangle$ and $|j_2m_2\rangle$, respectively. Thus, we have

$$J_1^2|j_1m_1\rangle = j_1(j_1+1)\hbar^2|j_1m_1\rangle; \qquad J_{1z}|j_1m_1\rangle = m_1\ \hbar|j_1m_1\rangle$$

and

$$J_2^2|j_2m_2\rangle = j_2(j_2+1)\hbar^2\ |j_2m_2\rangle; \qquad J_{2z}|j_2m_2\rangle = m_2\ \hbar|j_2m_2\rangle$$

where

$$m_1 = -j_1, -j_1+1, \ldots, j_1; \qquad m_2 = -j_2, -j_2+1, \ldots, j_2$$

Since the J_1 and J_2 are not interacting, we have

$$[J_1, J_2] = 0$$

and therefore[5]

$$[J_1^2, J_2^2] = 0$$

and thus, the operators J_1^2, J_{1z}, J_2^2 and J_{2z} form a complete set with common eigenkets $|j_1m_1, j_2m_2\rangle$, which is nothing but product of $|j_1m_1\rangle$ and $|j_2m_2\rangle$. That is

$$|j_1m_1, j_2m_2\rangle = |j_1m_1\rangle\ |j_2m_2\rangle$$

which for the given values of j_1 and j_2 can also be written as

$$|m_1m_2\rangle \equiv |j_1m_1, j_2m_2\rangle$$

Since m_1 and m_2 can have $(2J_1+1)$ and $(2J_2+1)$ orientations, respectively. Thus, we would have $(2J_1+1)(2J_2+1)$ values for $|m_1m_2\rangle$ eigenkets.

---

[5]We have

$$[J_1^2, J_2^2] = J_1[J_1, J_2^2] + [J_1, J_2^2]J_1 = -J_1[J_2^2, J_1] - [J_2^2, J_1]J_1$$
$$= -J_1J_2[J_2, J_1] - J_1[J_2, J_1]J_2 - J_2[J_2, J_1]J_1 - [J_2, J_1]J_2J_1 = 0$$

9.1 Clebsch-Gordan coefficients

For the total angular momentum $\vec{J} = \vec{J}_1 + \vec{J}_2$, we have

$$\vec{J} \times \vec{J} = i\,\hbar\,\vec{J}$$

It also follows that

$$[J^2, J_z] = 0 \qquad [J^2, J_1^2] = 0 \qquad [J^2, J_2^2] = 0$$

The orthonormal eigenkets of J^2 and J_z are $|jm\rangle$ such that

$$J^2|jm\rangle = j(j+1)\hbar^2|jm\rangle \qquad J_z|jm\rangle = m\hbar|jm\rangle$$

As J^2 commutes with J_z, J_1^2 and J_2^2, they form another complete set with common eigenkets $|j_1j_2jm\rangle$, which for the given values of j_1 and j_2 can also be written as

$$|jm\rangle \equiv |j_1j_2jm\rangle$$

Now, one set of eigenkets can be expressed as linear combination of other set of eigenkets as

$$|jm\rangle = \sum_{m_1,m_2} C_{jmm_1m_2}\,|m_1m_2\rangle \tag{7.22}$$

The coefficients $C_{jmm_1m_2}$ are known as the 'Clebsch-Gordan coefficients' or Wigner coefficients. Multiplying equation (7.22) from left by $\langle m_1'm_2'|$, we get

$$\langle m_1'm_2'|jm\rangle = \sum_{m_1,m_2} C_{jmm_1m_2}\langle m_1'm_2'|m_1m_2\rangle = C_{jmm_1'm_2'}$$

Thus,

$$C_{jmm_1m_2} = \langle m_1m_2|jm\rangle$$

and therefore,

$$|jm\rangle = \sum_{m_1,m_2} |m_1m_2\rangle\langle m_1m_2|jm\rangle \tag{7.23}$$

On the other side, second set of eigenkets can be expressed as linear combination of the first eigenkets as

$$|m_1m_2\rangle = \sum_{jm} C_{jmm_1m_2}|jm\rangle$$

The coefficients $C_{jmm_1m_2}$ are again known as the Clebsch-Gordan coefficients. Multiplying this equation from left by $\langle j'm'|$, we get

$$\langle j'm'|m_1m_2\rangle = \sum_{jm} C_{jmm_1m_2}\langle j'm'|jm\rangle C_{j'm'm_1m_2}$$

Thus,

$$C_{jmm_1m_2} = \langle jm|m_1m_2\rangle$$

and therefore,

$$|m_1m_2\rangle = \sum_{jm} |jm\rangle\langle jm|m_1m_2\rangle \tag{7.24}$$

9.2 Orthogonality relations for Clebsch-Gordan coefficients

Now, multiplying equation (7.24) from left by $\langle m_1' m_2'|$, we get

$$\langle m_1' m_2' | m_1 m_2 \rangle = \sum_{jm} \langle m_1' m_2' | jm \rangle \langle jm | m_1 m_2 \rangle$$

Taking complex conjugate of this equation, we have

$$\langle m_1 m_2 | m_1' m_2' \rangle = \sum_{jm} \langle jm | m_1' m_2' \rangle \langle m_1 m_2 | jm \rangle$$

and therefore, we get

$$\sum_{jm} \langle jm | m_1' m_2' \rangle \langle m_1 m_2 | jm \rangle = \delta_{m_1 m_1'} \, \delta_{m_2 m_2'}$$

Thus, for $m_1 = m_1'$ and $m_2 = m_2'$, we have

$$\sum_{jm} \langle jm | m_1 m_2 \rangle \langle m_1 m_2 | jm \rangle = 1$$

or

$$\sum_{jm} |\langle m_1 m_2 | jm \rangle|^2 = 1 \tag{7.25}$$

Now, multiplying equation (7.23) from left by $\langle j'm'|$, we get

$$\langle j'm' | jm \rangle = \sum_{m_1, m_2} \langle j'm' | m_1 m_2 \rangle \langle m_1 m_2 | jm \rangle$$

Taking complex conjugate of this equation, we have

$$\langle jm | j'm' \rangle = \sum_{m_1, m_2} \langle m_1 m_2 | j'm' \rangle \langle jm | m_1 m_2 \rangle$$

and therefore, we have

$$\sum_{m_1, m_2} \langle m_1 m_2 | j'm' \rangle \langle jm | m_1 m_2 \rangle = \delta_{jj'} \, \delta_{mm'}$$

Thus, for $m = m'$ and $j = j'$, we have

$$\sum_{m_1, m_2} \langle m_1 m_2 | jm \rangle \langle jm | m_1 m_2 \rangle = 1$$

or

$$\sum_{m_1, m_2} |\langle m_1 m_2 | jm \rangle|^2 = 1 \tag{7.26}$$

9.3 Selection rules for Clebsch-Gordan coefficients

Multiplying equation (7.23) from left by J_z, we get

$$J_z|jm\rangle = \sum_{m_1,m_2}(J_{1z} + J_{2z})|m_1m_2\rangle\langle m_1m_2|jm\rangle$$

$$m\hbar|jm\rangle = \sum_{m_1,m_2}(m_1 + m_2)\ \hbar\ |m_1m_2\rangle\langle m_1m_2|jm\rangle$$

Using (7.23) here, we get

$$m|m_1m_2\rangle\langle m_1m_2|jm\rangle = \sum_{m_1,m_2}(m_1 + m_2)|m_1m_2\rangle\langle m_1m_2|jm\rangle$$

$$\sum_{m_1,m_2}(m_1 + m_2 - m)|m_1m_2\rangle\langle m_1m_2|jm\rangle = 0$$

This equation is valid only when the coefficient of each term vanishes separately. This gives one of the rules of vector model of atom as the following.

$$m = m_1 + m_2 \tag{7.27}$$

9.4 Recursion relations for Clebsch-Gordan coefficients

Multiplying equation (7.23) from left by J_- and replacing J_- by $J_{1-} + J_{2-}$, we get

$$J_-\ |jm\rangle = \sum_{m_1',m_2'}(J_{1-} + J_{2-})\ |m_1'm_2'\rangle\langle m_1'm_2'|jm\rangle$$

Using equation (7.21), we get

$$[j(j+1) - m(m-1)]^{1/2}|j, m-1\rangle$$

$$= \sum_{m_1',m_2'}[j_1(j_1+1) - m_1'(m_1'-1)]^{1/2}|m_1'-1, m_2'\rangle\langle m_1'm_2'|jm\rangle$$

$$+ \sum_{m_1',m_2'}[j_2(j_2+1) - m_2'(m_2'-1)]^{1/2}|m_1', m_2'-1\rangle\langle m_1'm_2'|jm\rangle$$

Multiplying this equation from left by $\langle m_1m_2|$, we get

$$[j(j+1) - m(m-1)]^{1/2}\langle m_1m_2|j, m-1\rangle$$

$$= \sum_{m_1',m_2'}[j_1(j_1+1) - m_1'(m_1'-1)]^{1/2}\langle m_1m_2|m_1'-1, m_2'\rangle\langle m_1'm_2'|jm\rangle$$

$$+ \sum_{m_1',m_2'}[j_2(j_2+1) - m_2'(m_2'-1)]^{1/2}\langle m_1m_2|m_1', m_2'-1\rangle\langle m_1'm_2'|jm\rangle$$

This equation gives

$$\begin{aligned}[j(j+1)-m(m-1)]^{1/2}\langle m_1 m_2|j,m-1\rangle \\ = [j_1(j_1+1)-m_1(m_1+1)]^{1/2}\langle m_1+1,m_2|jm\rangle \\ +[j_2(j_2+1)-m_2(m_2+1)]^{1/2}\langle m_1,m_2+1|jm\rangle\end{aligned} \tag{7.28}$$

Multiplying equation (7.23) from left by J_+ and replacing J_+ by $J_{1+} + J_{2+}$, we get

$$J_+\ |jm\rangle = \sum_{m_1',m_2'}(J_{1+}+J_{2+})\ |m_1'm_2'\rangle\langle m_1'm_2'|jm\rangle$$

Using equation (7.19), we get

$$[j(j+1)-m(m+1)]^{1/2}|j,m+1\rangle$$

$$= \sum_{m_1',m_2'}[j_1(j_1+1)-m_1'(m_1'+1)]^{1/2}|m_1'+1,m_2'\rangle\langle m_1'm_2'|jm\rangle$$

$$+ \sum_{m_1',m_2'}[j_2(j_2+1)-m_2'(m_2'+1)]^{1/2}|m_1',m_2'+1\rangle\langle m_1'm_2'|jm\rangle$$

Multiplying this equation from left by $\langle m_1 m_2|$, we get

$$[j(j+1)-m(m+1)]^{1/2}\langle m_1 m_2|j,m+1\rangle$$

$$= \sum_{m_1',m_2'}[j_1(j_1+1)-m_1'(m_1'+1)]^{1/2}\langle m_1 m_2|m_1'+1,m_2'\rangle\langle m_1'm_2'|jm\rangle$$

$$+ \sum_{m_1',m_2'}[j_2(j_2+1)-m_2'(m_2'+1)]^{1/2}\langle m_1 m_2|m_1',m_2'+1\rangle\langle m_1'm_2'|jm\rangle$$

This equation gives

$$\begin{aligned}[j(j+1)-m(m+1)]^{1/2}\langle m_1 m_2|j,m+1\rangle \\ = [j_1(j_1+1)-m_1(m_1-1)]^{1/2}\langle m_1-1,m_2|jm\rangle \\ +[j_2(j_2+1)-m_2(m_2-1)]^{1/2}\langle m_1,m_2-1|jm\rangle\end{aligned} \tag{7.29}$$

The relations (7.28) and (7.29) are the recursion relations and they along with the equations (7.25) and (7.26) play vital role for computation of Clebsch-Gordan coefficients.

Let us change m_1 by $-m_1$, m_2 by $-m_2$ and m by $-m$ simultaneously in equation (7.28), so that we have

$$[j(j+1)-m(m+1)]^{1/2}\langle -m_1,-m_2|j,-m-1\rangle$$
$$=[j_1(j_1+1)-m_1(m_1-1)]^{1/2}\hbar\,\langle -m_1+1,-m_2|j,-m\rangle$$
$$+[j_2(j_2+1)-m_2(m_2-1)]^{1/2}\hbar\,\langle -m_1,-m_2+1|j,-m\rangle$$

This equation can rearranged as

$$[j(j+1)-m(m+1)]^{1/2}\langle -m_1,-m_2|j,-(m+1)\rangle$$
$$=[j_1(j_1+1)-m_1(m_1-1)]^{1/2}\hbar\,\langle -(m_1-1),-m_2|j,-m\rangle$$
$$+[j_2(j_2+1)-m_2(m_2-1)]^{1/2}\hbar\,\langle -m_1,-(m_2-1)|j,-m\rangle$$

Comparison of this equation with equation (7.29) gives

$$\langle m_1m_2|jm\rangle=\langle -m_1,-m_2|j,-m\rangle \tag{7.30}$$

It shows an important result that when we change sign of m_1,m_2 and m simultaneously in a Clebsch-Gordan coefficient, the value of the coefficient does not change.

10. Reduction of the matrix of Clebsch-Gordan coefficients

For the given values of j_1 and j_2, there are $(2j_1+1)(2j_2+1)$ combinations for m_1 and m_2 and equal number of combinations for j and m. Thus, a matrix of Clebsch-Gordan coefficients $\langle m_1m_2|jm\rangle$ has $(2j_1+1)(2j_2+1)$ rows and $(2j_1+1)(2j_2+1)$ columns.

For example, for $j_1 = 1/2$ and $j_2 = 1/2$. The value of m_1 can be $-1/2$, $1/2$ and that of m_2 as $-1/2$, $1/2$. The possible values of m are 1, 0, -1, whereas the possible values of j are 0, 1. The matrix of Clebsch-Gordan coefficients has 16 elements (denoted by zero as well as $\times$) and is shown in Table 1. Here, the zero (0) indicates that the value is zero, whereas the symbol $\times$ indicates that there is some value which may however be zero.

Table 1. Matrix of Clebsch-Gordan coefficients

		$jm\rangle$			
m_1	m_2	$\|11\rangle$	$\|10\rangle$	$\|00\rangle$	$\|1-1\rangle$
1/2	1/2	$\times$	0	0	0
1/2	−1/2	0	$\times$	$\times$	0
−1/2	1/2	0	$\times$	$\times$	0
−1/2	−1/2	0	0	0	$\times$

For $j_1 = 1$ and $j_2 = 1/2$, value of m_1 can be -1, 0, 1 and that of m_2 as $-1/2$, 1/2. The possible values of m are 3/2, 1/2, $-1/2$, $-3/2$ whereas the possible values of j are 1/2, 3/2. The matrix of Clebsch-Gordan coefficients has 36 elements (denoted by zero as well as $\times$) and is shown in Table 2.

For $j_1 = 1$ and $j_2 = 1$, value of m_1 can be -1, 0, 1 and that of m_2 as -1, 0, 1. The possible values of m are 2, 1, 0, -1, -2 whereas the possible values of j are 0, 1, 2. The matrix of Clebsch-Gordan coefficients has 81 elements (denoted by zero as well as $\times$) and is shown in Table 3.

Table 2. Matrix of Clebsch-Gordan coefficients

		$\|jm\rangle$					
m_1	m_2	$\|\frac{3}{2}\frac{3}{2}\rangle$	$\|\frac{3}{2}\frac{1}{2}\rangle$	$\|\frac{1}{2}\frac{1}{2}\rangle$	$\|\frac{3}{2},-\frac{1}{2}\rangle$	$\|\frac{1}{2},-\frac{1}{2}\rangle$	$\|\frac{3}{2},-\frac{3}{2}\rangle$
1	1/2	$\times$	0	0	0	0	0
1	$-1/2$	0	$\times$	$\times$	0	0	0
0	1/2	0	$\times$	$\times$	0	0	0
0	$-1/2$	0	0	0	$\times$	$\times$	0
-1	1/2	0	0	0	$\times$	$\times$	0
-1	$-1/2$	0	0	0	0	0	$\times$

Table 3. Matrix of Clebsch-Gordan coefficients

		$\|jm\rangle$								
m_1	m_2	$\|22\rangle$	$\|21\rangle$	$\|11\rangle$	$\|20\rangle$	$\|10\rangle$	$\|00\rangle$	$\|2,-1\rangle$	$\|1,-1\rangle$	$\|2,-2\rangle$
1	1	$\times$	0	0	0	0	0	0	0	0
1	0	0	$\times$	$\times$	0	0	0	0	0	0
0	1	0	$\times$	$\times$	0	0	0	0	0	0
1	-1	0	0	0	$\times$	$\times$	$\times$	0	0	0
0	0	0	0	0	$\times$	$\times$	$\times$	0	0	0
-1	1	0	0	0	$\times$	$\times$	$\times$	0	0	0
0	-1	0	0	0	0	0	0	$\times$	$\times$	0
-1	0	0	0	0	0	0	0	$\times$	$\times$	0
-1	-1	0	0	0	0	0	0	0	0	$\times$

Following the selection rules (7.27), a large number of elements are obviously zero. It shows that this matrix of Clebsch-Gordan coefficients can be decomposed into square sub-matrices of various orders. For $j_1 = 1/2$ and $j_2 = 1/2$, there are two matrices of order 1×1 and one matrix of order 2×2. For $j_1 = 1$ and $j_2 = 1/2$, there are two matrices of order 1×1 and two matrices of order 2×2. For $j_1 = 1$ and $j_2 = 1$, there are two matrices of order 1×1, two matrices of order 2×2 and one matrix of order 3×3. For the above cases, the values of m_1, m_2, j and m can be summarized as in Table 4 (for $m_1 = \frac{1}{2}$ and $m_2 = \frac{1}{2}$), Table 5 (for

$m_1 = 1$ and $m_2 = \frac{1}{2}$) and Table 6 (for $m_1 = 1$ and $m_2 = 1$).

Table 4.

m_1	m_2	m	j
$\frac{1}{2}$	$\frac{1}{2}$	1	1
$\frac{1}{2}$	$-\frac{1}{2}$	0	1
$-\frac{1}{2}$	$\frac{1}{2}$		0
$-\frac{1}{2}$	$-\frac{1}{2}$	-1	1

Table 5.

m_1	m_2	m	j
1	$\frac{1}{2}$	$\frac{3}{2}$	$\frac{3}{2}$
1	$-\frac{1}{2}$	$\frac{1}{2}$	$\frac{3}{2}$
0	$\frac{1}{2}$		$\frac{1}{2}$
0	$-\frac{1}{2}$	$-\frac{1}{2}$	$\frac{3}{2}$
-1	$\frac{1}{2}$		$\frac{1}{2}$
-1	$-\frac{1}{2}$	$-\frac{3}{2}$	$\frac{3}{2}$

Table 6.

m_1	m_2	m	j
1	1	2	2
1	0	1	2
0	1		1
1	-1	0	2
0	0		1
-1	1		0
0	-1	-1	2
-1	0		1
-1	-1	-2	2

For given values of j_1 and j_2, m_1 can have the value from $-j_1$ to j_1 and m_2 from $-j_2$ to j_2 in the steps of unity. Since $m = m_1 + m_2$, the possible value of m are $(j_1 + j_2)$, $j_1 + j_2 - 1$, $j_1 + j_2 - 2$, ..., $-(j_1 + j_2)$. The sub-matrices are given in Table 7.[6]

Since the smallest value of m_1 can be $-j_1$ and that of m_2 can be $-j_2$, the smallest value $j_1 + j_2 - k$ of j occurs when

$$j_1 - k = -j_1 \qquad \text{or} \qquad j_2 - k = -j_2$$

That is when $k = 2j_1$ or $k = 2j_2$. The smallest value of j is then

$$j_1 + j_2 - k = j_1 + j_2 - 2j_1 = j_2 - j_1$$

or

$$j_1 + j_2 - k = j_1 + j_2 - 2j_2 = j_1 - j_2$$

Since j must always be positive, the smallest value of j is $|j_1 - j_2|$. Thus, the permitted values of j vary from $(j_1 + j_2)$ to $|j_1 - j_2|$.

The number of these sub-matrices is $(2j_1 + 2j_2 + 1)$. When $(2j_1 + 1)(2j_2 + 1)$ can be expressed as $n(n + 1)$, where n is a positive integer, then there are pairs of sub-matrices of orders 1×1, 2×2, ..., $n \times n$. When $(2j_1 + 1)(2j_2 + 1)$ can be expressed as n^2, where n is a positive integer, then there are pairs of sub-matrices of orders 1×1, 2×2, ..., $(n - 1) \times (n - 1)$ and one sub-matrix is of order $n \times n$.

[6]For preparation of this table, we first take the values of m ranging from $(j_1 + j_2)$ to $-(j_1 + j_2)$ in the steps of unity. Thus, there are $(2j_1 + 2j_2 + 1)$ sub-matrices. For a given m, we take all the possible values of j and finally, we take combinations of the values of m_1 and m_2, such that $m = m_1 + m_2$.

11. Computation of Clebsch-Gordan coefficients

Now, the task for consideration is to compute the elements of these square sub-matrices of various orders.

There are two sub-matrices of order 1×1. For one sub-matrix, we $m_1 = j_1$, $m_2 = j_2$, $m = (j_1 + j_2)$ and $j = (j_1 + j_2)$ and for the other sub-matrix, we $m_1 = -j_1$, $m_2 = -j_2$, $m = -(j_1 + j_2)$ and $j = j_1 + j_2$. For the first sub-matrix, from equation (7.25), we have

$$\sum_{jm} |\langle m_1 m_2 | jm \rangle|^2 = 1$$

$$|\langle j_1, j_2 | (j_1 + j_2), (j_1 + j_2) \rangle|^2 = 1$$

$$\langle j_1, j_2 | (j_1 + j_2), (j_1 + j_2) \rangle = 1$$

Table 7. Sub-matrices

m_1	m_2	m	j
j_1	j_2	$j_1 + j_2$	$j_1 + j_2$
j_1 $j_1 - 1$	$j_2 - 1$ j_2	$j_1 + j_2 - 1$	$j_1 + j_2$ $j_1 + j_2 - 1$
j_1 $j_1 - 1$ $j_1 - 2$	$j_2 - 2$ $j_2 - 1$ j_2	$j_1 + j_2 - 2$	$j_1 + j_2$ $j_1 + j_2 - 1$ $j_1 + j_2 - 2$
j_1 $j_1 - 1$ $j_1 - 2$ $j_1 - 3$	$j_2 - 3$ $j_2 - 2$ $j_2 - 1$ j_2	$j_1 + j_2 - 3$	$j_1 + j_2$ $j_1 + j_2 - 1$ $j_1 + j_2 - 2$ $j_1 + j_2 - 3$
⋮	⋮	⋮	⋮
j_1 $j_1 - 1$ $j_1 - 2$ ⋮ $j_1 - k$	$j_2 - k$ $j_2 - k + 1$ $j_2 - k + 2$ ⋮ j_2	$j_1 + j_2 - k$	$j_1 + j_2$ $j_1 + j_2 - 1$ $j_1 + j_2 - 2$ ⋮ $j_1 + j_2 - k$
⋮	⋮	⋮	⋮
$-j_1 + 2$ $-j_1 + 1$ $-j_1$	$-j_2$ $-j_2 + 1$ $-j_2 + 2$	$-(j_1 + j_2 - 2)$	$(j_1 + j_2)$ $(j_1 + j_2 - 1)$ $(j_1 + j_2 - 2)$
$-j_1 + 1$ $-j_1$	$-j_2$ $-j_2 + 1$	$-(j_1 + j_2 - 1)$	$(j_1 + j_2)$ $(j_1 + j_2 - 1)$
$-j_1$	$-j_2$	$-(j_1 + j_2)$	$(j_1 + j_2)$

For the second sub-matrix, using equation (7.30), we have

$$\langle -j_1, -j_2|(j_1+j_2), -(j_1+j_2)\rangle = 1$$

There may be two sub-matrices of order 2×2. For the sub-matrix with $m = (j_1 + j_2 - 1)$, we put $m_1 = j_1$, $m_2 = j_2 - 1$, $j = j_1 + j_2$ and $m = j_1 + j_2$ in equation (7.28) to get

$$\Big[(j_1+j_2)(j_1+j_2+1) - (j_1+j_2)(j_1+j_2-1)\Big]^{1/2}\langle j_1, (j_2-1)|(j_1+j_2),$$

$$(j_1+j_2-1)\rangle = 0 + \Big[j_2(j_2+1) - (j_2-1)j_2\Big]^{1/2}\langle j_1, j_2|(j_1+j_2), (j_1+j_2)\rangle$$

$$\sqrt{2}(j_1+j_2)^{1/2}\langle j_1, (j_2-1)|(j_1+j_2), (j_1+j_2-1)\rangle$$

$$= \sqrt{2j_2}\langle j_1, j_2|(j_1+j_2), (j_1+j_2)\rangle$$

Using

$$\langle j_1, j_2|(j_1+j_2), (j_1+j_2)\rangle = 1$$

in this equation, we have

$$\langle j_1, (j_2-1)|(j_1+j_2), (j_1+j_2-1)\rangle = \left(\frac{j_2}{j_1+j_2}\right)^{1/2} \qquad (7.31)$$

From equation (7.25), we have

$$\sum_{jm} |\langle m_1 m_2|jm\rangle|^2 = 1$$

$$|\langle j_1, (j_2-1)|(j_1+j_2), (j_1+j_2-1)\rangle|^2$$
$$+|\langle j_1, (j_2-1)|(j_1+j_2-1), (j_1+j_2-1)\rangle|^2 = 1$$

Using equation (7.31) here, we get

$$\left(\frac{j_2}{j_1+j_2}\right) + |\langle j_1, (j_2-1)|(j_1+j_2-1), (j_1+j_2-1)\rangle|^2 = 1$$

$$|\langle j_1, (j_2-1)|(j_1+j_2-1), (j_1+j_2-1)\rangle|^2 = \left(\frac{j_1}{j_1+j_2}\right)$$

$$\langle j_1, (j_2-1)|(j_1+j_2-1), (j_1+j_2-1)\rangle = \left(\frac{j_1}{j_1+j_2}\right)^{1/2}$$

Now, on putting $m_1 = j_1 - 1$, $m_2 = j_2$, $j = j_1 + j_2$ and $m = j_1 + j_2$ in equation (7.28) to get

$$\Big[(j_1+j_2)(j_1+j_2+1) - (j_1+j_2)(j_1+j_2-1)\Big]^{1/2} \langle (j_1-1), j_2 | (j_1+j_2),$$

$$(j_1+j_2-1)\rangle = \Big[j_1(j_1+1) - (j_1-1)j_1\Big]^{1/2} \langle j_1, j_2 | (j_1+j_2), (j_1+j_2)\rangle + 0$$

$$\sqrt{2}\,(j_1+j_2)^{1/2} \langle (j_1-1), j_2 | (j_1+j_2), (j_1+j_2-1)\rangle$$

$$= \sqrt{2j_1}\,\langle j_1, j_2 | (j_1+j_2), (j_1+j_2)\rangle$$

Using

$$\langle j_1, j_2 | (j_1+j_2), (j_1+j_2)\rangle = 1$$

we have

$$\langle (j_1-1), j_2 | (j_1+j_2), (j_1+j_2-1)\rangle = \Big(\frac{j_1}{j_1+j_2}\Big)^{1/2} \qquad (7.32)$$

From equation (7.25), we have

$$\sum_{jm} |\langle m_1 m_2 | jm\rangle|^2 = 1$$

$$|\langle (j_1-1), j_2 | (j_1+j_2), (j_1+j_2-1)\rangle|^2 + |\langle (j_1-1), j_2 | (j_1+j_2-1), (j_1+j_2-1)\rangle|^2 = 1$$

Using equation (7.32) here, we get

$$\Big(\frac{j_1}{j_1+j_2}\Big) + |\langle (j_1-1), j_2 | (j_1+j_2-1), (j_1+j_2-1)\rangle|^2 = 1$$

$$|\langle (j_1-1), j_2 | (j_1+j_2-1), (j_1+j_2-1)\rangle|^2 = \Big(\frac{j_2}{j_1+j_2}\Big)$$

$$\langle (j_1-1), j_2 | (j_1+j_2-1), (j_1+j_2-1)\rangle = -\Big(\frac{j_2}{j_1+j_2}\Big)^{1/2}$$

It is a convention to take negative sign for the Clebsch-Gordan coefficient with lower value of m_1 and lower value of j.

The Clebsch-Gordan coefficients of the 2 × 2 matrix are as given in the following table.

		$jm\rangle$	
m_1	m_2	$\|(j_1+j_2), (j_1+j_2-1)\rangle$	$\|(j_1+j_2-1), (j_1+j_2-1)\rangle$
j_1	$j_2 - 1$	$\left(\frac{j_2}{j_1+j_2}\right)^{1/2}$	$\left(\frac{j_1}{j_1+j_2}\right)^{1/2}$
$j_1 - 1$	j_2	$\left(\frac{j_1}{j_1+j_2}\right)^{1/2}$	$-\left(\frac{j_2}{j_1+j_2}\right)^{1/2}$

Exercise 6. Compute Clebsch-Gordan coefficients for $j_1 = 1/2$ and $j_2 = 1/2$.

Solution: The sub-matrices for the Clebsch-Gordan coefficients are as the following.

m_1	m_2	m	j
1/2	1/2	1	1
1/2	−1/2	0	1
−1/2	1/2		0
−1/2	−1/2	−1	1

For the sub-matrices of order 1×1 with $m = 1$, we have

$$\sum_{jm} |\langle m_1 m_2 | jm \rangle|^2 = 1$$

$$\left|\left\langle \frac{1}{2}, \frac{1}{2} \middle| 1, 1 \right\rangle\right|^2 = 1 \qquad \text{and thus} \qquad \left\langle \frac{1}{2}, \frac{1}{2} \middle| 1, 1 \right\rangle = 1 \ .$$

For the second sub-matrix of order 1×1 with $m = -1$, using equation (7.30), we have

$$\left\langle -\frac{1}{2}, -\frac{1}{2} \middle| 1, -1 \right\rangle = \left\langle \frac{1}{2}, \frac{1}{2} \middle| 1, 1 \right\rangle = 1 \tag{7.33}$$

For the sub-matrices of order 2×2, we put $m_1 = 1/2$, $m_2 = -1/2$, $j = 1$ and $m = 1$ in equation (7.28) to get

$$\sqrt{2} \left\langle \frac{1}{2}, -\frac{1}{2} \middle| 1, 0 \right\rangle = 0 + \left\langle \frac{1}{2}, \frac{1}{2} \middle| 1, 1 \right\rangle$$

Using equation (7.33), we get

$$\left\langle \frac{1}{2}, -\frac{1}{2} \middle| 1, 0 \right\rangle = \frac{1}{\sqrt{2}} \tag{7.34}$$

From equation (7.25), we have

$$\sum_{jm} |\langle m_1 m_2 | jm \rangle|^2 = 1$$

$$\left|\left\langle \frac{1}{2}, -\frac{1}{2} \middle| 1, 0 \right\rangle\right|^2 + \left|\left\langle \frac{1}{2}, -\frac{1}{2} \middle| 0, 0 \right\rangle\right|^2 = 1$$

Using equation (7.34), we get

$$\left\langle \frac{1}{2}, -\frac{1}{2} \middle| 0, 0 \right\rangle = \frac{1}{\sqrt{2}}$$

Now, we put $m_1 = -1/2$, $m_2 = 1/2$, $j = 1$ and $m = 1$ in equation (7.28) to get

$$\sqrt{2}\left\langle -\frac{1}{2},\frac{1}{2}\middle|1,0\right\rangle = \left\langle \frac{1}{2},\frac{1}{2}\middle|1,1\right\rangle + 0$$

Using equation (7.33), we get

$$\left\langle -\frac{1}{2},\frac{1}{2}\middle|1,0\right\rangle = \frac{1}{\sqrt{2}} \tag{7.35}$$

From equation (7.25), we have

$$\sum_{jm} |\langle m_1 m_2 | jm\rangle|^2 = 1$$

$$\left|\left\langle -\frac{1}{2},\frac{1}{2}\middle|1,0\right\rangle\right|^2 + \left|\left\langle -\frac{1}{2},\frac{1}{2}\middle|0,0\right\rangle\right|^2 = 1$$

Using equation (7.35), we get

$$\left\langle -\frac{1}{2},\frac{1}{2}\middle|0,0\right\rangle = -\frac{1}{\sqrt{2}}$$

As a convention, negative sign is taken for the Clebsch-Gordan coefficients with lower value of m_1 and lower value of j_1. Thus, the Clebsch-Gordan coefficients for $j_1 = 1/2$ and $j_2 = 1/2$ are as in the following table.

		$jm >$			
m_1	m_2	$\|11>$	$\|10>$	$\|00>$	$\|1-1>$
$1/2$	$1/2$	1	0	0	0
$1/2$	$-1/2$	0	$1/\sqrt{2}$	$1/\sqrt{2}$	0
$-1/2$	$1/2$	0	$1/\sqrt{2}$	$-1/\sqrt{2}$	0
$-1/2$	$-1/2$	0	0	0	1

Exercise 7. Compute Clebsch-Gordan coefficients for $j_1 = 1$ and $j_2 = 1/2$.

Solution: The sub-matrices for the Clebsch-Gordan coefficients are as the following.

m_1	m_2	m	j
1	$1/2$	$3/2$	$3/2$
1	$-1/2$	$1/2$	$3/2$
0	$1/2$		$1/2$
0	$-1/2$	$-1/2$	$3/2$
-1	$1/2$		$1/2$
-1	$-1/2$	$-3/2$	$3/2$

For the sub-matrices of order 1×1 with $m = 3/2$, we have

$$\sum_{jm} |\langle m_1 m_2 | jm \rangle|^2 = 1$$

$$\left|\left\langle 1, \frac{1}{2} \middle| \frac{3}{2}, \frac{3}{2} \right\rangle\right|^2 = 1 \qquad \text{and thus} \qquad \left\langle 1, \frac{1}{2} \middle| \frac{3}{2}, \frac{3}{2} \right\rangle = 1$$

For the second sub-matrix of order 1×1 with $m = -3/2$,, using equation (7.30), we have

$$\left\langle -1, -\frac{1}{2} \middle| \frac{3}{2}, -\frac{3}{2} \right\rangle = \left\langle 1, \frac{1}{2} \middle| \frac{3}{2}, \frac{3}{2} \right\rangle = 1 \tag{7.36}$$

For the sub-matrices of order 2×2 with $m = 1/2$, we put $m_1 = 1$, $m_2 = -1/2$, $j = 3/2$ and $m = 3/2$ in equation (7.28) to get

$$\sqrt{3}\left\langle 1, -\frac{1}{2} \middle| \frac{3}{2}, \frac{1}{2} \right\rangle = 0 + \left\langle 1, \frac{1}{2} \middle| \frac{3}{2}, \frac{3}{2} \right\rangle$$

Using equation (7.36), we get

$$\left\langle 1, -\frac{1}{2} \middle| \frac{3}{2}, \frac{1}{2} \right\rangle = \frac{1}{\sqrt{3}} \tag{7.37}$$

From equation (7.25), we have

$$\sum_{jm} |\langle m_1 m_2 | jm \rangle|^2 = 1$$

$$\left|\left\langle 1, -\frac{1}{2} \middle| \frac{3}{2}, \frac{1}{2} \right\rangle\right|^2 + \left|\left\langle 1, -\frac{1}{2} \middle| \frac{1}{2}, \frac{1}{2} \right\rangle\right|^2 = 1$$

Using equation (7.37), we get

$$\left\langle 1, -\frac{1}{2} \middle| \frac{1}{2}, \frac{1}{2} \right\rangle = \sqrt{\frac{2}{3}}$$

Now, we put $m_1 = 0$, $m_2 = 1/2$, $j = 3/2$ and $m = 3/2$ in equation (7.28) to get

$$\sqrt{3}\left\langle 0, \frac{1}{2} \middle| \frac{3}{2}, \frac{1}{2} \right\rangle = \sqrt{2}\left\langle 1, \frac{1}{2} \middle| \frac{3}{2}, \frac{3}{2} \right\rangle + 0$$

Using equation (7.36), we get

$$\left\langle 0, \frac{1}{2} \middle| \frac{3}{2}, \frac{1}{2} \right\rangle = \sqrt{\frac{2}{3}}$$

From equation (7.25), we have

$$\sum_{jm} |\langle m_1 m_2 | jm \rangle|^2 = 1$$

$$\left|\left\langle 0, \frac{1}{2} \middle| \frac{3}{2}, 1/2 \right\rangle\right|^2 + \left|\left\langle 0, \frac{1}{2} \middle| \frac{1}{2}, \frac{1}{2} \right\rangle\right|^2 = 1$$

Using equation (7.37), we get

$$\left\langle 0, \frac{1}{2} \middle| \frac{1}{2}, \frac{1}{2} \right\rangle = -\frac{1}{\sqrt{3}}$$

As a convention, negative sign is taken for the Clebsch-Gordan coefficients with lower value of m_1 and lower value of j_1.

For the elements of second 2 × 2 matrix, we use the relation (7.30) and thus,

$$\left\langle 0, -\frac{1}{2} \middle| \frac{1}{2}, -\frac{1}{2} \right\rangle = \left\langle 0, \frac{1}{2} \middle| \frac{1}{2}, \frac{1}{2} \right\rangle = \frac{1}{\sqrt{3}}$$

$$\left\langle 0, -\frac{1}{2} \middle| \frac{3}{2}, -\frac{1}{2} \right\rangle = \left\langle 0, \frac{1}{2} \middle| \frac{3}{2}, \frac{1}{2} \right\rangle = \sqrt{\frac{2}{3}}$$

$$\left\langle -1, \frac{1}{2} \middle| \frac{3}{2}, -\frac{1}{2} \right\rangle = \left\langle 1, -\frac{1}{2} \middle| \frac{3}{2}, \frac{1}{2} \right\rangle = \frac{1}{\sqrt{3}}$$

$$\left\langle -1, \frac{1}{2} \middle| \frac{1}{2}, -\frac{1}{2} \right\rangle = \left\langle 1, -\frac{1}{2} \middle| \frac{1}{2}, \frac{1}{2} \right\rangle = -\sqrt{\frac{2}{3}}$$

As a convention, negative sign is taken for the lower value of m_1. Thus, the Clebsch-Gordan coefficients for $j_1 = 1$ and $j_2 = 1/2$ are as in the following table.

		$\lvert jm\rangle$					
m_1	m_2	$\lvert\frac{3}{2}\frac{3}{2}\rangle$	$\lvert\frac{3}{2}\frac{1}{2}\rangle$	$\lvert\frac{1}{2}\frac{1}{2}\rangle$	$\lvert\frac{3}{2},\text{-}\frac{1}{2}\rangle$	$\lvert\frac{1}{2},\text{-}\frac{1}{2}\rangle$	$\lvert\frac{3}{2},\text{-}\frac{3}{2}\rangle$
1	1/2	1	0	0	0	0	0
1	$-1/2$	0	$1/\sqrt{3}$	$\sqrt{2/3}$	0	0	0
0	1/2	0	$\sqrt{2/3}$	$-1/\sqrt{3}$	0	0	0
0	$-1/2$	0	0	0	$\sqrt{2/3}$	$1/\sqrt{3}$	0
-1	1/2	0	0	0	$1/\sqrt{3}$	$-\sqrt{2/3}$	0
-1	$-1/2$	0	0	0	0	0	1

Exercise 8. Compute Clebsch-Gordan coefficients for $j_1 = 1$ and $j_2 = 1$.

Solution: The sub-matrices for the Clebsch-Gordan coefficients for $j_1 = 1$ and $j_2 = 1$ are as the following.

m_1	m_2	m	j
1	1	2	2
1	0	1	2
0	1		1
1	−1	0	2
0	0		1
−1	1		0
0	−1	−1	2
−1	0		1
−1	−1	−2	2

For the sub-matrices of order 1×1 with $m = 2$; we have

$$\sum_{jm} |\langle m_1 m_2 | jm\rangle|^2 = 1$$

$$|\langle 1, 1|2, 2\rangle|^2 = 1 \qquad \text{and thus} \qquad \langle 1, 1|2, 2\rangle = 1 \tag{7.38}$$

For the second sub-matrix of order 1×1 with $m = -2$, using equation (7.30), we have

$$\langle -1, -1|2, -2\rangle = \langle 1, 1|2, 2\rangle = 1$$

For the sub-matrices of order 2×2 with $m = 1$, we put $m_1 = 1$, $m_2 = 0$, $j = 2$ and $m = 2$ in equation (7.28) to get

$$2\ \langle 1, 0|2, 1\rangle = 0 + \sqrt{2}\ \langle 1, 1|2, 2\rangle$$

Using equation (7.38), we get

$$\langle 1, 0|2, 1\rangle = 1/\sqrt{2} \tag{7.39}$$

From equation (7.25), we have

$$\sum_{jm} |\langle m_1 m_2 | jm\rangle|^2 = 1$$

$$|\langle 1, 0|2, 1\rangle|^2 + |\langle 1, 0|1, 1\rangle|^2 = 1$$

Using equation (7.39), we get

$$\langle 1, 0|1, 1\rangle = 1/\sqrt{2} \tag{7.40}$$

Now, we put $m_1 = 0$, $m_2 = 1$, $j = 2$ and $m = 2$ in equation (7.28) to get

$$2\ \langle 0, 1|2, 1\rangle = \sqrt{2}\ \langle 1, 1|2, 2\rangle + 0$$

Using equation (7.38), we get

$$\langle 0, 1|2, 1\rangle = 1/\sqrt{2} \tag{7.41}$$

From equation (7.25), we have

$$\sum_{jm} |\langle m_1 m_2|jm\rangle|^2 = 1$$

$$|\langle 0, 1|2, 1\rangle|^2 + |\langle 0, 1|1, 1\rangle|^2 = 1$$

Using equation (7.40), we get

$$\langle 0, 1|1, 1\rangle = -1/\sqrt{2} \tag{7.42}$$

As a convention, negative sign is taken for the Clebsch-Gordan coefficients with lower value of m_1 and lower value of j_1.

For the elements of second 2×2 matrix, we use the relation (7.30) and thus,

$$\langle 0, -1|2, -1\rangle = \langle 0, 1|2, 1\rangle = 1/\sqrt{2}$$
$$\langle 0, -1|1, -1\rangle = \langle 0, 1|1, 1\rangle = 1/\sqrt{2}$$
$$\langle -1, 0|2, -1\rangle = \langle 1, 0|2, 1\rangle = 1/\sqrt{2}$$
$$\langle -1, 0|1, -1\rangle = \langle 1, 0|1, 1\rangle = -1/\sqrt{2}$$

As a convention, negative sign is taken for the Clebsch-Gordan coefficients with lower value of m_1 and lower value of j_1.

For the sub-matrices of order 3×3, we put $m_1 = 1$, $m_2 = -1$, $j = 2$ and $m = 1$ in equation (7.28) to get

$$\sqrt{6}\ \langle 1, -1|2, 0\rangle = 0 + \sqrt{2}\ \langle 1, 0|2, 1\rangle$$

Using equation (7.39), we get

$$\langle 1, -1|2, 0\rangle = 1/\sqrt{6} \tag{7.43}$$

Now, we put $m_1 = 1$, $m_2 = -1$, $j = 1$ and $m = 1$ in equation (7.28) to get

$$\sqrt{2}\ \langle 1, -1|1, 0\rangle = 0 + \sqrt{2}\ \langle 1, 0|1, 1\rangle$$

Using equation (7.42) without convention, we get

$$\langle 1, -1|1, 0\rangle = 1/\sqrt{2} \tag{7.44}$$

From equation (7.25), we have

$$\sum_{jm} |\langle m_1 m_2|jm\rangle|^2 = 1$$

$$|\langle 1, -1|2, 0\rangle|^2 + |\langle 1, -1|1, 0\rangle|^2 + |\langle 1, -1|0, 0\rangle|^2 = 1$$

Using equations (7.43) and (7.44) here, we get

$$\langle 1, -1|0, 0\rangle = 1/\sqrt{3}$$

Now, we put $m_1 = -1$, $m_2 = 1$, $j = 2$ and $m = 1$ in equation (7.28) to get

$$\sqrt{6}\ \langle -1, 1|2, 0\rangle = \sqrt{2}\ \langle 0, 1|2, 1\rangle + 0$$

Using equation (7.41), we get

$$\langle -1, 1|2, 0\rangle = 1/\sqrt{6} \tag{7.45}$$

Now, we put $m_1 = -1$, $m_2 = 1$, $j = 1$ and $m = 1$ in equation (7.28) to get

$$\sqrt{2}\ \langle -1, 1|1, 0\rangle = \sqrt{2}\ \langle 0, 1|1, 1\rangle + 0$$

Using equation (7.42) with convention, we get

$$\langle -1, 1|1, 0\rangle = -1/\sqrt{2} \tag{7.46}$$

From equation (7.25), we have

$$\sum_{jm} |\langle m_1 m_2|jm\rangle|^2 = 1$$

$$|\langle -1, 1|2, 0\rangle|^2 + |\langle -1, 1|1, 0\rangle|^2 + |\langle -1, 1|0, 0\rangle|^2 = 1$$

Using equations (7.45) (7.46), we get

$$\langle\ \ 1, 1|0, 0\rangle = 1/\sqrt{3}$$

Now, we put $m_1 = 0$, $m_2 = 0$, $j = 2$ and $m = 1$ in equation (7.28) to get

$$\sqrt{6}\ \langle 0, 0|2, 0\rangle = \sqrt{2}\ \langle 1, 0|2, 1\rangle + \sqrt{2}\ \langle 0, 1|2, 1\rangle$$

Using equations (7.39) and (7.41), we get

$$\langle 0, 0|2, 0\rangle = \sqrt{2/3} \tag{7.47}$$

Now, we put $m_1 = 0$, $m_2 = 0$, $j = 1$ and $m = 1$ in equation (7.28) to get

$$\sqrt{2}\ \langle 0, 0|1, 0\rangle = \sqrt{2}\ \langle 1, 0|1, 1\rangle + \sqrt{2}\ \langle 0, 1|1, 1\rangle$$

Using equations (7.40) and (7.42) here, we get

$$\langle 0, 0|1, 0\rangle = 0 \tag{7.48}$$

From equation (7.25), we have

$$\sum_{jm} |\langle m_1 m_2 | jm\rangle|^2 = 1$$

$$|\langle 0,0|2,0\rangle|^2 + |\langle 0,0|1,0\rangle|^2 + |\langle 0,0|0,0\rangle|^2 = 1$$

Using equations (7.47) (7.48), we get

$$\langle 0,0|0,0\rangle = -1/\sqrt{3}$$

As a convention, it is taken negative. Thus, the Clebsch-Gordan coefficients for $j_1 = 1$ and $j_2 = 1/2$ are as in the following table.

		$\|jm\rangle$								
m_1	m_2	$\|22\rangle$	$\|21\rangle$	$\|11\rangle$	$\|20\rangle$	$\|10\rangle$	$\|00\rangle$	$\|2,-1\rangle$	$\|1,-1\rangle$	$\|2,-2$
1	1	1	0	0	0	0	0	0	0	0
1	0	0	$\frac{1}{\sqrt{2}}$	$\frac{1}{\sqrt{2}}$	0	0	0	0	0	0
0	1	0	$\frac{1}{\sqrt{2}}$	$-\frac{1}{\sqrt{2}}$	0	0	0	0	0	0
1	−1	0	0	0	$\frac{1}{\sqrt{6}}$	$\frac{1}{\sqrt{2}}$	$\frac{1}{\sqrt{3}}$	0	0	0
0	0	0	0	0	$\sqrt{\frac{2}{3}}$	0	$-\frac{1}{\sqrt{3}}$	0	0	0
−1	1	0	0	0	$\frac{1}{\sqrt{6}}$	$\frac{1}{\sqrt{2}}$	$\frac{1}{\sqrt{3}}$	0	0	0
0	−1	0	0	0	0	0	0	$\frac{1}{\sqrt{2}}$	$-\frac{1}{\sqrt{2}}$	0
−1	0	0	0	0	0	0	0	$\frac{1}{\sqrt{2}}$	$-\frac{1}{\sqrt{2}}$	0
−1	−1	0	0	0	0	0	0	0	0	1

12. Problems and questions

1. For the components L_x, L_y, L_z for angular momentum, show that

$$[L_x,\ L_y] = i\,\hbar\ L_z \qquad [L_y,\ L_z] = i\,\hbar\ L_x \qquad [L_z,\ L_x] = i\,\hbar\ L_y$$

2. For the angular momentum operator $\vec{L}$ show that $\vec{L} \times \vec{L} = i\,\hbar\,\vec{L}$.

3. Define raising and lowering operators. For the raising operator $L_+ (\equiv L_x + i\ L_y)$ and lowering operator $L_- (\equiv L_x - i\ L_y)$, prove the following relations:

$$[L^2,\ L_+] = 0 \qquad [L^2,\ L_-] = 0 \qquad [L_z,\ L_+] = \hbar\ L_+$$
$$[L_z,\ L_-] = -\hbar\ L_- \qquad [L_x,\ L_+] = -\hbar L_z \qquad [L_x,\ L_-] = \hbar L_z$$

4. Obtain expressions for matrices for J^2 and J_z for $j = 1$.

5. Obtain expressions for matrices for J_+, J_-, J_x and J_y for $j = 1$.

6. Derive the following relations for Pauli's spin matrices:

$$\sigma_x \, \sigma_y = i \, \sigma_z \qquad \sigma_y \, \sigma_z = i \, \sigma_x \qquad \sigma_z \, \sigma_y = i \, \sigma_y$$

$$\sigma_x \, \sigma_y + \sigma_y \, \sigma_x = \sigma_y \, \sigma_z + \sigma_z \, \sigma_y = \sigma_z \, \sigma_x + \sigma_x \, \sigma_z = 0$$

7. Obtain spin vectors for a system with one-half spin.

8. Define the Clebsch-Gordan coefficients and derive the selection rules for them.

9. Define the Clebsch-Gordan coefficients and derive the orthogonality relations for them.

10. Define the Clebsch-Gordan coefficients and derive the recursion relations for them. Show that

$$\langle m_1 m_2 | jm \rangle = \langle -m_1, -m_2 | j, -m \rangle$$

11. Show that tne matrix of Clebsch-Gordan coefficients can be decomposed into a number of square matrices of various orders. For $j_1 = 1$ and $J_2 = 2$, find the orders of various sub-matrices.

12. Calculate Clebsch-Gordan coefficients for $j_1 = 1/2$ and $J_2 = 1/2$.

13. Calculate Clebsch-Gordan coefficients for $j_1 = 1$ and $J_2 = 1/2$.

14. Calculate Clebsch-Gordan coefficients for $j_1 = 1$ and $J_2 = 1$.

15. Write notes on the following

    (i) Raising and lowering operators

    (ii) Spin vectors for a system with one-half spin

    (ii) Clebsch-Gordan coefficients

Time-independent Perturbation

In the preceding chapters, we considered some systems for which the Schrödinger equation can be solved exactly. For example, we solved Schrödinger equation for a harmonic oscillator, hydrogen atom, etc. For those systems, potential energy was expressed in a suitable form. But, in a large number of cases, exact solution of Schrödinger equation may not be possible. Then we solve the Schrödinger equation by using various approximate methods, such as the time-independent perturbation method, variation method, WKB approximation, time-dependent perturbation method. In this chapter, we shall discuss about the time-independent perturbation method. Other methods will be discussed in the following chapters.

1. Basic concepts

When the Hamiltonian of a system is slightly different from that which can be solved exactly and the Hamiltonian does not depend on time, the time-independent perturbation method can be applied conveniently. Here, the Hamiltonian H is expressed as

$$H = H^0 + H'$$

where H^0 is known as the unperturbed Hamiltonian, whose nondegenerate eigenvalues E_n^0, $n = 1, 2, \ldots$ and corresponding wave functions ψ_n^0 can be obtained accurately. So that

$$H^0 \psi_n^0 = E_n^0 \psi_n^0, \qquad n = 1, 2, \ldots \tag{8.1}$$

The time-independent operator H' is a small perturbation from H^0. The orthonormal wave functions ψ_n^0, $n = 1, 2, \ldots$ form a basis. Now, our aim is to solve the Schrödinger equation

$$H\psi_n = E_n \psi_n \tag{8.2}$$

For applying time-independent perturbation method, we express the Hamiltonian H as

$$H = H^0 + \lambda H'$$

where the parametre λ can very from 0 to 1. That is, for $\lambda = 0$, the Schrödinger equation can be solved exactly. But, our aim is to solve the Schrödinger equation with $\lambda = 1$. For that, let us express E_n and ψ_n in terms of λ as

$$E_n = E_n^0 + \lambda E_n^{(1)} + \lambda^2 E_n^{(2)} + \ldots \tag{8.3}$$

$$\psi_n = \psi_n^0 + \lambda \psi_n^{(1)} + \lambda^2 \psi_n^{(2)} + \ldots \tag{8.4}$$

where E_n^0 and ψ_n^0, known as the zeroth order terms, represent the unperturbed part and $E_n^{(1)}$, $E_n^{(2)}$, ... as well as $\psi_n^{(1)}$, $\psi_n^{(2)}$, etc. represent various order corrections. Using equations (8.3) and (8.4) in (8.2), we get

$$\left(H^0 + \lambda H'\right)\left(\psi_n^{(0)} + \lambda \psi_n^{(1)} + \lambda^2 \psi_n^{(2)} + \ldots\right)$$

$$= \left(E_n^{(0)} + \lambda E_n^{(1)} + \lambda^2 E_n^{(2)} + \ldots\right)\left(\psi_n^0 + \lambda \psi_n^{(1)} + \lambda^2 \psi_n^{(2)} + \ldots\right)$$

This equation can be rearranged as

$$\left(H^0 - E_n^0\right)\psi_n^0 + \lambda\left(H'\psi_n^0 + H^0\psi_n^{(1)} - E_n^{(1)}\psi_n^0 - E_n^0\psi_n^{(1)}\right)$$

$$+\lambda^2\left(H'\psi_n^{(1)} + H^0\psi_n^{(2)} - E_n^{(2)}\psi_n^0 - E_n^{(1)}\psi_n^{(1)} - E_n^0\psi_n^{(2)}\right) + \ldots = 0 \tag{8.5}$$

Since λ is a free-parametre, equation (8.5) can be satisfied only when each of the coefficients of various powers of λ is zero separately. Thus, we have

$$H^0\ \psi_n^0 = E_n^0\ \psi_n^0 \tag{8.6}$$

$$H'\ \psi_n^0 + H^0\ \psi_n^{(1)} = E_n^{(1)}\ \psi_n^0 + E_n^0\ \psi_n^{(1)} \tag{8.7}$$

$$H'\ \psi_n^{(1)} + H^0\ \psi_n^{(2)} = E_n^{(2)}\ \psi_n^0 + E_n^{(1)}\ \psi_n^{(1)} + E_n^0\ \psi_n^{(2)} \tag{8.8}$$

Here, we are interested only up to the second order. Since equation (8.6) refers to the zeroth order perturbation, as per our expectation, it is identical to equation (8.1).

2. Nondegenerate energy levels

For applying the time-independent perturbation theory, we have first to specify whether the perturbed energy level is nondegenerate or degenerate, as the treatment depends on the situation. Let us first consider the perturbation of nondegenerate levels. For convenience, we shall use the wave function itself to label the state.

2.1 First-order correction to energy

Multiplying equation (8.7) from left by $\langle\psi_n^0|$, we get

$$\langle\psi_n^0|H'|\psi_n^0\rangle + \langle\psi_n^0|H^0|\psi_n^{(1)}\rangle = E_n^{(1)}\langle\psi_n^0|\psi_n^0\rangle + E_n^0\langle\psi_n^0|\psi_n^{(1)}\rangle \tag{8.9}$$

Since H^0 is Hermitian, we have

$$\langle\psi_n^0|H^0|\psi_n^{(1)}\rangle = \langle\psi_n^{(1)}|H^0|\psi_n^0\rangle^* = \langle\psi_n^{(1)}E_n^0\psi_n^0\rangle^* = E_n^0\langle\psi_n^0|\psi_n^{(1)}\rangle \tag{8.10}$$

Using equation (8.10) in (8.9), we have

$$\langle\psi_n^0|H'|\psi_n^0\rangle + E_n^0\langle\psi_n^0|\psi_n^{(1)}\rangle = E_n^{(1)} + E_n^0\langle\psi_n^0|\psi_n^{(1)}\rangle$$

$$E_n^{(1)} = \langle\psi_n^0|H'|\psi_n^0\rangle$$

This relation can also be expressed as

$$E_n^{(1)} = \langle n|H'|n\rangle \tag{8.11}$$

It is often called a matrix element. The first-order correction to the energy is therefore the average value of the perturbation over the corresponding unperturbed states of the system.

2.2 First-order correction to wave function

The first-order correction to the wave function is expressed as linear combination of the unperturbed wave functions of the system.

$$\psi_n^{(1)} = \sum_{l=1}^{\infty} a_l\psi_l^0 \tag{8.12}$$

Using equation (8.12) in (8.7) and multiplying from left by $\langle\psi_m^0|$, we get

$$\langle\psi_m^0|H'|\psi_n^0\rangle + \langle\psi_m^0|H^0|\psi_n^{(1)}\rangle = \langle\psi_m^0|E_n^{(1)}|\psi_n^0\rangle + \langle\psi_m^0|E_n^0|\psi_n^{(1)}\rangle$$

$$\langle\psi_m^0|H'|\psi_n^0\rangle + \left\langle\psi_m^0|H^0|\sum_{l=1}^{\infty} a_l\psi_l^0\right\rangle = E_n^{(1)}\langle\psi_m^0|\psi_n^0\rangle + \left\langle\psi_m^0|E_n^0|\sum_{l=1}^{\infty} a_l\psi_l^0\right\rangle$$

$$\langle\psi_m^0|H'|\psi_n^0\rangle + \sum_{l=1}^{\infty} a_l E_l^0\left\langle\psi_m^0|\psi_l^0\right\rangle = 0 + \sum_{l=1}^{\infty} a_l E_n^0\left\langle\psi_m^0|\psi_l^0\right\rangle$$

$$\langle\psi_m^0|H'|\psi_n^0\rangle + a_m E_m^0 = a_m E_n^0$$

$$a_m = \frac{\langle\psi_m^0|H'|\psi_n^0\rangle}{E_n^0 - E_m^0} = \frac{\langle m|H'|n\rangle}{E_n^0 - E_m^0} \tag{8.13}$$

All the a's, except a_n, in equation (8.12) can be calculated with the help of equation (8.13). From the normalization condition $\langle\psi_n|\psi_n\rangle = 1$, the coefficient a_n is found to be zero.[1] Thus, we have

$$\psi_n^{(1)} = {\sum_m}' \frac{\langle m|H'|n\rangle}{E_n^0 - E_m^0}\psi_m^0 \tag{8.14}$$

Here, the prime on the sum shows that the state $m = n$ is to be excluded from the summation. Thus, the energy and wave function corrected to first-order are

$$E_n = E_n^0 + \langle n|H'|n\rangle$$

and

$$\psi_n = \psi_n^0 + {\sum_m}' \frac{\langle m|H'|n\rangle}{E_n^0 - E_m^0}\psi_m^0$$

---

[1]We have

$$\psi_n = \psi_n^0 + \lambda\psi_n^{(1)} = \psi_n^0 + \lambda\sum_{l=1}^{\infty} a_l\psi_l^0$$

Now,

$$\langle\psi_n|\psi_n\rangle = \langle\psi_n^0|\psi_n^0\rangle + \lambda\left\langle\psi_n^0|\sum_{l=1}^{\infty} a_l\psi_l^0\right\rangle + \lambda\left\langle\sum_{l=1}^{\infty} a_l\psi_l^0|\psi_n^0\right\rangle$$

For normalization of ψ_n, we have

$$1 = 1 + \lambda a_n + \lambda a_n$$

giving $a_n = 0$

2.3 Second-order correction to energy

For getting the second-order correction to energy, multiply equation (8.8) from left by $\langle\psi_n^0|$ and use the Hermitian nature of H^0 for getting

$$\langle\psi_n^0|H'|\psi_n^{(1)}\rangle + \langle\psi_n^0|H^0|\psi_n^{(2)}\rangle = \langle\psi_n^0|E_n^{(2)}|\psi_n^0\rangle + \langle\psi_n^0|E_n^{(1)}|\psi_n^{(1)}\rangle$$

$$+\langle\psi_n^0|E_n^0|\psi_n^{(2)}\rangle \qquad (8.15)$$

Since H^0 is Hermitian, we have

$$\langle\psi_n^0|H^0|\psi_n^{(2)}\rangle = \langle\psi_n^{(2)}|H^0|\psi_n^0\rangle^* = \langle\psi_n^{(2)}|E_n^0|\psi_n^0\rangle^* = \langle\psi_n^0|E_n^0|\psi_n^{(2)}\rangle \qquad (8.16)$$

Using equations (8.16) in (8.15), we get

$$\langle\psi_n^0|H'|\psi_n^{(1)}\rangle = E_n^{(2)}\langle\psi_n^0|\psi_n^0\rangle + E_n^{(1)}\langle\psi_n^0|\psi_n^{(1)}\rangle$$

$$= E_n^{(2)} + E_n^{(1)}\left\langle\psi_n^0\middle|\sum_m{}' a_m\psi_m^0\right\rangle = E_n^{(2)}$$

Thus, we have

$$E_n^{(2)} = \langle\psi_n^0|H'|\psi_n^{(1)}\rangle \qquad (8.17)$$

Using the value of $\psi_n^{(1)}$ from equation (8.14) in (8.17), we get

$$E_n^{(2)} = \left\langle\psi_n^0\middle|H'\middle|\sum_m{}' \frac{\langle m|H'|n\rangle}{E_n^0 - E_m^0}|\psi_m^0\right\rangle = \sum_m{}' \frac{\langle m|H'|n\rangle\langle n|H'|m\rangle}{E_n^0 - E_m^0}$$

If H' is Hermitian $\langle n|H'|m\rangle = \langle m|H'|n\rangle^*$, and therefore, we have

$$E_n^{(2)} = \sum_m{}' \frac{\left|\langle m|H'|n\rangle\right|^2}{E_n^0 - E_m^0} \qquad (8.18)$$

Since $|\langle m|H'|n\rangle|^2$ is always positive, the sign of the correction depends on the denominator. Thus, the second-order correction in energy is positive when $E_n^0 > E_m^0$ and negative when $E_n^0 < E_m^0$.

2.4 Second-order correction to wave function

The second-order correction to the wave function $\psi_n^{(2)}$ is expressed as linear combination of the unperturbed wave functions of the system.

$$\psi_n^{(2)} = \sum_k b_k\psi_k^0 \qquad (8.19)$$

Using equation (8.19) in (8.8) and multiplying from left by $\langle\psi_l^0|$, we get

$$\langle\psi_l^0|H'|\psi_n^{(1)}\rangle + \langle\psi_l^0|H^0|\psi_n^{(2)}\rangle = E_n^{(2)}\langle\psi_l^0|\psi_n^0\rangle + E_n^{(1)}\langle\psi_l^0|\psi_n^{(1)}\rangle$$

$$+E_n^0\langle\psi_l^0|\psi_n^{(2)}\rangle$$

$$\left\langle\psi_l^0|H'|\sum_m{}' a_m\psi_m^0\right\rangle + \left\langle\psi_l^0|H^0|\sum_k b_k\psi_k^0\right\rangle = E_n^{(2)}\langle\psi_l^0|\psi_n^0\rangle$$

$$+E_n^{(1)}\left\langle\psi_l^0|\sum_m{}' a_m\psi_m^0\right\rangle + E_n^0\left\langle\psi_l^0|\sum_k b_k\psi_k^0\right\rangle$$

$$\sum_m{}' a_m\langle l|H'|m\rangle + \sum_k b_k\left\langle l|H^0|k\right\rangle = 0 + \sum_m{}' a_m E_n^{(1)}\langle l|m\rangle$$

$$+\sum_k b_k E_n^0\langle l|k\rangle$$

On rearranging, we get

$$\sum_m{}' a_m\langle l|H'|m\rangle + b_l E_l^0 = E_n^{(1)}a_l + E_n^0 b_l$$

$$b_l(E_l^0 - E_n^0) = E_n^{(1)}a_l - \sum_m{}' a_m\langle l|H'|m\rangle$$

Substituting values of the a's from equation (8.11) and $E_n^{(1)}$ from equation (8.13) here, we get

$$b_l = \frac{\langle n|H'|n\rangle\langle l|H'|n\rangle}{(E_l^0 - E_n^0)(E_n^0 - E_l^0)} - \sum_m{}' \frac{\langle m|H'|n\rangle\langle l|H'|m\rangle}{(E_n^0 - E_m^0)(E_l^0 - E_n^0)}$$

$$= \sum_m{}' \frac{\langle m|H'|n\rangle\langle l|H'|m\rangle}{(E_n^0 - E_m^0)(E_n^0 - E_l^0)} - \frac{\langle n|H'|n\rangle\langle l|H'|n\rangle}{(E_n^0 - E_l^0)^2} \tag{8.20}$$

From the normalization condition of the wave function shows that the coefficient b_n is zero. Thus, the energy and wave function corrected to second-order are

$$E_n = E_n^0 + \langle n|H'|n\rangle + \sum_m{}' \frac{\left|\langle n|H'|m\rangle\right|^2}{E_n^0 - E_m^0}$$

and

$$\psi_n = \psi_n^0 + \sum_m{}' \frac{\langle m|H'|n\rangle}{E_n^0 - E_m^0}\psi_m^0$$

$$+ \sum_l{}' \Big[\sum_m{}' \frac{\langle m|H'|n\rangle \langle l|H'|m\rangle}{(E_n^0 - E_m^0)(E_n^0 - E_l^0)} - \frac{\langle n|H'|n\rangle \langle l|H'|n\rangle}{(E_n^0 - E_l^0)^2}\Big]\psi_l^0$$

The prime on the summation shows that the states $m = n$ or $l = n$, as the case may be, are to be excluded from the summation.

3. Anharmonic oscillator: First-order correction

As an example of time-independent perturbation theory, let us calculate the first-order correction for the energy of motion of an anharmonic oscillator. Suppose, a particle of mass m is moving in a one-dimensional potential

$$V(x) = \frac{1}{2}\,\omega^2 x^2 + bx^4$$

where ω is angular frequency of oscillation and b a constant small parametre. When b is zero, it is the case of a harmonic oscillator. Thus, inclusion of the term bx^4 in the potential changes from a harmonic oscillator to an anharmonic oscillator. Now, we would like to calculate the first-order correction to the nth state, which is expressed as (equation 8.11)

$$E^{(1)} = \left\langle n|bx^4|n\right\rangle$$

where $|n\rangle$ is the eigen ket for the nth state of a harmonic oscillator. The treatment of the problem becomes simple when we apply operator method (discussed in the chapter IV) for a linear harmonic oscillator where the creation ($a^\dagger$) and annihilation (a) operators were expressed as

$$a = \left(\frac{m\omega}{2\hbar}\right)^{1/2} x + i\left(\frac{1}{2m\hbar\omega}\right)^{1/2} p_x \tag{8.21}$$

and

$$a^\dagger = \left(\frac{m\omega}{2\hbar}\right)^{1/2} x - i\left(\frac{1}{2m\hbar\omega}\right)^{1/2} p_x \tag{8.22}$$

so that

$$aa^\dagger = \frac{H}{\hbar\omega} + \frac{1}{2} \tag{8.23}$$

$$a^\dagger a = \frac{H}{\hbar\omega} - \frac{1}{2} \tag{8.24}$$

On multiplying equation (8.23) by $|n\rangle$ from right, we get

$$aa^\dagger|n\rangle = \left(\frac{H}{\hbar\omega} + \frac{1}{2}\right)|n\rangle = \frac{1}{\hbar\omega}H|n\rangle + \frac{1}{2}|n\rangle$$

$$= \frac{1}{\hbar\omega}E_n|n\rangle + \frac{1}{2}|n\rangle = \frac{1}{\hbar\omega}\left(n + \frac{1}{2}\right)\hbar\omega + \frac{1}{2}|n\rangle$$

$$= (n+1)|n\rangle \tag{8.25}$$

Now, multiplying equation (8.24) by $|n\rangle$ from right we get

$$a^\dagger a|n\rangle = \left(\frac{H}{\hbar\omega} - \frac{1}{2}\right)|n\rangle = \frac{1}{\hbar\omega}H|n\rangle - \frac{1}{2}|n\rangle$$

$$= \frac{1}{\hbar\omega}E_n|n\rangle - \frac{1}{2}|n\rangle = \frac{1}{\hbar\omega}\left(n + \frac{1}{2}\right)\hbar\omega - \frac{1}{2}|n\rangle$$

$$= n|n\rangle \tag{8.26}$$

From equations (8.25) and (8.26),.we have

$$(aa^\dagger - a^\dagger a)|n\rangle = |n\rangle \tag{8.27}$$

Equations (8.25), (8.26) and (8.27) can be satisfied when

$$a^\dagger\,|n\rangle = \sqrt{n+1}\,|n+1\rangle \qquad\qquad a\,|n\rangle = \sqrt{n}\,|n-1\rangle \tag{8.28}$$

The sum of the creation $(a^\dagger)$ and annihilation (a) operators is

$$a + a^\dagger = \left(\frac{m\omega}{2\hbar}\right)^{1/2}x + i\left(\frac{1}{2m\hbar\omega}\right)^{1/2}p_x + \left(\frac{m\omega}{2\hbar}\right)^{1/2}x - i\left(\frac{1}{2m\hbar\omega}\right)^{1/2}p_x$$

$$= \left(\frac{2m\omega}{\hbar}\right)^{1/2}x$$

Thus,

$$x = \left(\frac{\hbar}{2m\omega}\right)^{\frac{1}{2}}(a + a^\dagger)$$

and therefore

$$E^{(1)} = b\left(\frac{\hbar}{2m\omega}\right)^2\langle n|(a+a^\dagger)(a+a^\dagger)(a+a^\dagger)(a+a^\dagger)|n\rangle \tag{8.29}$$

We have

$$(a+a^\dagger)(a+a^\dagger) = aa + a^\dagger a + aa^\dagger + a^\dagger a^\dagger$$

and thus,

$$(aa + a^\dagger a + aa^\dagger + a^\dagger a^\dagger)(aa + a^\dagger a + aa^\dagger + a^\dagger a^\dagger) = aaaa + aaa^\dagger a$$
$$+aaaa^\dagger + aaa^\dagger a^\dagger + a^\dagger aaa + a^\dagger aa^\dagger a + a^\dagger aaa^\dagger + a^\dagger aa^\dagger a^\dagger$$
$$+aa^\dagger aa + aa^\dagger a^\dagger a + aa^\dagger aa^\dagger + aa^\dagger a^\dagger a^\dagger + a^\dagger a^\dagger aa + a^\dagger a^\dagger a^\dagger a$$
$$+a^\dagger a^\dagger aa^\dagger + a^\dagger a^\dagger a^\dagger a^\dagger \tag{8.30}$$

Using equation (8.30) in (8.29), we get

$$E^{(1)} = b\left(\frac{\hbar}{2m\omega}\right)^2 \langle n|(aaaa + aaa^\dagger a + aaaa^\dagger + aaa^\dagger a^\dagger + a^\dagger aaa$$
$$+a^\dagger aa^\dagger a + a^\dagger aaa^\dagger + a^\dagger aa^\dagger a^\dagger + aa^\dagger aa + aa^\dagger a^\dagger a + aa^\dagger aa^\dagger$$
$$+aa^\dagger a^\dagger a^\dagger + a^\dagger a^\dagger aa + a^\dagger a^\dagger a^\dagger a + a^\dagger a^\dagger aa^\dagger + a^\dagger a^\dagger a^\dagger a^\dagger)|n\rangle \tag{8.31}$$

There are 16 terms on the right side of equation (8.31). Application of equations (8.28) says that for the terms where a and $a^\dagger$ are not equal in number (*i.e.*, two each of them), the term will vanish. Thus, the terms containing $aaaa$, $aaa^\dagger a$, $aaaa^\dagger$, $a^\dagger aaa$, $a^\dagger aa^\dagger a^\dagger$, $aa^\dagger aa$, $aa^\dagger a^\dagger a^\dagger$, $a^\dagger a^\dagger a^\dagger a$, $a^\dagger a^\dagger aa^\dagger$ and $a^\dagger a^\dagger a^\dagger a^\dagger$ will vanish and we have

$$E^{(1)} = b\left(\frac{\hbar}{2m\omega}\right)^2 \langle n|(aaa^\dagger a^\dagger + a^\dagger aa^\dagger a + a^\dagger aaa^\dagger + aa^\dagger a^\dagger a$$
$$+aa^\dagger aa^\dagger + a^\dagger a^\dagger aa)|n\rangle \tag{8.32}$$

(i) For the ground state $n = 0$ energy, we have

$$\langle 0|aaa^\dagger a^\dagger|0\rangle = \langle 0|aaa^\dagger|1\rangle = \sqrt{2}\langle 0|aa|2\rangle = 2\langle 0|a|1\rangle = 2\langle 0|0\rangle = 2$$

$$\langle 0|a^\dagger aa^\dagger a|0\rangle = 0$$

$$\langle 0|a^\dagger aaa^\dagger|0\rangle = \langle 0|a^\dagger aa|1\rangle = \langle 0|a^\dagger a|0\rangle = 0$$

$$\langle 0|aa^\dagger a^\dagger a|0\rangle = 0$$

$$\langle 0|aa^\dagger aa^\dagger|0\rangle = \langle 0|aa^\dagger a|1\rangle = \langle 0|aa^\dagger|0\rangle = \langle 0|a|1\rangle = \langle 0|0\rangle = 1$$

$$\langle 0|a^\dagger a^\dagger aa|0\rangle = 0$$

Using these expressions, we have

$$E^{(1)} = b\left(\frac{\hbar}{2m\omega}\right)^2 (2+1) = 3b\left(\frac{\hbar}{2m\omega}\right)^2$$

Then, the first-order corrected energy of the ground state is

$$E_0 = \frac{1}{2}\hbar\omega + 3b\left(\frac{\hbar}{2m\omega}\right)^2$$

(ii) For the $n = 1$ state, we have

$$
\begin{aligned}
\langle 1|aaa^\dagger a^\dagger|1\rangle &= \langle 1|\sqrt{2}\sqrt{3}\sqrt{3}\sqrt{2}|1\rangle = 6 \\
\langle 1|a^\dagger aa^\dagger a|1\rangle &= \langle 1|1111|1\rangle = 1 \\
\langle 1|a^\dagger aaa^\dagger|1\rangle &= \langle 1|\sqrt{2}\sqrt{2}11|1\rangle = 2 \\
\langle 1|aa^\dagger a^\dagger a|1\rangle &= \langle 1|11\sqrt{2}\sqrt{2}|1\rangle = 2 \\
\langle 1|aa^\dagger aa^\dagger|1\rangle &= \langle 1|\sqrt{2}\sqrt{2}\sqrt{2}\sqrt{2}|1\rangle = 4 \\
\langle 1|a^\dagger a^\dagger aa|1\rangle &= \langle 1|a^\dagger a^\dagger a|0\rangle = 0
\end{aligned}
\tag{8.33}
$$

Using these expressions, we have

$$
E^{(1)} = b\left(\frac{\hbar}{2m\omega}\right)^2 (6 + 1 + 2 + 2 + 4 + 0) = 15b\left(\frac{\hbar}{2m\omega}\right)^2
$$

Then, the first-order corrected energy of the first excited state is

$$
E_1 = \frac{3}{2}\hbar\omega + 15b\left(\frac{\hbar}{2m\omega}\right)^2
$$

(iii) For $n > 1$ state, we have

$$
\begin{aligned}
\langle n|aaa^\dagger a^\dagger|n\rangle &= \langle n|\sqrt{n+1}\sqrt{n+2}\sqrt{n+2}\sqrt{n+1}|n\rangle = (n+1)(n+2) \\
\langle n|a^\dagger uu^\dagger u|n\rangle &- \langle n|\sqrt{n}\sqrt{n}\sqrt{n}\sqrt{n}|n\rangle = n^2 \\
\langle n|a^\dagger aaa^\dagger|n\rangle &= \langle n|\sqrt{n+1}\sqrt{n+1}\sqrt{n}\sqrt{n}|n\rangle = n(n+1) \\
\langle n|aa^\dagger a^\dagger a|n\rangle &= \langle n|\sqrt{n}\sqrt{n}\sqrt{n+1}\sqrt{n+1}|n\rangle = n(n+1) \\
\langle n|aa^\dagger aa^\dagger|n\rangle &= \langle n|\sqrt{n+1}\sqrt{n+1}\sqrt{n+1}\sqrt{n+1}|n\rangle = (n+1)^2 \\
\langle n|a^\dagger a^\dagger aa|n\rangle &= \langle n|\sqrt{n}\sqrt{n-1}\sqrt{n-1}\sqrt{n}|n\rangle = n(n-1)
\end{aligned}
$$

Using these expressions, we have

$$
\begin{aligned}
E^{(1)} = b\left(\frac{\hbar}{2m\omega}\right)^2 &[(n+1)(n+2) + n^2 + n(n+1) + n(n+1) \\
&+(n+1)^2 + n(n-1)] = b\left(\frac{\hbar}{2m\omega}\right)^2 [6n^2 + 6n + 3]
\end{aligned}
$$

Then, the first-order corrected energy of the n-th state is

$$E_n = \left(n + \frac{1}{2}\right)\hbar\omega + b\left(\frac{\hbar}{2m\omega}\right)^2 [6n^2 + 6n + 3]$$

Exercise 1. The ground state wave function ψ_0^0 of a harmonic oscillator of mass m oscillating with angular frequency ω in a potential field $V(x) = m\omega^2 x^2/2$ is

$$\psi_0^0 = \left(\frac{m\omega}{\pi\hbar}\right)^{1/4} \exp\left(-\frac{m\omega x^2}{2\hbar}\right)$$

Calculate the first-order correction to the ground state when potential field is changed by bx^4, where b is a constant.

Solution: Here, the perturbed Hamiltonian H' is bx^4. The first-order correction to the ground state energy is

$$E_0^{(1)} = \langle\psi_0^0|H'|\psi_0^0\rangle = \left(\frac{m\omega}{\pi\hbar}\right)^{1/2} b\int_{-\infty}^{\infty} x^4 \exp\left(-\frac{m\omega x^2}{\hbar}\right) dx$$

$$= \left(\frac{m\omega}{\pi\hbar}\right)^{1/2} 2b\int_0^{\infty} x^4 \exp\left(-\frac{m\omega x^2}{\hbar}\right) dx \tag{8.34}$$

Using

$$\int_0^{\infty} x^n \exp(-ax^2)\, dx = \frac{\Gamma\{(n+1)/2\}}{2a^{(n+1)/2}}$$

we get

$$E_0^{(1)} = \left(\frac{m\omega}{\pi\hbar}\right)^{1/2} b\frac{3\sqrt{\pi}}{4}\left(\frac{\hbar}{m\omega}\right)^{5/2} = \frac{3b\hbar^2}{4m^2\omega^2}$$

Exercise 2. A simple harmonic oscillator of mass m and angular frequency ω is perturbed by an additional potential bx^3. Calculate the second-order correction to the ground state energy of the oscillator.

Solution: The second-order correction to the ground state energy is

$$E_0^{(2)} = \sum_m{}' \frac{|\langle 0|bx^3|m\rangle|^2}{E_0^0 - E_m^0}$$

In terms of the creation ($a^\dagger$) and annihilation (a) operators, we have

$$x = \left(\frac{\hbar}{2m\omega}\right)^{1/2} (a + a^\dagger)$$

Thus,

$$\sum_m{}' \frac{|\langle 0|x^3|m\rangle|^2}{E_0^0 - E_m^0} = \left(\frac{\hbar}{2m\omega}\right)^3 \sum_m{}' \frac{|\langle 0|(a+a^\dagger)(a+a^\dagger)(a+a^\dagger)|m\rangle|^2}{E_0^0 - E_m^0}$$

$$= \left(\frac{\hbar}{2m\omega}\right)^3 \times$$

$$\sum_m{}' \frac{|\langle 0|(aaa + aaa^\dagger + aa^\dagger a + aa^\dagger a^\dagger + a^\dagger aa + a^\dagger aa^\dagger + a^\dagger a^\dagger a + a^\dagger a^\dagger a^\dagger)|m\rangle|^2}{E_0^0 - E_m^0}$$

$$= \left(\frac{\hbar}{2m\omega}\right)^3 \left[\frac{|\langle 0|(aaa)|3\rangle|^2}{-3\hbar\omega} + \frac{|\langle 0|(aaa^\dagger + aa^\dagger a)|1\rangle|^2}{-\hbar\omega}\right]$$

the other contributions vanish. Now, we have

$\langle 0|aaa|3\rangle = \sqrt{6}$ $\qquad \langle 0|aaa^\dagger|1\rangle = 2 \qquad$ $\langle 0|aa^\dagger a|1\rangle = 1$

Therefore, we have

$$E_0^{(2)} = b^2\left(\frac{\hbar}{2m\omega}\right)^3\left(\frac{6}{-3\hbar\omega} + \frac{9}{-\hbar\omega}\right) = -\frac{11b^2\hbar^2}{8m^3\omega^4}$$

Exercise 3. A particle of mass m and charge e oscillates along x-axis in one-dimensional potential with an angular frequency ω. If an electric field E is applied along the x-axis, calculate the first order and second-order corrections to the energy of the nth state.

Solution: The potential energy due to the electric field E is $-eEx$. In terms of the creation ($a^\dagger$) and annihilation (a) operators, we have

$$x = \left(\frac{\hbar}{2m\omega}\right)^{1/2}(a + a^\dagger)$$

The first-order correction to the energy of the nth state is

$$E_n^{(1)} = \langle n| - eEx|n\rangle = -eE\left(\frac{\hbar}{2m\omega}\right)^{1/2}\langle n|(a + a^\dagger)|n\rangle = 0$$

Thus, the first-order correction to the energy is zero. The second-order correction to the ground state energy is

$$E_n^{(2)} = \sum_m{}' \frac{|\langle n| - eEx|m\rangle|^2}{E_n^0 - E_m^0} = e^2E^2\left(\frac{\hbar}{2m\omega}\right)\sum_m{}' \frac{|\langle n|(a + a^\dagger)|m\rangle|^2}{E_n^0 - E_m^0}$$

$$= e^2E^2\left(\frac{\hbar}{2m\omega}\right)\left[\frac{|\langle n|a|n+1\rangle|^2}{E_n^0 - E_{n+1}^0} + \frac{|\langle n|a^\dagger|n-1\rangle|^2}{E_n^0 - E_{n-1}^0}\right]$$

$$= e^2E^2\left(\frac{\hbar}{2m\omega}\right)\left[\frac{(\sqrt{n+1})^2}{-\hbar\omega} + \frac{(\sqrt{n})^2}{\hbar\omega}\right] = -\frac{e^2E^2}{2m\omega^2}$$

4. Ground state of helium

In the helium atom, two electrons revolve around a nucleus which has two protons and two neutrons. The nucleus considered as situated on the origin of the coordinate system and having a charge Ze. Let the two electrons 1 and 2 are at the position vectors $\vec{r_1}$ and $\vec{r_2}$, respectively.

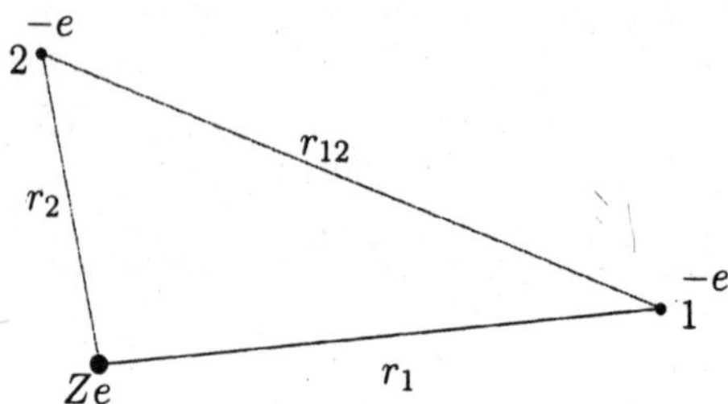

Figure 1: Helium atom having two electrons revolving around the nucleus

On neglecting the motion of the nucleus, the Hamiltonian of the system can be expressed as

$$H = \left(-\frac{\hbar^2}{2m}\nabla_1^2 - \frac{Ze^2}{r_1}\right) + \left(-\frac{\hbar^2}{2m}\nabla_2^2 - \frac{Ze^2}{r_2}\right) + \frac{e^2}{r_{12}} \tag{8.35}$$

where ∇_1 and ∇_2 refer to the coordinates of the electrons 1 and 2, respectively. The last term e^2/r_{12} is the electrostatic repulsion between the electrons. Denoting

$$H(1) = -\frac{\hbar^2}{2m}\nabla_1^2 - \frac{Ze^2}{r_1} \qquad \text{and} \qquad H(2) = -\frac{\hbar^2}{2m}\nabla_2^2 - \frac{Ze^2}{r_2}$$

equation (8.35) can be expressed as

$$H = H(1) + H(2) + \frac{e^2}{r_{12}}$$

The Hamiltonians $H(1)$ and $H(2)$ are for hydrogen-like atom and can be solved exactly, as discussed earlier. Schrödinger equation for hydrogen-like atom in the ground state is

$$\left(-\frac{\hbar^2}{2m}\nabla_1^2 - \frac{Ze^2}{r_1}\right)\psi_1^0 = E_1^0\psi_1^0$$

where

$$E_1^0 = -Z^2 W_H \qquad \psi_1^0 = \left(\frac{Z^3}{\pi a_0^3}\right)^{\frac{1}{2}} \exp\left(-\frac{Zr_1}{a_0}\right)$$

where $W_H = me^4/2\hbar^2 = 13.6$ eV. Further,

$$\left(-\frac{\hbar^2}{2m}\nabla_2^2 - \frac{Ze^2}{r_2}\right)\psi_2^0 = E_2^0\psi_2^0$$

where

$$E_2^0 = -Z^2 W_H \qquad \psi_2^0 = \left(\frac{Z^3}{\pi a_0^3}\right)^{\frac{1}{2}} \exp\left(-\frac{Zr_2}{a_0}\right)$$

Here, $a_0 \equiv \hbar^2/me^2$ is the Bohr radius. The wave function ψ^0 and energy eigenvalue E^0 of the operator $H(1) + H(2)$ can be expressed as

$$\psi^0 = \psi_1^0\psi_2^0 \qquad E^0 = E_1^0 + E_2^0 = -2Z^2 W_H$$

Thus, $H(1)+H(2)$ can be taken as unperturbed Hamiltonian and e^2/r_{12} as the perturbing Hamiltonian. The first order correction to the ground state energy is

$$E^{(1)} = \iint \psi_l^{0*}\psi_2^{0*}\ \frac{e^2}{r_{12}}\ \psi_l^0\psi_2^0 d\tau_1 d\tau_2$$

Substituting ψ_l^0 and ψ_2^0 in this equation, we have

$$E^{(1)} = \frac{Z^6 e^2}{\pi^2 a_0^6}\iint \frac{1}{r_{12}}\exp\left[-\frac{2Z}{a_0}(r_1 + r_2)\right] d\tau_1 d\tau_2 \tag{8.36}$$

where

$$d\tau_1 = r_1^2\ \sin\theta_1\ dr_1\ d\theta_1\ d\phi_1 \qquad d\tau_2 = r_2^2\ \sin\theta_2\ dr_2\ d\theta_2\ d\phi_2$$

The term $1/r_{12}$ can be expressed as

$$\frac{1}{r_{12}} = \sum_{l=0}^{\infty}\sum_{m=-l}^{l} \frac{4\pi}{2l+1}\frac{r_<^l}{r_>^{l+1}} Y_{lm}^*(\theta_1, \phi_1)\ Y_{lm}(\theta_2, \phi_2) \tag{8.37}$$

where $r_<$ denotes smaller of r_1 and r_2, and $r_>$ denotes larger of r_1 and r_2. Using equation (8.37) and the identity $4\pi Y_{00}^* Y_{00} = 1$ in (8.36), we get

$$E^{(1)} = \frac{Z^6 e^2}{\pi^2 a_0^6}\iint \sum_{l=0}^{\infty}\sum_{m=-l}^{l} \frac{4\pi}{2l+1}\frac{r_<^l}{r_>^{l+1}} Y_{lm}^*(\theta_1, \phi_1)\ Y_{lm}(\theta_2, \phi_2)\ 4\pi\ Y_{00}^*\ Y_{00}$$

$$\times \exp\Big[-\frac{2Z}{a_0}(r_1+r_2)\Big] r_1^2 \sin\theta_1 \, dr_1 \, d\theta_1 \, d\phi_1 \, r_2^2 \sin\theta_2 \, dr_2 \, d\theta_2 \, d\phi_2$$

Using the relations

$$\int_0^{2\pi}\int_0^{\pi} Y_{lm}^*(\theta_1,\phi_1) Y_{00}^*(\theta_1,\phi_1) \sin\theta_1 \, d\theta_1 \, d\phi_1 = \delta_{l0}\delta_{m0}$$

$$\int_0^{2\pi}\int_0^{\pi} Y_{00}^*(\theta_2,\phi_2) Y_{lm}^*(\theta_2,\phi_2) \sin\theta_2 \, d\theta_2 \, d\phi_2 = \delta_{l0}\delta_{m0}$$

These relations show that except $l=0$ and $m=0$, all the integrals are zero. Thus, we have

$$E^{(1)} = \frac{16Z^6 e^2}{a_0^6} \iint \frac{1}{r_>} \exp\Big[-\frac{2Z}{a_0}(r_1+r_2)\Big] r_1^2 \, r_2^2 \, dr_1 \, dr_2 \tag{8.38}$$

We shall first integrate over r_1 and then over r_2. In the range $0 \le r_1 \le r_2$, we have $r_> = r_2$ and in the range $r_2 \le r_1 \le \infty$, we have $r_> = r_1$. Then, we have

$$E^{(1)} = \frac{16Z^6 e^2}{a_0^6}\int_0^\infty \exp\Big(-\frac{2Zr_2}{a_0}\Big) r_2 \Big[\int_0^{r_2} \exp\Big(-\frac{2Zr_1}{a_0}\Big) r_1^2 \, dr_1\Big] dr_2$$

$$+\frac{16Z^6 e^2}{a_0^6}\int_0^\infty \exp\Big(-\frac{2Zr_2}{a_0}\Big) r_2^2 \Big[\int_{r_2}^{\infty} \exp\Big(-\frac{2Zr_1}{a_0}\Big) r_1 \, dr_1\Big] dr_2 \tag{8.39}$$

Using the expressions

$$\int x e^{ax} \, dx = \Big(\frac{x}{a} - \frac{1}{a^2}\Big) e^{ax}$$

$$\int x^2 e^{ax} \, dx = \Big(\frac{x^2}{a} - \frac{2x}{a^2} + \frac{2}{a^3}\Big) e^{ax}$$

we have

$$\int_0^{r_2} \exp\Big(-\frac{2Zr_1}{a_0}\Big) r_1^2 dr_1 = \Big[\Big(-\frac{a_0}{2Z} r_1^2 - \frac{a_0^2}{2Z^2} r_1 - \frac{a_0^3}{4Z^3}\Big) \exp\Big(-\frac{2Zr_1}{a_0}\Big)\Big]_0^{r_2}$$

$$= \Big(-\frac{a_0}{2Z} r_2^2 - \frac{a_0^2}{2Z^2} r_2 - \frac{a_0^3}{4Z^3}\Big) \exp\Big(-\frac{2Zr_2}{a_0}\Big) + \frac{a_0^3}{4Z^3} \tag{8.40}$$

$$\int_{r_2}^{\infty} \exp\Big(-\frac{2Zr_1}{a_0}\Big) r_1 \, dr_1 = \Big[\Big(-\frac{a_0}{2Z} r_1 - \frac{a_0^2}{4Z^2}\Big) \exp\Big(-\frac{2Zr_1}{a_0}\Big)\Big]_{r_2}^{\infty}$$

$$= \Big(-\frac{a_0}{2Z} r_2 - \frac{a_0^2}{4Z^2}\Big) \exp\Big(-\frac{2Zr_2}{a_0}\Big) \tag{8.41}$$

Using equations (8.40) and (8.41) in (8.39), we get

$$E^{(1)} = \frac{16Z^6e^2}{a_0^6}\int_0^{\infty}\Big(-\frac{a_0}{Z}\,r_2^3 - \frac{a_0^2}{2Z^2}\,r_2^2 - \frac{a_0^3}{4Z^3}r_2\Big)\exp\Big(-\frac{4Zr_2}{a_0}\Big)\,dr_2$$

$$+\frac{4Z^3e^2}{a_0^3}\int_0^{\infty} r_2\,\exp\Big(-\frac{2Zr_2}{a_0}\Big)\,dr_2$$

Using

$$\int_0^{\infty} x^n\exp(-ax)\,dx = \frac{n!}{a^{n+1}}$$

here, we get

$$E^{(1)} = \frac{16Z^6e^2}{a_0^6}\Big[-\frac{a_0}{Z}\,\frac{3!a_0^4}{256Z^4} + \frac{a_0^2}{2Z^2}\,\frac{2!a_0^3}{64Z^3} - \frac{a_0^3}{4Z^3}\,\frac{1!a_0^2}{16Z^2}\Big] + \frac{4Z^3e^2}{a_0^3}\,\frac{1!a_0^2}{4Z^2}$$

$$= \frac{5}{8}\,\frac{Ze^2}{a_0} = \frac{5}{4}\,ZW_H \tag{8.42}$$

Thus, the energy corrected to the first-order is

$$E = -2Z^2W_H + \frac{5}{4}ZW_H = -\Big[-2Z^2 - \frac{5}{4}Z\Big]W_H$$

The ratio of $E^{(1)}$ and E^0 is

$$\frac{E^{(1)}}{E^0} = \frac{5ZW_H/4}{2Z^2W_H} = \frac{5}{8Z}$$

showing that $E^{(1)}$ is about 31% of E^0 for helium.

5. Effect of electric field on the ground state of hydrogen atom

When an atom is placed in an external electric field, its energy levels are shifted. Such shifting of energy levels is known as the Stark effect, which was observed by Stark in hydrogen atom in 1913. Here, we shall consider a uniform electric field.

Let an electric field $\vec{E}$ is applied along z-axis (Figure 2) on a hydrogen atom situated at the origin of the coordinate system. If $\vec{r}$ be the position vector of electron, the electric dipole moment $\vec{\mu} = -e\,\vec{r}$ interacts with the field $\vec{E}$ producing an additional potential energy to the electron

$$H' = -\vec{\mu}\,.\,\vec{E} = erE\cos\theta = eEz \tag{8.43}$$

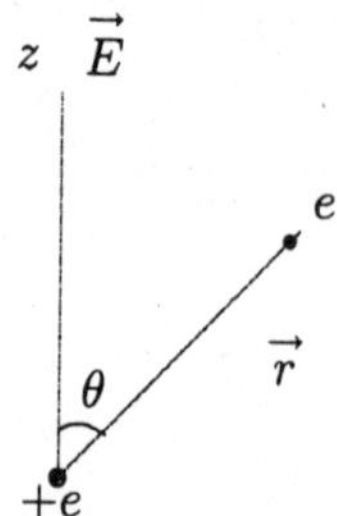

Figure 2: Electric field $\vec{E}$ is applied along z-axis on a hydrogen atom situated at the origin of the coordinate system

where θ is the angle between the position vector and the z-axis. Since this energy is very small, it can be treated as a perturbation part of the Hamiltonian

$$H^0 = -\frac{\hbar^2}{2\mu}\nabla^2 - \frac{e^2}{r}$$

The ground state of hydrogen atom is nondegenerate and the first-order correction to the energy is

$$E^{(1)} = \langle \psi_{100} | eEr\cos\theta | \psi_{100} \rangle \tag{8.44}$$

where the ground state wave function ψ_{100} is

$$\psi_{100} = \frac{1}{\sqrt{\pi a_0^3}} \exp\left(-\frac{r}{a_0}\right)$$

The θ part of the integral in equation (8.44) vanishes as

$$\int_0^\pi \cos\theta\ \sin\theta\ \mathrm{d}\theta = -\int_0^\pi \cos\theta\ \mathrm{d}(\cos\theta) = 0$$

This shows that the ground state of hydrogen atom will not show a first-order Stark effect. Let us go for the second-order Stark effect.

For calculating the second-order Stark effect, we have to use equation (8.18) where the sum goes over an infinite number of states. That problem of summation over an infinite number of states can be avoided with the help of a method developed by Dalgarno and Lewis. In this method instead of using equation (8.18), we go back to equation (8.17) and introduce an operator F (scalar function) such that

$$F\psi_1^0 = \psi_1^{(1)} \tag{8.45}$$

With the use of equations (8.45), equations (8.17) and (8.8), respectively, now reduce to

$$E_1^{(2)} = \langle \psi_1^0 \, | H' | \psi_1^{(1)} \rangle = \langle \psi_1^0 \, | H' F | \, \psi_1^0 \rangle$$

and

$$H' \psi_1^0 + H^0 \psi_1^{(1)} = E_1^{(1)} \psi_1^0 + E_1^0 \psi_1^{(1)}$$

$$H' \psi_1^0 + H^0 F \psi_1^0 = E_1^{(1)} \psi_1^0 + F E_1^0 \psi_1^0$$

Since $E_1^{(1)}$ is zero and $E_1^0 \psi_1^0 = H^0 \psi_1^0$, we have

$$H' \psi_1^0 + H^0 F \psi_1^0 = F H^0 \psi_1^0$$

$$(F H^0 - H^0 F) \psi_1^0 = H' \psi_1^0 \tag{8.46}$$

Since the scalar function F commutes with $-e^2/r$ in H^0, use of the relation

$$\nabla^2 (F \psi_1^0) = F \nabla^2 \psi_1^0 + \psi_1^0 \nabla^2 F + 2 \nabla \psi_1^0 \cdot \nabla F$$

in equation (8.46) gives

$$\left[F \left(-\frac{\hbar^2}{2\mu} \nabla^2 - \frac{e^2}{r} \right) - \left(-\frac{\hbar^2}{2\mu} \nabla^2 - \frac{e^2}{r} \right) F \right] \psi_1^0 = H' \psi_1^0$$

$$-\frac{\hbar^2}{2\mu} F \nabla^2 \psi_1^0 - F \frac{e^2}{r} \psi_1^0 + \frac{\hbar^2}{2\mu} \nabla^2 F \psi_1^0 + \frac{e^2}{r} F \psi_1^0 = H' \psi_1^0$$

$$-\frac{\hbar^2}{2\mu} F \nabla^2 \psi_1^0 + \frac{\hbar^2}{2\mu} \left[F \nabla^2 \psi_1^0 + \psi_1^0 \nabla^2 F + 2 \nabla \psi_1^0 . \nabla F \right] = H' \psi_1^0$$

$$\frac{\hbar^2}{2\mu} (\psi_1^0 \nabla^2 F + 2 \nabla \psi_1^0 \cdot \nabla F) = H' \psi_1^0 \tag{8.47}$$

This gives a differential equation that determines the operator F. We have

$$\psi_1^0 = \frac{1}{\sqrt{\pi a_0^3}} \exp\left(-\frac{r}{a_0} \right)$$

$$\nabla \psi_1^0 = \frac{\partial \psi_1^0}{\partial r} \, \hat{e}_r = -\frac{1}{a_0} \psi_1^0 \, \hat{e}_r$$

Thus, equation (8.47) gives

$$\psi_1^0 \nabla^2 F + 2\Big(-\frac{1}{a_0}\psi_1^0\Big)\frac{\partial F}{\partial r} = \frac{2\mu}{\hbar^2}\ \mathrm{e}Er\ \cos\theta\psi_1^0$$

$$\nabla^2 F - \frac{2}{a_0}\frac{\partial F}{\partial r} = \frac{2\mu}{\hbar^2}\ \mathrm{e}Er\ \cos\theta$$

$$\frac{1}{r^2}\frac{\partial}{\partial r}\left(r^2\frac{\partial F}{\partial r}\right) + \frac{1}{r^2\ \sin\theta}\frac{\partial}{\partial\theta}\left(\sin\theta\frac{\partial F}{\partial\theta}\right) + \frac{1}{r^2\ \sin^2\theta}\frac{\partial^2 F}{\partial\phi^2} - \frac{2}{a_0}\frac{\partial F}{\partial r} = \frac{2\mu}{\hbar^2}\mathrm{e}Er\ \cos\theta$$

Let the solution of this equation be of the form $f(r)\cos\theta$, so that we have

$$\frac{\cos\theta}{r^2}\frac{\partial}{\partial r}\left(r^2\frac{\partial f}{\partial r}\right) + \frac{f}{r^2\ \sin\theta}\ \frac{\partial}{\partial\theta}\left(\sin\theta\frac{\partial\cos\theta}{\partial\theta}\right) + 0 - \frac{2}{a_0}\cos\theta\ \frac{\partial f}{\partial r} = \frac{2\mu}{\hbar^2}\mathrm{e}Er\cos\theta$$

$$\frac{\cos\theta}{r^2}\frac{\partial}{\partial r}\left(r^2\frac{\partial f}{\partial r}\right) - \frac{2f\cos\theta}{r^2} - \frac{2}{a_0}\cos\theta\frac{\partial f}{\partial r} = \frac{2\mu}{\hbar^2}\mathrm{e}Er\ \cos\theta$$

$$\frac{1}{r^2}\frac{\partial}{\partial r}\left(r^2\frac{\partial f}{\partial r}\right) - \frac{2f}{r^2} - \frac{2}{a_0}\frac{\partial f}{\partial r} = \frac{2\mu}{\hbar^2}\mathrm{e}Er$$

Solution of this equation is[2]

$$f(r) = -\frac{\mu}{\hbar^2}\ \mathrm{e}Ea_0\Big(\frac{r^2}{2} + a_0 r\Big)$$

---

[2]We have

$$\frac{\partial f}{\partial r} = -\frac{\mu}{\hbar^2}\ \mathrm{e}Ea_0\Big(r + a_0\Big) \qquad r^2\ \frac{\partial f}{\partial r} = -\frac{\mu}{\hbar^2}\ \mathrm{e}Ea_0\Big(r^3 + a_0 r^2\Big)$$

$$\frac{\partial}{\partial r}\left(r^2\ \frac{\partial f}{\partial r}\right) = -\frac{\mu}{\hbar^2}\ \mathrm{e}Ea_0\Big(3r^2 + 2a_0 r\Big), \qquad \frac{1}{r^2}\frac{\partial}{\partial r}\left(r^2\ \frac{\partial f}{\partial r}\right) = -\frac{\mu}{\hbar^2}\ \mathrm{e}Ea_0\Big(3 + \frac{2a_0}{r}\Big)$$

Therefore, we have

$$\frac{1}{r^2}\frac{\partial}{\partial r}\left(r^2\frac{\partial f}{\partial r}\right) - \frac{2f}{r^2} - \frac{2}{a_0}\frac{\partial f}{\partial r} = -\frac{\mu}{\hbar^2}\ \mathrm{e}Ea_0\Big(3 + \frac{2a_0}{r}\Big) + \frac{2\mu}{\hbar^2 r^2}\ \mathrm{e}Ea_0\Big(\frac{r^2}{2} + a_0 r\Big) + \frac{2\mu}{a_0\hbar^2}\ \mathrm{e}Ea_0\Big(r + a_0\Big) = \frac{2\mu}{\hbar^2}\ \mathrm{e}Er$$

Thus, we have

$$F = f(r)\cos\theta = -\frac{\mu}{\hbar^2}\, eEa_0\Big(\frac{r^2}{2} + a_0 r\Big)\ \cos\theta$$

The second-order correction is then

$$E_1^{(2)} = \langle \psi_1^0 \mid H' \mid \psi_1^{(1)} \rangle = \langle \psi_1^0 \mid H'F \mid \psi_1^0 \rangle$$

$$= \int_0^\infty\int_0^\pi\int_0^{2\pi} \frac{1}{\pi a_0^3}\exp\Big(-\frac{2r}{a_0}\Big) eEr\cos\theta\Big[-\frac{\mu}{\hbar^2}\, eEa_0\Big(\frac{r^2}{2} + a_0 r\Big)\cos\theta\Big]$$

$$\times r^2 \sin\theta\ dr d\theta\ d\phi$$

$$= -\frac{\mu e^2 E^2}{\hbar^2 \pi a_0^2}\int_0^\pi\int_0^{2\pi}\cos^2\theta\sin\theta d\theta d\phi \int_0^\infty \Big[\frac{r^5}{2} + a_0 r^4\Big]\exp\Big(-\frac{2r}{a_0}\Big)dr$$

$$= -\frac{4\mu e^2 E^2}{3\hbar^2 a_0^2}\int_0^\infty \Big[\frac{r^5}{2} + a_0 r^4\Big]\exp\Big(-\frac{2r}{a_0}\Big)dr$$

Using the integral

$$\int_0^\infty x^n e^{-ax} dx = \frac{n!}{a^{n+1}}$$

we have

$$E_1^{(2)} = -\frac{9\mu}{4h^2}\, e^2 E^2 a_0^4 = -\frac{9}{4}a_0^3 E^2$$

where $a_0 = \hbar^2/\mu e^2$. The ground state energy corrected to the second-order correction is therefore

$$E_1 = E_1^0 - \frac{9}{4}a_0^3 E^2$$

Since the correction is proportional to the square of the electric field E, it is often known as the quadratic Stark effect.

Physically, the correction term $-9a_0^3E^2/4$ is due to the distortion of the charge distribution by the electric field $\vec{E}$. Thus, we may say that the atom is polarized by the electric field. The extent of this polarization is measured by the polarizability α which is defined as the ratio of the induced electric dipole moment $\vec{\mu}$ to the applied field $\vec{E}$

$$\vec{\mu} = \alpha\,\vec{E}$$

The energy of an induced dipole in an applied field is given by $-\alpha E^2/2$. Equating this energy with the energy due to perturbation calculation, we have

$$-\frac{1}{2}\alpha E^2 = -\frac{9}{4}\, a_0^3 E^2 \qquad \text{or} \qquad \alpha = \frac{9}{2}a_0^3$$

Hence, the polarizability α for hydrogen atom in the ground state is $9a_0^3/2$.

6. Degenerate energy levels

For nondegenerate case, the unperturbed wave function ψ_n^0 is a unique one. In case of degeneracy of an energy level, a linear combination of the degenerate wave functions can be taken as the unperturbed wave function. Let us consider a simple case of two-fold non-degeneracy of the level with energy E_n^0. Let ψ_n^0 and ψ_l^0 be two eigenfunctions corresponding to the eigenvalues $E_n^0 = E_l^0$ and a linear combination of them be

$$\phi = c_n\psi_n^0 + c_l\psi_l^0 \tag{8.48}$$

where c_n and c_l are constants.

6.1 First-order correction

Replacing ψ_n^0 in equation (8.7) by ϕ, we get

$$H'|(c_n\psi_n^0 + c_l\psi_l^0)\rangle + H^0|\psi_n^{(1)}\rangle = E_n^{(1)}|(c_n\psi_n^0 + c_l\psi_l^0)\rangle + E_n^0|\psi_n^{(1)}\rangle \tag{8.49}$$

Multiplying equation (8.49) from left by $\langle\psi_n^0|$, we get

$$c_n\langle\psi_n^0|H'|\psi_n^0\rangle + c_l\langle\psi_n^0|H'|\psi_l^0\rangle + \langle\psi_n^0|H^0|\psi_n^{(1)}\rangle = c_nE_n^{(1)}\langle\psi_n^0|\psi_n^0\rangle$$

$$+E_n^{(1)}c_l\langle\psi_n^0|\psi_l^0\rangle + E_n^0\langle\psi_n^0|\psi_n^{(1)}\rangle \tag{8.50}$$

Since H^0 is Hermitian, we have

$$\langle\psi_n^0|H^0|\psi_n^{(1)}\rangle = \langle\psi_n^{(1)}|H^0|\psi_n^0\rangle^* = E_n^0\langle\psi_n^{(1)}|\psi_n^0\rangle^* = E_n^0\langle\psi_n^0|\psi_n^{(1)}\rangle$$

and equation (8.50) reduces to

$$c_nH'_{nn} + c_lH'_{nl} = c_nE_n^{(1)}$$

$$(H'_{nn} - E_n^{(1)})\, c_n + H'_{nl}c_l = 0 \tag{8.51}$$

Operating equation (8.49) from left by $\langle\psi_l^0|$, we get

$$c_n\langle\psi_l^0|H'|\psi_n^0\rangle + c_l\langle\psi_l^0|H'|\psi_l^0\rangle + \langle\psi_l^0|H^0|\psi_n^{(1)}\rangle = c_n E_n^{(1)}\langle\psi_l^0|\psi_n^0\rangle$$

$$+c_l E_n^{(1)}\langle\psi_l^0|\psi_l^0\rangle + E_n^0\langle\psi_l^0|\psi_n^{(1)}\rangle$$

Since H^0 is Hermitian, we have

$$\langle\psi_n^0|H^0|\psi_n^{(1)}\rangle = \langle\psi_n^{(1)}|H^0|\psi_n^0\rangle^* = E_n^0\langle\psi_n^{(1)}|\psi_n^0\rangle^* = E_n^0\langle\psi_n^0|\psi_n^{(1)}\rangle$$

Thus, we have

$$c_n H'_{ln} + c_l H'_{ll} = c_l E_n^{(1)}$$

$$(H'_{ln}c_n + (H'_{ll} - E_n^{(1)})c_l = 0 \tag{8.52}$$

Equations (8.51) and (8.52) together from a set of simultaneous linear equations for the coefficients c_n and c_l. A nontrivial solution of these equations exist when the determinant of the coefficients vanishes.

$$\begin{vmatrix} H'_{nn} - E_n^{(1)} & H'_{nl} \\ H'_{ln} & H'_{ll} - E_n^{(1)} \end{vmatrix} = 0$$

This secular determinant can be expressed as $|H' - E_n^{(1)}\ I| = 0$, where I is a unit matrix and H' is a matrix

$$H' \equiv \begin{bmatrix} H'_{nn} & H'_{nl} \\ H'_{ln} & H'_{ll} \end{bmatrix}$$

This secular equation can be solved as the following.

$$[H'_{nn} - E_n^{(1)}][H'_{ll} - E_n^{(1)}] - |H'_{nl}|^2 = 0$$

$$E_n^{(1)^2} - E_n^{(1)}[H'_{ll} + H'_{nn}] + [H'_{nn}H'_{ll} - |H'_{nl}|^2] = 0$$

There are two roots

$$E_{n\pm}^{(1)} = \frac{1}{2}(H'_{nn} + (H'_{ll}) \pm \frac{1}{2}\left[(H'_{nn} + H'_{ll})^2 - 4(H'_{nn}H'_{ll} - |H'_{nl}|^2)\right]^{1/2}$$

$$= \frac{1}{2}(H'_{nn} + (H'_{ll}) \pm \frac{1}{2}\left[(H'_{nn} - H'_{ll})^2 + 4|H'_{nl}|^2\right]^{1/2} \tag{8.53}$$

Hence, the corrected energies are

$$E_n = E_n^0 + E_{n+}^{(1)} \qquad \text{and} \qquad E_l = E_n^0 + E_{n-}^{(1)}$$

Since the diagonal elements H'_{nn} and H'_{ll} of the Hermitian operator H' are real, therefore, both the energies E_n and E_l will be real. If $H'_{nn} = H'_{ll}$ and $H'_{nl} = 0$, we have $E_{n+} = E_{n-}$ and thus the degeneracy is not removed in the first order.

When the two roots of equation (8.53) are distinct, each (say E_n) can be used to calculate the ratio c_n/c_l either from equation (8.51) or (8.52). The normalization $c_n^2 + c_l^2 = 1$ allows us to calculate the values of c_n and c_l. These values of c_n and c_l give the desires linear combination associated with the level $E_n^0 + E_{n+}$. In a similar manner, the combination associated with the level $E_n^0 + E_{n-}$ can be obtained.

7. Effect of electric field on $n = 2$ state of hydrogen

The ground state ($n = 1$) of hydrogen atom had $l = 0$ and $m = 0$ and therefore, it was a nondegenerate one. But, the first excited state ($n = 2$) of hydrogen atom has $l = 0$ and 1, and for $l = 0$, we have $m = 0$ whereas for $l = 1$, we have $m = 1$, 0 and -1. Thus, the first excited state has four-fold degeneracy. The four degenerate states may be specified by the quantum numbers (nlm) as

$$|nlm\rangle : |200\rangle,\ |210\rangle,\ |211\rangle,\ |21,-1\rangle$$

The wave functions for these states are

$$\psi_{200} = \frac{1}{4\sqrt{2\pi a_0^3}}\left(2 - \frac{r}{a_0}\right)\exp\left(-\frac{r}{2a_0}\right)$$

$$\psi_{210} = \frac{1}{4\sqrt{2\pi a_0^5}}\, r \exp\left(-\frac{r}{2a_0}\right)\cos\theta$$

$$\psi_{211} = \frac{1}{8\sqrt{\pi a_0^5}}\, r \exp\left(-\frac{r}{2a_0}\right)\sin\theta\, e^{i\phi}$$

$$\psi_{21,-1} = \frac{1}{8\sqrt{\pi a_0^5}}\, r \exp\left(-\frac{r}{2a_0}\right)\sin\theta\, e^{-i\phi}$$

As earlier, let the electric field $\vec{E}$ be applied along the positive z-axis which interacts with the electric dipole giving the perturbing Hamiltonian (equation 8.43), $H' = eEz = erE\cos\theta$. For this four-fold degenerate state, application of the perturbation theory requires evaluation of sixteen matrix elements. Let us look into these matrix elements.

The θ-integral of the matrix elements $\langle 200|H'|200\rangle$ is

$$\int_0^{\pi} \cos\theta \ \sin\theta \mathrm{d}\theta$$

which is equal to zero. The θ-integral of the matrix elements $\langle 210|H'|210\rangle$ is

$$\int_0^{\pi} \cos^3\theta \ \sin\theta \mathrm{d}\theta$$

which is equal to zero. The θ-integral of the matrix elements $\langle 211|H'|211\rangle$ as well as $\langle 21,-1|H'|21,-1\rangle$ is

$$\int_0^{\pi} \cos\theta \ \sin^3\theta \mathrm{d}\theta$$

which is equal to zero. Thus, the four diagonal matrix elements are zero. The ϕ-integral of the matrix elements $\langle 200|H'|211\rangle$, $\langle 210|H'|211\rangle$, $\langle 21,-1|H'|200\rangle$ and $\langle 21,-1|H'|210\rangle$ is

$$\int_0^{2\pi} \mathrm{e}^{i\phi}\mathrm{d}\phi$$

which is equal to zero. The ϕ-integral of the matrix elements $\langle 211|H'|200\rangle$, $\langle 211|H'|210\rangle$, $\langle 200|H'|21,-1\rangle$ and $\langle 210|H'|21,-1\rangle$ is

$$\int_0^{2\pi} \mathrm{e}^{-i\phi}\mathrm{d}\phi$$

which is equal to zero. The ϕ-integral of the matrix elements $\langle 211|H'|21,-1\rangle$ is

$$\int_0^{2\pi} \mathrm{e}^{-2i\phi}\mathrm{d}\phi$$

which is equal to zero. The ϕ-integral of the matrix elements $\langle 21,-1|H'|211\rangle$ is

$$\int_0^{2\pi} \mathrm{e}^{2i\phi}\mathrm{d}\phi$$

which is equal to zero. Thus, 10 off-diagonal matrix elements are zero. The remaining two matrix elements $\langle 200|H'|210\rangle$ and $\langle 210|H'|200\rangle$ are the only nonvanishing ones, showing that the stats $|200\rangle$ and $|210\rangle$ are affected by the electric field whereas the states $|211\rangle$ and $|21,-1\rangle$ are not. Let us now evaluate these two non-vanishing matrix elements. We have

$$\langle 200|H'|210\rangle = \int_0^\infty \int_0^\pi \int_0^{2\pi} \frac{1}{4\sqrt{2\pi a_0^3}}\left(2-\frac{r}{a_0}\right)\exp\left(-\frac{r}{2a_0}\right) erE\cos\theta$$

$$\frac{1}{4\sqrt{2\pi a_0^5}}\, r\exp\left(-\frac{r}{2a_0}\right)\cos\theta r^2\ \sin\theta\ \mathrm{d}r\ \mathrm{d}\theta\ \mathrm{d}\phi$$

$$= \frac{eE}{32\pi a_0^4}\int_0^\infty \int_0^\pi \int_0^{2\pi}\left(2-\frac{r}{a_0}\right) r^4 \exp\left(-\frac{r}{a_0}\right)\cos^2\theta\sin\theta \mathrm{d}r\ \mathrm{d}\theta\ \mathrm{d}\phi$$

We have

$$\int_0^\pi \cos^2\theta\sin\theta\ \mathrm{d}\theta = \frac{2}{3}$$

and therefore, we have

$$\langle 200|H'|210\rangle = \frac{eE}{24a_0^4}\int_0^\infty \left(2r^4-\frac{r^5}{a_0}\right)\exp\left(-\frac{r}{a_0}\right)\mathrm{d}r$$

Using

$$\int_0^\infty x^n \mathrm{e}^{-ax}\mathrm{d}x = \frac{n!}{a^{n+1}}$$

we get

$$\langle 200|H'|210\rangle = -3eEa_0$$

Since ψ_{200} and ψ_{210} both are real values, we have

$$\langle 210|H'|200\rangle = -3eEa_0$$

The perturbation matrix is therefore

$$H' = \begin{bmatrix} 0 & -3eEa_0 & 0 & 0 \\ -3eEa_0 & 0 & 0 & 0 \\ 0 & 0 & 0 & 0 \\ 0 & 0 & 0 & 0 \end{bmatrix}$$

The secular determinant $|H' - E_2^{(1)}\ I| = 0$ is therefore

$$\begin{vmatrix} -E_2^{(1)} & -3eEa_0 & 0 & 0 \\ -3eEa_0 & -E_2^{(1)} & 0 & 0 \\ 0 & 0 & -E_2^{(1)} & 0 \\ 0 & 0 & 0 & -E_2^{(1)} \end{vmatrix} = 0$$

On expansion of this determinant, we have

$$[-E_2^{(1)}][-E_2^{(1)}][E_2^{(1)^2} - (3eEa_0)^2] = 0$$

$$E_2^{(1)}\ E_2^{(1)}\left(E_2^{(1)} - 3eEa_0\right)\left(E_2^{(1)} + 3eEa_0\right) = 0$$

The four roots of this polynomial equation are $3eEa_0$, $-3eEa_0$, 0 and 0. The effect of the electric field is in the states $|200\rangle$ and $|210\rangle$ whereas the states $|211\rangle$ and $|21,-1\rangle$ remain unaffected. Thus, the four-fold degeneracy is removed partially. We have three eigenvalues: E_2^0, $E_2^0 + 3eEa_0$ and $E_2^0 - 3eEa_0$. The eigenstate $|211\rangle$ and $|21,-1\rangle$ obviously correspond to the eigenvalue E_2^0.

Eigenstates corresponding to the eigenvalues can be calculated with the help of equation (8.51) or (8.52). Let us use equation (8.51) for $E_2^{(1)} = 3eEa_0$.

$$(H'_{nn} - E_n^{(1)})\ c_n + H'_{nl}c_l = 0$$

$$(0 - 3eEa_0)\ c_n - 3eEa_0c_l = 0$$

giving $c_n/c_l = -1$. The condition $c_n^2 + c_l2 = 1$, gives $c_n = 1/\sqrt{2}$ and $c_l = -1/\sqrt{2}$. Thus, the eigenstate is (equation 8.48)

$$\phi = c_n\psi_n^0 + c_l\psi_l^0 = \frac{1}{\sqrt{2}}|200\rangle - \frac{1}{\sqrt{2}}|210\rangle = \frac{1}{\sqrt{2}}\left(|200\rangle - |210\rangle\right)$$

Now, we use equation (8.51) for $E_2^{(1)} = -3eEa_0$.

$$(H'_{nn} - E_n^{(1)})\ c_n + H'_{nl}c_l = 0$$

$$(0 + 3eEa_0)\ c_n - 3eEa_0c_l = 0$$

giving $c_n/c_l = 1$. The condition $c_n^2 + c_l2 = 1$, gives $c_n = c_l = 1/\sqrt{2}$. Thus, the eigenstate is (equation 8.48)

$$\phi = c_n\psi_n^0 + c_l\psi_l^0 = \frac{1}{\sqrt{2}}|200\rangle + \frac{1}{\sqrt{2}}|210\rangle = \frac{1}{\sqrt{2}}\left(|200\rangle + |210\rangle\right)$$

The energies along with the eigenstates of the $n = 2$ state of hydrogen atom placed in an electric field along positive z-direction is shown in Figure 3.

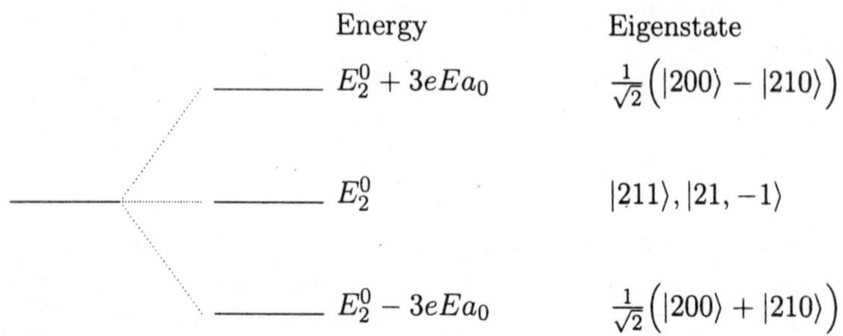

Figure 3: Energies and eigenfunctions of the first excited $n = 2$ state of hydrogen atom placed in an electric field.

8. Spin-orbit interaction

An electron in an atom is assumed to revolve in a potential field, produced by the nucleus and other electrons of the atom. This potential field for convenience is generally assumed to be spherically symmetric, $V = V(r)$. Besides this revolution, the electron has its spin motion. These orbital and spin motions of an electron interact and introduce a term in the energy of the system. This phenomenon is known as the spin-orbit interaction. The spin-orbit Hamiltonian H_{so} is expressed as

$$H_{so} = \xi(r)\, \vec{L} \cdot \vec{S}$$

where $\vec{L}$ and $\vec{S}$ are orbital and spin angular momenta and

$$\xi(r) = \frac{1}{2m^2c^2}\,\frac{1}{r}\,\frac{\mathrm{d}V}{\mathrm{d}r}$$

The radial average of the function $\xi(r)\hbar^2$ can be written as $hc\,\zeta$ where ζ is known as the spin-orbit coupling constant. Then we have

$$hc\,\zeta = \langle nlm|\xi|nlm\rangle\hbar^2$$

Following this definition, ζ is a wave number and $hc\,\zeta$ is an energy. For an electron in a hydrogenic atom, the potential is

$$V(r) = -\frac{Ze^2}{r} \qquad \text{or} \qquad \frac{\mathrm{d}V}{\mathrm{d}r} = \frac{Ze^2}{r^2}$$

and thus, we have

$$H_{so} = \xi(r)\, \vec{L} \cdot \vec{S} = \frac{Ze^2}{2m^2c^2}\, \frac{1}{r^3}\, \vec{L} \cdot \vec{S}$$

The expectation value of $1/r^3$ for hydrogenic wave function is

$$\left\langle nlm | \frac{1}{r^3} | nlm \right\rangle = \frac{Z^3}{n^3 a_0^3 l(l+\frac{1}{2})(l+1)}$$

where a_0 is the Bohr radius. The spin-orbit coupling constant for a hydrogenic atom is therefore

$$hc\, \zeta_{nl} = \langle nlm|\xi|nlm\rangle \hbar^2 = \left\langle nlm | \frac{Ze^2}{2m^2c^2}\, \frac{1}{r^3} | nlm \right\rangle \hbar^2$$
$$= \frac{Ze^2}{2m^2c^2}\, \frac{Z^3\hbar^2}{n^3 a_0^3 l(l+\frac{1}{2})(l+1)} = \frac{Z^4 e^2 \hbar^2}{2m^2c^2 a_0^3 n^3 l(l+\frac{1}{2})(l+1)}$$

We generally express this expression in terms of the fine structure constant $\alpha = e^2/\hbar c$ and Rydberg constant $R = 2\pi^2 me^4/ch^3$ as

$$\zeta_{nl} = \frac{Z^4 \alpha^2 R}{n^3 l[l+(\frac{1}{2})](l+1)}$$

The presence of the factor Z^4 in the numerator shows that the spin-orbit interaction is more predominant in heavy elements. Let us now calculate the effect of spin-orbit interaction on the energy levels. Since H_{so} is very weak in comparison to the energy level separations of the atom, it can be treated as a perturbation on the unperturbed Hamiltonian H^0 which has degenerate eigenvalues for a given n and l. However, it is possible to avoid working with degenerate perturbation theory by using the basis $|lsjm\rangle$ as L^2, S^2, J^2 and J_z commute. The first order correction to the energy of the state $|lsjm\rangle$ is

$$E_{so} = \langle lsjm \mid H_{so} \mid lsjm\rangle = \langle lsjm \mid \xi(r)\, \vec{L} \cdot \vec{S} \mid lsjm\rangle$$

The matrix elements of $\vec{L} \cdot \vec{S}$ can be evaluated easily by writing

$$J^2 = L^2 + S^2 + 2\, \vec{L} \cdot \vec{S} \qquad \text{or} \qquad \vec{L} \cdot \vec{S} = \frac{1}{2}(J^2 - L^2 - S^2)$$

Thus, we have

$$\langle lsjm|\, \vec{L} \cdot \vec{S}\, |lsjm\rangle = \frac{1}{2}\hbar^2[j(j+1) - l(l+1) - s(s+1)]$$

and the interaction energy is

$$E_{so} = \frac{1}{2}hc\zeta_{nl}\,[j(j+1) - l(l+1) - s(s+1)]$$

$$= \frac{chZ^4\alpha^2 R}{2}\,\frac{j(j+1) - l(l+1) - s(s+1)}{n^3 l[l + (\frac{1}{2})](l+1)}$$

The spin-orbit interaction energy E_{so} is independent of m_j and thus each level has $(2j+1)$-fold degeneracy.

9. Problems and questions

1. Describe the principle of time-dependent perturbation theory.

2. Derive expressions for the first-order correction to energy as well as to the wave function in the time-independent perturbation theory.

3. Derive expressions for the second-order correction to energy as well as to the wave function in the time-independent perturbation theory.

4. Explain why the hydrogen atom in the ground state does not show a first-order Stark effect.

5. Explain the quadratic Stark effect

6. Derive an expression for polarizability for hydrogen atom in the ground state.

7. What is spin-orbit interaction? Obtain an expression for spin-orbit coupling constant. Show that it is very large in heavy elements.

8. Obtain an expression for the second-order correction to the energy of the ground state and show that is is always negative.

9. An oscillator of mass m and angular frequency ω is subjected to a potential

$$V(x) = \frac{1}{2}\,m\omega^2 x^2 + bx$$

   b is a constant and $bx << (m\omega^2 x^2/2)$. Calculate the first and second order corrections to the energy of $n = 1$ state of the oscillator.

10. An oscillator of mass m and angular frequency ω is subjected to a potential

$$V(x) = \frac{1}{2}\, m\omega^2 x^2 + bx^3$$

b is a constant and $bx^3 << (m\omega^2 x^2/2)$. Calculate the first and second order corrections to the energy of $n = 1$ state of the oscillator.

11. Write notes on the following

(i) Time-dependent perturbation theory

(ii) Stark effect

(iii) Quadratic Stark effect

Variation Method

For application of the perturbation method in the preceding chapter, we had to know a closely related problem whose exact solution could be obtained. The system of interest may not necessarily fulfill this requirement of exact solution of a closely related problem. Then, one of the other options it to use the variation method, where one has to make a guess about the wave function and then to apply the variational principle to improve the guessed wave function and to obtain an upper limit for the ground state energy. The accuracy of results in this method however depends considerably on the ability of the guess. Though this variation method is generally applied to obtain the ground state energy and the wave function, but it can be extended to the excited states, as well.

1. Variational principle

The variation method is based on the philosophy to take a trial wave function ϕ and to calculate the expectation value $\langle H \rangle$ of the Hamiltonian H. That is $\langle H \rangle = \langle \phi | H | \phi \rangle$. According to the variational principle, the ground state energy E_1 of the system satisfies the condition

$$\langle H \rangle \geq E_1 \tag{9.1}$$

It shows that the lowest expectation value of H would be taken as the ground state energy of the system. When the trial function approaches to the actual (not known) ground state wave function, the value of $\langle H \rangle$ approaches to E_1. Thus, when the trial function is the ground state wave function, we have $\langle H \rangle = E_1$. Hence, in the variation method, we go on modifying the trial function as long as the expectation value $\langle H \rangle$ goes on decreasing, and the final lowest value of $\langle H \rangle$ is taken as the ground state energy. In case, the wave function ϕ is not normalized, we take

$$\langle H \rangle = \frac{\langle \phi | H | \phi \rangle}{\langle \phi | \phi \rangle} \tag{9.2}$$

Proof of this theorem (9.1) can be obtained in the following manner. Suppose, Ψ_1, Ψ_2, ... are true (but not known) wave functions of the Hamiltonian H with energies E_1, E_2, ..., respectively, so that $H\Psi_n = E_n\Psi_n$. Let the trial function ϕ be expressed as linear combination of these true wave functions[1]

$$\phi = \sum_i a_i|\Psi_i\rangle$$

Thus, we have

$$\langle H\rangle = \langle\phi|H|\phi\rangle = \sum_i\sum_j a_i^* a_j \langle\Psi_i|H|\Psi_j\rangle$$

$$= \sum_i\sum_j a_i^* a_j E_j \langle\Psi_i|\Psi_j\rangle = \sum_i |a_i|^2 E_i$$

Keeping in view that $E_i \geq E_1$, let E_i differs from E_1 by ΔE_i, so that $E_i = E_1 + \Delta E_i$, and we have

$$\langle H\rangle = \sum_i |a_i|^2(E_1 + \Delta E_i) = E_1 \sum_i |a_i|^2 + \sum_i |a_i|^2 \Delta E_i$$

Since $\sum |a_i|^2 = 1$ and ΔE_i is positive, we have

$$\langle H\rangle \geq E_1$$

It proves the required condition. In practice, in a trial function, we take one or more adjustable parameters, and by adjusting them, the value of $\langle H\rangle$ is calculated till it decreases. The minimum value of $\langle H\rangle$ is taken as the energy of the ground state.

Exercise 1. Optimize the trial function $\phi = \mathrm{e}^{-\alpha r}$, where α is a constant, and calculate the ground state energy of hydrogen atom.

Solution: Electron in a hydrogen atom revolves in a spherically symmetric potential and the Hamiltonian of hydrogen atom is

$$H = -\frac{\hbar^2}{2m}\nabla^2 - \frac{e^2}{r}$$

[1]For normalized wave function ϕ, we have

$$1 = \langle\phi|\phi\rangle = \sum_i\sum_j a_i^* a_j \langle\Psi_i|\Psi_j\rangle = \sum_i a_i^* a_i = \sum_i |a_i|^2$$

Since the trial function $\phi = \mathrm{e}^{-\alpha r}$ depends only on r, ∇^2 in spherical polar coordinates contains only the radial derivative

$$\nabla^2 \equiv \frac{1}{r^2}\frac{\mathrm{d}}{\mathrm{d}r}\left(r^2\frac{\mathrm{d}}{\mathrm{d}r}\right) = \frac{\mathrm{d}^2}{\mathrm{d}r^2} + \frac{2}{r}\frac{\mathrm{d}}{\mathrm{d}r}$$

and the Hamiltonian is

$$H = -\frac{\hbar^2}{2m}\frac{\mathrm{d}^2}{\mathrm{d}r^2} - \frac{\hbar^2}{mr}\frac{\mathrm{d}}{\mathrm{d}r} - \frac{e^2}{r}$$

Thus, we have

$$\langle\phi|H|\phi\rangle = -\frac{\hbar^2}{2m}\left\langle\phi\left|\frac{\mathrm{d}^2}{\mathrm{d}r^2}\right|\phi\right\rangle - \frac{\hbar^2}{m}\left\langle\phi\left|\frac{1}{r}\frac{\mathrm{d}}{\mathrm{d}r}\right|\phi\right\rangle - \left\langle\phi\left|\frac{e^2}{r}\right|\phi\right\rangle$$

The angular part of the volume element in spherical polar coordinates contributes 4π. Now, we have

$$\left\langle\phi\left|\frac{\mathrm{d}^2}{\mathrm{d}r^2}\right|\phi\right\rangle = 4\pi\alpha^2\int_0^\infty r^2\mathrm{e}^{-2\alpha r}\,\mathrm{d}r = \frac{\pi}{\alpha}$$

$$\left\langle\phi\left|\frac{1}{r}\frac{\mathrm{d}}{\mathrm{d}r}\right|\phi\right\rangle = -4\pi\alpha\int_0^\infty r\mathrm{e}^{-2\alpha r}\,\mathrm{d}r = -\frac{\pi}{\alpha}$$

$$\left\langle\phi\left|\frac{e^2}{r}\right|\phi\right\rangle = 4\pi e^2\int_0^\infty r\mathrm{e}^{-2\alpha r}\,\mathrm{d}r = \frac{\pi e^2}{\alpha^2}$$

$$\langle\phi|\phi\rangle = 4\pi\int_0^\infty r^2\mathrm{e}^{-2\alpha r}\,\mathrm{d}r = \frac{\pi}{\alpha^3}$$

On substituting these values, we get

$$\langle\phi|H|\phi\rangle = -\frac{\hbar^2}{2m}\frac{\pi}{\alpha} + \frac{\hbar^2}{m}\frac{\pi}{\alpha} - \frac{\pi e^2}{\alpha^2} = \frac{\hbar^2}{2m}\frac{\pi}{\alpha} - \frac{\pi e^2}{\alpha^2}$$

Thus, we have

$$\langle H\rangle = \frac{\langle\phi|H|\phi\rangle}{\langle\phi|\phi\rangle} = \frac{\hbar^2\alpha^2}{2m} - \alpha e^2$$

Exercise 2. Optimize the trial function $\phi = A\mathrm{e}^{-\alpha x^2}$, where A and α are constants, and calculate the ground state energy of one-dimensional harmonic oscillator of mass m and frequency ω.

Solution: The Hamiltonian of one-dimensional harmonic oscillator is

$$H = -\frac{\hbar^2}{2m}\frac{\mathrm{d}^2}{\mathrm{d}x^2} + \frac{1}{2}m\omega^2x^2$$

Thus, we have

$$\langle\phi|H|\phi\rangle = -\frac{\hbar^2}{2m}\left\langle\phi\left|\frac{d^2}{dx^2}\right|\phi\right\rangle + \frac{1}{2}\, m\omega^2\left\langle\phi\left|x^2\right|\phi\right\rangle$$

Now, we have

$$\left\langle\phi\left|\frac{d^2}{dx^2}\right|\phi\right\rangle = -2\alpha|A|^2\int_{-\infty}^{\infty} e^{-2\alpha x^2}dx + 4\alpha^2|A|^2\int_{-\infty}^{\infty} x^2 e^{-2\alpha x^2}dx$$

$$= -2\alpha|A|^2\left(\frac{\pi}{2\alpha}\right)^{1/2} + 4\alpha^2|A|^2\frac{1}{4\alpha}\left(\frac{\pi}{2\alpha}\right)^{1/2} = -\alpha|A|^2\left(\frac{\pi}{2\alpha}\right)^{1/2}$$

$$\left\langle\phi\left|x^2\right|\phi\right\rangle = |A|^2\int_{-\infty}^{\infty} x^2 e^{-2\alpha x^2}\, dx = |A|^2\frac{1}{4\alpha}\left(\frac{\pi}{2\alpha}\right)^{1/2}$$

$$\langle\phi|\phi\rangle = |A|^2\int_{-\infty}^{\infty} e^{-2\alpha x^2}\, dx = |A|^2\left(\frac{\pi}{2\alpha}\right)^{1/2}$$

On substituting these values, we get

$$\langle\phi|H|\phi\rangle = \frac{\hbar^2}{2m}\,\alpha|A|^2\left(\frac{\pi}{2\alpha}\right)^{1/2} + \frac{1}{2}\, m\omega^2|A|^2\frac{1}{4\alpha}\left(\frac{\pi}{2\alpha}\right)^{1/2}$$

Thus, we have

$$\langle H\rangle = \frac{\langle\phi|H|\phi\rangle}{\langle\phi|\phi\rangle} = \frac{\hbar^2\alpha}{2m} + \frac{m\omega^2}{8\alpha}$$

Here, α is a free parameter. On minimizing with respect to α, we get

$$0 = \frac{\partial\langle H\rangle}{\partial\alpha} = \frac{\hbar^2}{2m} - \frac{m\omega^2}{8\alpha^2}$$

It gives $\alpha = m\omega/2\hbar$. Using this value of α, we get

$$\langle H\rangle_{\text{min}} = \frac{\hbar^2}{2m}\frac{m\omega}{2\hbar} + \frac{m\omega^2}{8}\frac{2\hbar}{m\omega} = \frac{\hbar\omega}{2}$$

Since this value is the same as obtained earlier for one-dimensional harmonic oscillator. It shows that the trial wave function is the exact one.

2. Rayleigh-Ritz method

Here, we consider a trial wave function, which is a linear combination of basis functions Ψ_i, $i = 1, 2, \ldots, n$

$$\phi = \sum_{i=1}^{n} a_i|\Psi_i\rangle \tag{9.3}$$

where a_i's are real or complex coefficients. This method is due to Rayleigh and Ritz. Using equation (9.3) in (9.2), we get

$$E = \langle H \rangle = \frac{\langle \phi | H | \phi \rangle}{\langle \phi | \phi \rangle}$$

$$= \frac{\sum_i \sum_j a_i^* a_j \langle \Psi_i | H | \Psi_j \rangle}{\sum_i \sum_j a_i^* a_j \langle \Psi_i | \Psi_j \rangle} = \frac{\sum_i \sum_j a_i^* a_j H_{ij}}{\sum_i \sum_j a_i^* a_j S_{ij}} \tag{9.4}$$

where

$$H_{ij} = \langle \Psi_i | H | \Psi_j \rangle \qquad \text{and} \qquad S_{ij} = \langle \Psi_i | \Psi_j \rangle$$

For finding out the minimum value of E or $\langle H \rangle$, for convenience, we rearrange equation (9.4) in the following form.

$$E \sum_i \sum_j a_i^* a_j S_{ij} = \sum_i \sum_j a_i^* a_j H_{ij} \tag{9.5}$$

Now, we differentiate this equation (9.5) with respect to a_k^*

$$\frac{\partial E}{\partial a_k^*} \sum_i \sum_j a_i^* a_j S_{ij} + E \sum_j a_j S_{kj} = \sum_j a_j H_{kj}$$

For the minimum value of energy E, we set

$$\frac{\partial E}{\partial a_k^*} = 0, \qquad k = 1, 2, \ldots, n$$

and obtain

$$\sum_{j=1}^{n} a_j (H_{kj} - E S_{kj}) = 0, \qquad k = 1, 2, \ldots, n \tag{9.6}$$

Now, we differentiate this equation (9.5) with respect to a_k

$$\frac{\partial E}{\partial a_k} \sum_i \sum_j a_i^* a_j S_{ij} + E \sum_i a_i^* S_{ik} = \sum_i a_i^* H_{ik}$$

For the minimum value of energy E, we set

$$\frac{\partial E}{\partial a_k} = 0, \qquad k = 1, 2, \ldots, n$$

and obtain

$$\sum_{i=1}^{n} a_i^* (H_{ik} - E S_{ik}) = 0, \qquad k = 1, 2, \ldots, n \tag{9.7}$$

Equation (9.7) is Hermitian adjoint of (9.6).

Equation (9.6) is a set of n simultaneous homogeneous linear equations in variables $a_1, a_2, \ldots, a_n$ and for a non-trivial solution of equation (9.6), we have

$$\begin{vmatrix} H_{11} - ES_{11} & H_{12} - ES_{12} & \ldots & H_{1n} - ES_{1n} \\ H_{21} - ES_{21} & H_{22} - ES_{22} & \ldots & H_{2n} - ES_{2n} \\ \vdots & \vdots & \vdots & \vdots \\ H_{n1} - ES_{n1} & H_{n2} - ES_{n2} & \ldots & H_{nn} - ES_{nn} \end{vmatrix} = 0$$

Expansion of this determinant gives nth degree polynomial whose solution gives n values for E (*i.e.*, n roots of the polynomial). The lowest value of E is the one closest to the ground state energy with the selected trial wave function.

3. Variation method for excited states

This variation method can be easily applied to the excited states as well, where we have to select trial wave functions which are orthogonal to the wave functions of the lower states. Suppose a trial wave function of the $(n+1)$th state be ϕ_{n+1}. It may be the first excited state, so that $n = 1$. Let the normalized wave functions of the lower states be Ψ_i, $i = 1, 2, \ldots, n$. These wave functions must obviously be orthogonal to ϕ_{n+1}, and we can have[2]

$$\phi_{n+1} = \sum_{i=n+1} a_i|\Psi_i\rangle$$

Thus, the average value of the Hamiltonian is

$$\langle H\rangle = \langle\phi_{n+1}|H|\phi_{n+1}\rangle = \sum_{i=n+1}\sum_{j=n+1} a_i^* a_j \langle\Psi_i^*|H|\Psi_j\rangle$$

$$= \sum_{i=n+1}\sum_{j=n+1} a_i^* a_j \langle\Psi_i^* E_j \Psi_j\rangle$$

---

[2]If we express ϕ_{n+1} as

$$\phi_{n+1} = \sum_{i=1} a_i|\Psi_i\rangle$$

where $a_i = \langle\Psi_i|\phi_{n+1}\rangle$. Since $|\Psi_i\rangle$, $i = 1, 2, \ldots, n$ are orthogonal to ϕ_{n+1}, we have $a_i = 0$ for $i = 1, 2, \ldots, n$.

$$= \sum_{i=n+1} \sum_{j=n+1} a_i^* a_j E_j \langle \Psi_i^* | \Psi_j \rangle = \sum_{i=n+1} a_i^* a_i E_i$$

$$= |a_{n+1}|^2 E_{n+1} + |a_{n+2}|^2 E_{n+2} + \ldots$$

As

$$\sum_{i=n+1} |a_i|^2 = 1 \qquad \text{and} \qquad E_{n+k} > E_{n+1}, \qquad k = 2, 3, \ldots$$

we have

$$\langle H \rangle \geq E_{n+1}$$

Thus, the lowest value of $\langle H \rangle$ in the excited state is the energy of the excited state.

Trial wave functions

The trial wave functions, orthogonal to the lower states, can be obtained in the following manner. We first obtain the ground state variational function ϕ_1. Let us express ϕ_1 as

$$\phi_1 = a_1 |\Psi_1\rangle + a_2 |\Psi_2\rangle + a_3 |\Psi_3\rangle + \ldots$$

where the expansion coefficients are

$$a_i = \langle \Psi_i | \phi_1 \rangle$$

Since Ψ_1 is orthogonal to ϕ_2, we have

$$\begin{aligned} \phi_2 &= a_2 \, |\Psi_2\rangle + a_3 \, |\Psi_3\rangle + \ldots \\ &= \phi_1 - a_1 \, |\Psi_1\rangle = \phi_1 - \langle \Psi_1 | \phi_1 \rangle \, |\Psi_1\rangle \end{aligned}$$

With ϕ_2 as a trial wave function, we get $\langle H \rangle \geq E_2$. Since Ψ_1 and Ψ_2 are orthogonal to ϕ_3, we have

$$\begin{aligned} \phi_3 &= a_3 \, |\Psi_3\rangle + a_4 \, |\Psi_4\rangle + \ldots = \phi_1 - a_1 \, |\Psi_1\rangle - a_2 \, |\Psi_2\rangle \\ &= \phi_1 - \langle \Psi_1 | \phi_1 \rangle \, |\Psi_1\rangle - \langle \Psi_2 | \phi_1 \rangle \, |\Psi_2\rangle \end{aligned}$$

With ϕ_3 as a trial wave function, we get $\langle H \rangle \geq E_3$. This procedure goes on step by step and we get

$$\phi_{n+1} = \phi_1 - \sum_{i=1}^{n} \langle \Psi_i | \phi_1 \rangle \, |\Psi_n\rangle$$

4. Ground state of helium

Helium atom consists of a nucleus having two protons and two neutrons. The nucleus is considered as situated on the origin of the coordinate system and it carries a charge Ze. Let the two electrons 1 and 2 are at the position vectors $\vec{r_1}$ and $\vec{r_2}$, respectively (Figure 1). On neglecting the motion of the nucleus, the Hamiltonian of the system is

$$H = \Big(-\frac{\hbar^2}{2m}\nabla_1^2 - \frac{Ze^2}{r_1}\Big) + \Big(-\frac{\hbar^2}{2m}\nabla_2^2 - \frac{Ze^2}{r_2}\Big) + \frac{e^2}{r_{12}} \tag{9.8}$$

where ∇_1 and ∇_2 refer to the coordinates of 1 and 2 electrons, respectively. The last term e^2/r_{12} is the electrostatic repulsion between the two electrons. Denoting

$$H(1) = -\frac{\hbar^2}{2m}\nabla_1^2 - \frac{Ze^2}{r_1} \qquad \text{and} \qquad H(2) = -\frac{\hbar^2}{2m}\nabla_2^2 - \frac{Ze^2}{r_2}$$

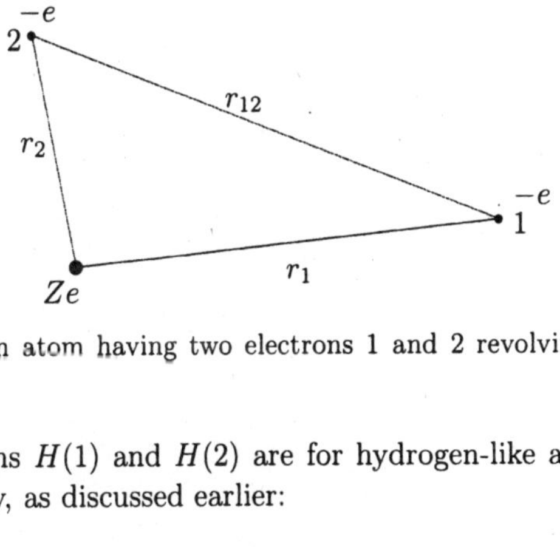

Figure 1: Helium atom having two electrons 1 and 2 revolving around the nucleus.

The Hamiltonians $H(1)$ and $H(2)$ are for hydrogen-like atom and can be solved exactly, as discussed earlier:

$$\Big(-\frac{\hbar^2}{2m}\nabla_1^2 - \frac{Ze^2}{r_1}\Big)\phi_1 = E_1\phi_1$$

where

$$E_1 = -\frac{Z^2me^4}{2\hbar^2} \qquad \phi_1 = \Big(\frac{Z^3}{\pi a_0^3}\Big)^{1/2}\exp\Big(-\frac{Zr_1}{a_0}\Big)$$

and

$$\Big(-\frac{\hbar^2}{2m}\nabla_2^2 - \frac{Ze^2}{r_2}\Big)\phi_2 = E_2\phi_2$$

where

$$E_2 = -\frac{Z^2 me^4}{2\hbar^2} \qquad\qquad \phi_2 = \left(\frac{Z^3}{\pi a_0^3}\right)^{1/2} \exp\left(-\frac{Zr_2}{a_0}\right)$$

Here, $a_0(\equiv \hbar^2/me^2)$ is the Bohr radius. The exact solution of the two hydrogenic Hamiltonian is

$$\phi = \phi_1(r_1)\phi_2(r_2) = \frac{Z^3}{\pi a_0^3} \exp\left[-\frac{Z}{a_0}(r_1 + r_2)\right] \tag{9.9}$$

Owing to the screening of attraction between one electron and the nucleus by the other electron, the nucleus charge experienced by the electron during attraction is not exactly $2e$, but is expected to be lesser. Hence, the effective atomic number Z' may be regarded as a free parameter and the function in equation (9.9) corresponding to the charge $Z'e$ may be used as a trial wave function. For convenience, we write the Hamiltonian (9.8) as

$$H = \left(-\frac{\hbar^2}{2m}\nabla_1^2 - \frac{Z'e^2}{r_1}\right) + \left(-\frac{\hbar^2}{2m}\nabla_2^2 - \frac{Z'e^2}{r_2}\right) + (Z' - Z)e^2\left(\frac{1}{r_1} + \frac{1}{r_2}\right) + \frac{e^2}{r_{12}}$$

Expectation value of H with the trial wave function (9.9), where Z is replaced by Z', is

$$\begin{aligned}
\langle H\rangle &= \left\langle \phi_1\phi_2 \left| -\frac{\hbar^2}{2m}\nabla_1^2 - \frac{Z'e^2}{r_1} \right| \phi_1\phi_2 \right\rangle + \left\langle \phi_1\phi_2 \left| -\frac{\hbar^2}{2m}\nabla_2^2 - \frac{Z'e^2}{r_2} \right| \phi_1\phi_2 \right\rangle \\
&\quad + (Z' - Z)\left\langle \phi_1\phi_2 \left| \frac{e^2}{r_1} \right| \phi_1\phi_2 \right\rangle + (Z' - Z)\left\langle \phi_1\phi_2 \left| \frac{e^2}{r_2} \right| \phi_1\phi_2 \right\rangle + \left\langle \phi_1\phi_2 \left| \frac{e^2}{r_{12}} \right| \phi_1\phi_2 \right\rangle \\
&= \left\langle \phi_1 \left| -\frac{\hbar^2}{2m}\nabla_1^2 - \frac{Z'e^2}{r_1} \right| \phi_1 \right\rangle \langle \phi_2 | \phi_2 \rangle + \left\langle \phi_2 \left| -\frac{\hbar^2}{2m}\nabla_2^2 - \frac{Z'e^2}{r_2} \right| \phi_2 \right\rangle \langle \phi_1 | \phi_1 \rangle \\
&\quad + (Z' - Z)\left\langle \phi_1 \left| \frac{e^2}{r_1} \right| \phi_1 \right\rangle \langle \phi_2 | \phi_2 \rangle + (Z' - Z)\left\langle \phi_2 \left| \frac{e^2}{r_2} \right| \phi_2 \right\rangle \langle \phi_1 | \phi_1 \rangle \\
&\quad + \left\langle \phi_1\phi_2 \left| \frac{e^2}{r_{12}} \right| \phi_1\phi_2 \right\rangle \\
&= \left\langle \phi_1 \left| -\frac{\hbar^2}{2m}\nabla_1^2 - \frac{Z'e^2}{r_1} \right| \phi_1 \right\rangle + \left\langle \phi_2 \left| -\frac{\hbar^2}{2m}\nabla_2^2 - \frac{Z'e^2}{r_2} \right| \phi_2 \right\rangle \\
&\quad + (Z' - Z)\left\langle \phi_1 \left| \frac{e^2}{r_1} \right| \phi_1 \right\rangle + (Z' - Z)\left\langle \phi_2 \left| \frac{e^2}{r_2} \right| \phi_2 \right\rangle + \left\langle \phi_1\phi_2 \left| \frac{e^2}{r_{12}} \right| \phi_1\phi_2 \right\rangle
\end{aligned} \tag{9.10}$$

The values of the first and second terms on the right-hand side are equal and each of them is $-Z^2W_H$, where $W_H = me^4/2\hbar^2$. Now,

$$\left\langle \phi_1 \left| \frac{e^2}{r_1} \right| \phi_1 \right\rangle = \frac{Z'^3 e^2}{\pi a_0^3} \int_\theta^{2\pi} \mathrm{d}\phi \int_0^\pi \sin\theta_1 \mathrm{d}\theta_1 \int_0^\infty r_1 \exp\left(-\frac{2Z' r_1}{a_0}\right) \mathrm{d}r_1$$

$$= \frac{Z'^3 e^2}{\pi a_0^3} 4\pi \left(\frac{a_0}{2Z'}\right)^2 = \frac{Z' e^2}{a_0} = 2Z' W_H$$

Similarly, we have

$$\left\langle \phi_2 \left| \frac{e^2}{r_2} \right| \phi_2 \right\rangle = 2Z'\, W_H$$

As we obtained in the last chapter (equation 8.42), we have

$$\left\langle \phi_1 \phi_2 \left| \frac{e^2}{r_{12}} \right| \phi_1 \phi_2 \right\rangle = \frac{5}{4} Z'\, W_H$$

Putting all these terms in equation (9.10), we get

$$\langle H \rangle = -2Z' W_H + 4(Z' - Z) Z'\, W_H + \frac{5}{4} Z'\, W_H \tag{9.11}$$

Minimizing $\langle H \rangle$ with respect to Z', we have

$$0 = -4Z' W_H + 8Z' W_H - 4Z W_H + \frac{5}{4} W_H$$

This equation gives

$$Z' = Z - \frac{5}{16}$$

For this value of Z', equation (9.11) gives

$$E = \langle H \rangle = -2\left(Z - \frac{5}{16}\right)^2 W_H$$

For $W_H = 13.6$ eV, the ground state energy for helium is -77.46 eV. The experimental value for ground state energy for helium is -78.98 eV.

5. Hydrogen molecule ion

Let us now evaluate the ground state energy of hydrogen molecule ion. It consists of an electron moving in the Coulomb filed of two electrons (Figure 2). The Hamiltonian is

$$H = -\frac{\hbar^2}{2m} \nabla^2 - e^2 \left(\frac{1}{r_1} + \frac{1}{r_2}\right) + \frac{e^2}{R}$$

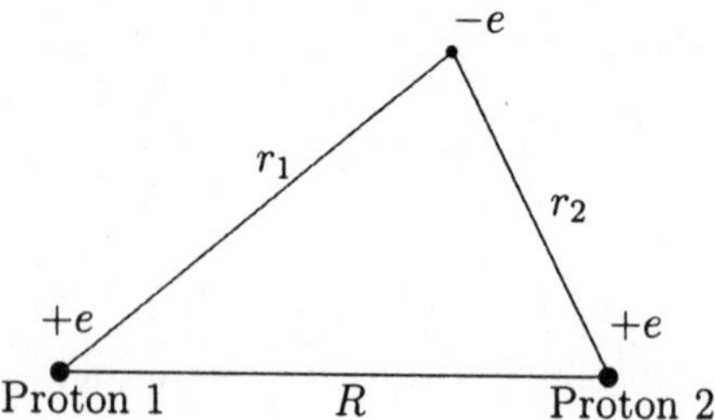

Figure 2: Hydrogen molecular ion having two protons and one electron.

The electron has the same probability of being associated with either of the protons, we can consider a trial wave function of the form

$$\phi = c_1\phi_1(r_1) + c_2\phi_2(r_2) \tag{9.12}$$

where c_1 and c_2 are free parameters (assumed to be real). Here, $\phi_1(r_1)$ is the atomic wave function when the electron is in the neighbourhood of proton 1 and $\phi_2(r_2)$ is the atomic wave function when the electron is in the neighbourhood of proton 2. The expressions for $\phi_1(r_1)$ and $\phi_2(r_2)$ are (ground state wave function for hydrogen atom)

$$\phi_1(r_1) = \frac{1}{\sqrt{\pi a_0^3}} \exp\Big(-\frac{r_1}{a_0}\Big)$$

$$\phi_2(r_2) = \frac{1}{\sqrt{\pi a_0^3}} \exp\Big(-\frac{r_2}{a_0}\Big)$$

This way of expressing the molecular wave functions as a linear combination of atomic wave functions, as in equation (9.12), is known as LACO method. The expectation value of H with the trial wave function ϕ is

$$\begin{aligned} E = \langle H \rangle &= \frac{\langle (c_1\phi_1 + c_2\phi_2)|H|(c_1\phi_1 + c_2\phi_2)\rangle}{\langle (c_1\phi_1 + c_2\phi_2)|(c_1\phi_1 + c_2\phi_2)\rangle} \\ &= \frac{c_1^2\langle \phi_1|H|\phi_1\rangle + c_1c_2\langle \phi_1|H|\phi_2\rangle + c_2c_1\langle \phi_2|H|\phi_1\rangle + c_2^2\langle \phi_2|H|\phi_2\rangle}{c_1^2 + c_2^2 + c_1c_2\langle \phi_1|\phi_2\rangle + c_2c_1\langle \phi_2|\phi_1\rangle} \end{aligned} \tag{9.13}$$

The integrals in equation (9.13) can be written in the form of the matrix elements

$$H_{11} = \langle \phi_1|H|\phi_1\rangle; \qquad H_{22} = \langle \phi_2|H|\phi_2\rangle$$
$$H_{12} = \langle \phi_1|H|\phi_2\rangle; \qquad H_{21} = \langle \phi_2|H|\phi_1\rangle$$

Here, H_{11} and H_{22} are equal and we denote them by α (*i.e.*, $H_{11} = H_{22} = \alpha$. Further, H_{12} and H_{21} are equal and we denote them by β (*i.e.*, $H_{12} = H_{21} = \beta$. Let us write

$$S_{ij} = \langle \phi_i | \phi_j \rangle$$

so that $S_{12} = S_{21} = S$ (say). Obviously, we have $S_{11} = S_{22} = 1$.

The integrals α are simply the energies of atomic wave functions ϕ_1 and ϕ_2 and are known as the Coulomb integrals. The integrals β and S are known as the resonance integral and the overlap integral. The overlap integral is a measure of amount by which $\phi_1(r_1)$ and $\phi_2(r_2)$ overlap with each other and it is equal to 0 or 1 depending on whether the two nuclei are infinitely separated from each other or are very close together.

The optimum values of the coefficients c_1 and c_2 can be obtained by minimizing $\langle H \rangle$ with respect to c_1 and c_2, separately. Thus, we have to solve the secular equations (equations 9.6 and 9.7) and which in the present case are

$$c_1(H_{11} - ES_{11}) + c_2(H_{12} - ES_{12}) = 0 \tag{9.14}$$

$$c_1(H_{12} - ES_{12}) + c_2(H_{22} - ES_{22}) = 0 \tag{9.15}$$

Remember that c_1 and c_2 are assumed to be real. Since $S_{11} = S_{22} = 1$ and $S_{12} = S_{21} = S$, for a nontrivial solution, we have

$$\begin{vmatrix} \alpha - E & \beta - ES \\ \beta - ES & \alpha - E \end{vmatrix} = 0$$

On expending this determinant, we have

$$E^2(1 - S^2) + 2E(\beta S - \alpha) + (\alpha^2 - \beta^2) = 0$$

This is a quadratic equation and its roots are

$$E_+ = \frac{\alpha + \beta}{1 + S} \qquad \text{and} \qquad E_- = \frac{\alpha - \beta}{1 - S}$$

Using E_+ and E_- in equation (9.14) or (9.15), we get

$$c_1 = c_2 \quad \text{for } E_+ \qquad \text{and} \qquad c_1 = -c_2 \quad \text{for } E_-$$

The wave functions are

$$\phi_+ = c_1(\phi_1 + \phi_2) \qquad \text{and} \qquad \phi_- = c_2(\phi_1 - \phi_2)$$

for E_+ and E_-, respectively. The normalization condition gives

$$\phi_+ = \frac{\phi_1 + \phi_2}{\sqrt{2+2S}} \qquad \text{and} \qquad \phi_- = \frac{\phi_1 - \phi_2}{\sqrt{2-2S}}$$

Now, the α is

$$\alpha = \left\langle \phi_1 \right| - \frac{\hbar^2}{2m}\nabla^2 - \frac{e^2}{r_1}\left|\phi_1\right\rangle - \left\langle \phi_1 \left| \frac{e^2}{r_2} \right| \phi_1 \right\rangle + \left\langle \phi_1 \left| \frac{e^2}{R} \right| \phi_1 \right\rangle$$

Since the nuclear repulsion term does not depend on the electronic coordinates, the last term gives e^2/R. The first term gives $-W_H$ as it is simply the ground state energy of hydrogen atom. Expressing

$$V_{aa} = \left\langle \phi_1 \left| \frac{e^2}{r_2} \right| \phi_1 \right\rangle$$

we obtain

$$\alpha = -W_H - V_{aa} + \frac{e^2}{R}$$

Now, the β is

$$\beta = \left\langle \phi_1 \right| - \frac{\hbar^2}{2m}\nabla^2 - \frac{e^2}{r_1}\left|\phi_2\right\rangle - \left\langle \phi_1 \left| \frac{e^2}{r_2} \right| \phi_2 \right\rangle + \left\langle \phi_1 \left| \frac{e^2}{R} \right| \phi_2 \right\rangle$$

It can be expressed as

$$\beta = \left\langle \phi_1 \right| - \frac{\hbar^2}{2m}\nabla^2 - \frac{e^2}{r_1}\left|\phi_1\right\rangle\left\langle \phi_1 | \phi_2 \right\rangle - \left\langle \phi_1 \left| \frac{e^2}{r_2} \right| \phi_2 \right\rangle + \left\langle \phi_1 \left| \frac{e^2}{R} \right| \phi_1 \right\rangle\left\langle \phi_1 | \phi_2 \right\rangle$$

Expressing

$$V_{ab} = \left\langle \phi_1 \left| \frac{e^2}{r_2} \right| \phi_2 \right\rangle$$

we obtain

$$\beta = -W_H S - V_{ab} + \frac{e^2}{R}S$$

Substituting the values of α and β, we get

$$E_+ = -W_H + \frac{e^2}{R} - \frac{V_{aa} + V_{ab}}{1+S}$$

and

$$E_- = -W_H + \frac{e^2}{R} - \frac{V_{aa} - V_{ab}}{1-S}$$

The integral V_{aa} and V_{ab} are both positive with $V_{aa} > V_{ab}$ and therefore $E_+ < E_-$. The lower energy orbital of the form $\phi_1 + \phi_2$ is known as the bonding orbital whereas the higher energy orbital of the form $\phi_1 - \phi_2$ is an antibonding orbital.

6. Ground state of deuteron

The deuteron has a nucleus comprising of one proton and one neutron of almost equal mass m. In the nucleus these two particles are held together by an attractive short-range force (nuclear force). This nuclear force between two nucleons can be expressed by the Yukawa potential

$$V(r) = -V_0 + \frac{e^{-r/\beta}}{r/\beta}$$

where V_0 and β are constants. The binding energy of deuteron is the ground energy of this two-particle system. In the center of mass system, the Hamiltonian for the s state is

$$H = -\frac{\hbar^2}{2\mu}\nabla^2 + V(r)$$

where μ is the proton-neutron reduced mass which is equal to $m/2$ and r is the proton-neutron separation. The ground state energy may be estimated with the help of a trial wave function

$$\phi = \exp\left(-\frac{\alpha r}{\beta}\right)$$

where α is a free parameter. Since the trial function depends only on r, ∇^2 in spherical polar coordinates contains only the radial derivative

$$\nabla^2 \equiv \frac{1}{r^2}\frac{d}{dr}\left(r^2\frac{d}{dr}\right) = \frac{d^2}{dr^2} + \frac{2}{r}\frac{d}{dr}$$

Thus,

$$\begin{aligned}\langle H\rangle &= \frac{\langle\phi|H|\phi\rangle}{\langle\phi|\phi\rangle}\\ &= \left[-\frac{\hbar^2}{2\mu}\left\langle\phi\left|\frac{d^2}{dr^2}\right|\phi\right\rangle + \frac{\hbar^2}{2\mu}\left\langle\phi\left|\frac{2}{r}\frac{d}{dr}\right|\phi\right\rangle + \langle\phi|V|\phi\rangle\right]\Big/\langle\phi|\phi\rangle \qquad (9.16)\end{aligned}$$

In equation (9.16), the integration of angular parts gives

$$\int_0^{2\pi} d\phi = 2\pi \qquad \text{and} \qquad \int_0^{\pi} \sin\theta \, d\theta = 2$$

Thus, the angular part contributes a factor of 4π in all these integrals. Using the integral

$$\int_0^{\infty} x^n \, e^{-ax} dx = \frac{n!}{a^{n+1}}$$

let us now perform integrations one by one as the following.

$$\langle\phi|\phi\rangle = 4\pi\int_0^\infty r^2\exp\Big(-\frac{2\alpha r}{\beta}\Big)\mathrm{d}r = \frac{\pi\beta^3}{\alpha^3}$$

$$\Big\langle\phi\Big|\frac{\mathrm{d}^2}{\mathrm{d}r^2}\Big|\phi\Big\rangle = 4\pi\frac{\alpha^2}{\beta^2}\int_0^\infty r^2\exp\Big(-\frac{2\alpha r}{\beta}\Big)\mathrm{d}r = \frac{\pi\beta}{\alpha}$$

$$\Big\langle\phi\Big|\frac{2}{r}\frac{\mathrm{d}}{\mathrm{d}r}\Big|\phi\Big\rangle = -8\pi\frac{\alpha}{\beta}\int_0^\infty r\exp\Big(-\frac{2\alpha r}{\beta}\Big)\mathrm{d}r = \frac{2\pi\beta}{\alpha}$$

$$\langle\phi|V|\phi\rangle = -4\pi V_0\beta\int_0^\infty r\exp\Big(-\frac{2\alpha+1}{\beta}r\Big)\mathrm{d}r = -\frac{4\pi V_0\beta^3}{(2\alpha+1)^2} \tag{9.17}$$

Substitution of these integrals in equation (9.16) gives

$$E = \langle H\rangle = \frac{\hbar^2\alpha^2}{2\mu\beta^2} - \frac{4V_0\alpha^3}{(2\alpha+1)^2} \tag{9.18}$$

Minimizing $\langle H\rangle$ with respect to α, we have

$$0 = \frac{\hbar^2\alpha}{\mu\beta^2} - \frac{4V_0\alpha^2(2\alpha+3)}{(2\alpha+1)^3} \tag{9.19}$$

$$\frac{(2\alpha+1)^3}{\alpha(2\alpha+3)} = \frac{4\mu V_0\beta^2}{\hbar^2} \tag{9.20}$$

It gives α corresponding to the minimum energy. Using equation (9.20) in (9.18), we get the minimum value of energy is

$$E_{\text{min}} = -\frac{\hbar^2}{2\mu\beta^2}\frac{\alpha^2(2\alpha-1)}{2\alpha+3} \tag{9.21}$$

A relation between β and V_0 is obtained by substituting the experimental binding energy (E_{min}) of 2.226 MeV in equation (9.21). The relation is checked for a range $\beta \equiv 1.4 \times 10^{-13}$ cm, which corresponds to a meson mass of 270 m_e (m_e is the mass of electron). The agreement is found satisfactory.

7. Problems and questions

1. State and prove the variational principle.
2. Explain why the variational method always gives an upper limit for the ground state energy.

3. Calculate the ground state energy of a harmonic oscillator of mass m oscillating with frequency ω by using the trial function

$$\phi = \begin{cases} \cos\left(\pi x/2a\right) & -a \le x \le a \\ 0 & |x| > a \end{cases}$$

4. Optimize the wave function $\phi = A\ x\ e^{-\alpha x}$, where A and α are constants, and calculate the ground state energy of a particle of mass m moving in the potential

$$V(x) = \begin{cases} kx & x > 0 \\ \infty & x < 0 \end{cases}$$

where k is a constant.

5. Discuss briefly the variational method for excited states.

6. Discuss the Rayleigh-Ritz method for obtaining the lowest value of energy.

7. Apply the variation method to ground state of helium.

8. Apply the variation method to hydrogen molecule ion.

9. Apply the variation method to ground state of deuteron.

10. Write notes on the following

    (i) Variational method for ground state

    (i) Variational method for excited states

WKB Approximation

The WKB method, also known as the WKB approximation, is a technique designed by Wentzel, Kramers and Brillouin (WKB), which plays a magnificent role for solving one-dimensional Schrödinger equation for a slowly varying potential. In this method, we expand the wave function in powers of $\hbar$. The coefficients of zero, first and second powers of $\hbar$, respectively, give the classical theory, quantum theory and higher-order results. It therefore shows a connection between the classical theory, quantum theory and quantum mechanics. It is also known as the classical approximation, as it produces the classical results when $\hbar$ is set to zero.

1. WKB approximation

One-dimensional Schrödinger equation for a particle moving in a region of constant potential V_0 is

$$\frac{\mathrm{d}^2\psi}{\mathrm{d}x^2} + k^2\psi = 0, \qquad \text{where} \qquad k^2 = \frac{2m}{\hbar^2}(E - V_0) \tag{10.1}$$

Solution of this equation is

$$\psi = \mathrm{e}^{\pm ikx}$$

When the given potential is not constant, but varies with x, the parameter k in equation (10.1) is a function of x as

$$k^2 = \frac{2m}{\hbar^2}[E - V(x)] \tag{10.2}$$

Now, there can be two situations: (i) $E > V(x)$, *i.e.*, the energy of the system is larger than the potential, and (ii) $E < V(x)$, *i.e.*, the energy of the system is smaller than the potential. Let us consider these cases one by one.

(i) Case of $E > V(x)$

For this case, the k is positive. Let us express the solution of equation (10.1) in the form (WKB approximation)

$$\psi = \exp\left[\frac{i}{\hbar}S(x)\right] \tag{10.3}$$

Using equation (10.3) in (10.1), we get[1]

$$\left(\frac{\mathrm{d}S}{\mathrm{d}x}\right)^2 - i\hbar\frac{\mathrm{d}^2S}{\mathrm{d}x^2} - k^2\hbar^2 = 0 \tag{10.4}$$

Solution of equation (10.4) gives a form of the function $S(x)$. In the WKB method, $S(x)$ is expressed in powers of $\hbar$ as

$$S(x) = S_0(x) + \hbar S_1(x) + \frac{\hbar^2}{2!}S_2(x) + \ldots \tag{10.5}$$

On differentiating equation (10.5), we have

$$\frac{\mathrm{d}S}{\mathrm{d}x} = \frac{\mathrm{d}S_0}{\mathrm{d}x} + \hbar\frac{\mathrm{d}S_1}{\mathrm{d}x} + \frac{\hbar^2}{2}\frac{\mathrm{d}S_2}{\mathrm{d}x} + \ldots \tag{10.6}$$

$$\frac{\mathrm{d}^2S}{\mathrm{d}x^2} = \frac{\mathrm{d}^2S_0}{\mathrm{d}x^2} + \hbar\frac{\mathrm{d}^2S_1}{\mathrm{d}x^2} + \frac{\hbar^2}{2}\frac{\mathrm{d}^2S_2}{\mathrm{d}x^2} + \ldots \tag{10.7}$$

Using equations (10.5), (10.6) and (10.7) in (10.4) and retaining terms up to $\hbar$, we get

$$\left(\frac{\mathrm{d}S_0}{\mathrm{d}x}\right)^2 - k^2\hbar^2 + \left(2\frac{\mathrm{d}S_0}{\mathrm{d}x}\frac{\mathrm{d}S_1}{\mathrm{d}x} - i\frac{\mathrm{d}^2S_0}{\mathrm{d}x^2}\right)\hbar = 0 \tag{10.8}$$

[1]Differentiating equation (10.3) with respect to x, we have

$$\frac{\mathrm{d}\psi}{\mathrm{d}x} = \frac{i}{\hbar}\frac{\mathrm{d}S}{\mathrm{d}x}\exp\left[\frac{i}{\hbar}S(x)\right]$$

Again differentiating with respect to x, we have

$$\frac{\mathrm{d}^2\psi}{\mathrm{d}x^2} = \left[-\frac{1}{\hbar^2}\left(\frac{\mathrm{d}S}{\mathrm{d}x}\right)^2\right]\exp\left[\frac{i}{\hbar}S(x)\right] + \frac{i}{\hbar}\frac{\mathrm{d}^2S}{\mathrm{d}x^2}\exp\left[\frac{i}{\hbar}S(x)\right]$$

Using these expressions in equation (10.1), we get

$$\left[-\frac{1}{\hbar^2}\left(\frac{\mathrm{d}S}{\mathrm{d}x}\right)^2\right]\exp\left[\frac{i}{\hbar}S(x)\right] + \frac{i}{\hbar}\frac{\mathrm{d}^2S}{\mathrm{d}x^2}\exp\left[\frac{i}{\hbar}S(x)\right] + k^2\exp\left[\frac{i}{\hbar}S(x)\right] = 0$$

On cancellation of common terms and rearranging, we get

$$\left(\frac{\mathrm{d}S}{\mathrm{d}x}\right)^2 - i\hbar\frac{\mathrm{d}^2S}{\mathrm{d}x^2} - k^2\hbar^2 = 0$$

The term $k^2\hbar^2$ is independent of $\hbar$, since $k^2\hbar^2 = 2m[E - V(x)]$. Since equation (10.8) is a polynomial in $\hbar$, it can only be satisfied when coefficient of each power of $\hbar$ vanishes separately. Then we have

$$\left(\frac{dS_0}{dx}\right)^2 - k^2\hbar^2 = 0 \quad \text{or} \quad \frac{dS_0}{dx} = \pm k\hbar \tag{10.9}$$

and

$$2\,\frac{dS_0}{dx}\frac{dS_1}{dx} - i\,\frac{d^2S_0}{dx^2} = 0 \tag{10.10}$$

For better results, one has to account also for the terms corresponding to the coefficients of $\hbar^2$. Integration of equation (10.9) gives

$$S_0(x) = \pm\int k\hbar\ dx = \pm\int [2m(E - V(x)]^{1/2}\ dx \tag{10.11}$$

Further, we have

$$\frac{dS_0}{dx} = \pm k\hbar \quad \text{and} \quad \frac{d^2S_0}{dx^2} = \pm\hbar\,\frac{dk}{dx} \tag{10.12}$$

Using equations (10.11) and (10.12) in (10.10), we get

$$\frac{dS_1}{dx} = \frac{i}{2k}\frac{dk}{dx}$$

Integration of this equation gives

$$S_1 = \frac{i}{2}\ln k \quad \text{or} \quad iS_1 = \ln k^{-1/2} \quad \text{or} \quad \exp(iS_1) = k^{-1/2} \tag{10.13}$$

Using equations (10.5) and (10.13) in (10.3) and retaining the terms up to $\hbar$, we have

$$\psi = A\exp\left(\frac{i}{\hbar}S\right) = A\exp\left(\frac{i}{\hbar}S_0\right)\exp(iS_1)$$

$$= \frac{A}{k^{\frac{1}{2}}}\exp\left(\pm\, i\int k\ dx\right) \tag{10.14}$$

where A is a constant and k is given by equation (10.2). A general solution is a linear combination of these two terms, one with each sign. As the wavenumber k is real, the solution is oscillatory. The probability density is

$$\psi\psi^* = \frac{|A|^2}{k}\exp\left(\pm\, i\int k\ dx\right)\exp\left(\mp\, i\int k\ dx\right) = \frac{|A|^2}{k}$$

Since $p = \hbar\, k$, the probability density is inversely proportional to the velocity. This is understandable as classically the time spent by a particle in a region is inversely proportional to its velocity.

(ii) Case of $E < V(x)$

Contrary to classical mechanics, in quantum mechanics a particle can even penetrate the regions where $E < V(x)$. The basic equation for this case is

$$\frac{\mathrm{d}^2\psi}{\mathrm{d}x^2} - \gamma^2\psi = 0, \quad \text{where} \quad \gamma^2 = \frac{2m}{\hbar^2}[V(x) - E] \tag{10.15}$$

Following the similar steps, as for the case $E > V(x)$ discussed above, equation (10.15) has a solution

$$\psi = \frac{B}{\gamma^{1/2}} \exp\Big(\pm\int\gamma\ \mathrm{d}x\Big) \tag{10.16}$$

where B is a constant. The plus sign gives an exponentially increasing term whereas the minus sign gives an exponentially decreasing term. A general solution is a linear combination of these two terms, one with each sign. In the region $E \approx V(x)$, both k as well as γ tend to zero and therefore, the wave function ψ becomes infinite. Consequently, the WKB method fails. The point $E = V(x)$ is known as the classical turning point, as at this point a classical particle would stop and would start to move in the reverse direction.

2. Connection formulas

On one side of a turning point, we have an exponential solution (equation 10.16) whereas on the other side, the solution is oscillatory (equation 10.14). Near the turning point, $E \approx V(x)$, the WKB approximation breaks down. Hence, to get a complete solution, the wave functions near the turning point must be connected with the WKB solutions in the regions $E > V(x)$ and $E < V(x)$. In most of the cases, near a turning point, one is not interested in the exact solutions, but only in the connection formulas.

Let us consider a turning point at $x = x_1$ (Figure 1). A smoothly varying potential $V(x)$ can be expanded in a Taylor series as

$$V(x) = V(x_1) + \left(\frac{\partial V}{\partial x}\right)_{x=x_1}(x - x_1)$$

where higher order terms are neglected. This assumption is the same as approximating the potential by a straight line.[2] As $E = V(x_1)$, the Schrödinger equation near $x = x_1$ is

$$\frac{d^2\psi}{dx^2} - \frac{2m}{\hbar^2}\left(\frac{\partial V}{\partial x}\right)_{x=x_1}(x - x_1)\psi = 0 \tag{10.17}$$

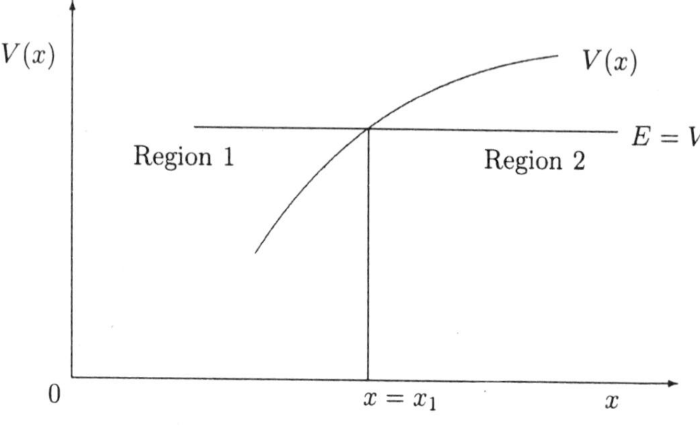

Figure 1: Turning point is at $x = x_1$

Now, the task is to work out independent solutions of equation (10.17) for the region $x < x_1$ (two solutions) and for the region $x > x_1$ (two solutions). The next step is to extend these solutions to the regions where WKB solutions are applicable and to obtain connection formulas. Such connection formulas were obtained by Kramers and are summarized in the following. Here, for each turning point, we have two connection formulas. We shall have two cases: in one case, the barrier would be right to the turning point x_1 and in the other case, the barrier would be left

[2]For straight line variation, we have

$$\left(\frac{\partial V}{\partial x}\right)_{x=x_1} = \frac{V(x) - V(x_1)}{x - x_1}$$

It gives

$$V(x) = V(x_1) + \left(\frac{\partial V}{\partial x}\right)_{x=x_1}(x - x_1)$$

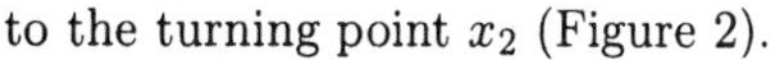

to the turning point x_2 (Figure 2).

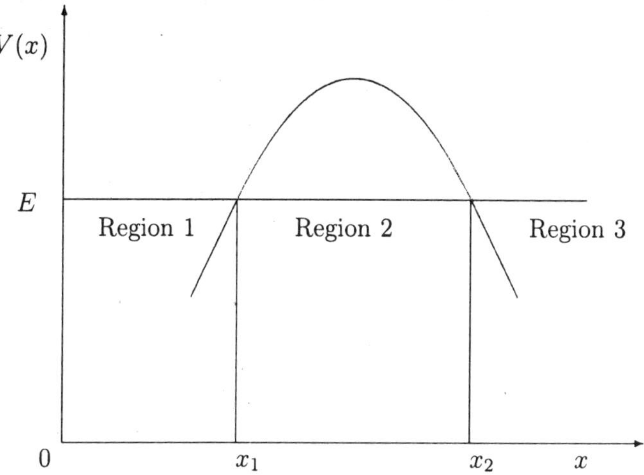

Figure 2: A particle of energy E penetrating a barrier

Barrier to the right of the turning point:

$$\frac{2}{k^{1/2}}\cos\left(\int_x^{x_1} k\, dx - \frac{\pi}{4}\right) \longleftarrow \frac{1}{\gamma^{1/2}}\exp\left(-\int_{x_1}^{x} \gamma\, dx\right) \tag{10.18}$$

$$\frac{1}{k^{1/2}}\sin\left(\int_x^{x_1} k\, dx - \frac{\pi}{4}\right) \longrightarrow -\frac{1}{\gamma^{1/2}}\exp\left(\int_{x_1}^{x} \gamma\, dx\right) \tag{10.19}$$

Barrier to the left of the turning point:

$$\frac{1}{\gamma^{1/2}}\exp\left(-\int_x^{x_2} \gamma\, dx\right) \longrightarrow \frac{2}{k^{1/2}}\cos\left(\int_{x_2}^{x} k\, dx - \frac{\pi}{4}\right) \tag{10.20}$$

$$-\frac{1}{\gamma^{1/2}}\exp\left(\int_x^{x_2} \gamma\, dx\right) \longleftarrow \frac{1}{k^{1/2}}\sin\left(\int_{x_2}^{x} k\, dx - \frac{\pi}{4}\right) \tag{10.21}$$

Here, in both the cases the cosine term is connected to the decreasing exponential whereas the sine term is connected to the increasing exponential.[3] Because of the negative sign, the increasing exponential term tends to minus infinity. For accurate results, it is advisable to use the above connection relations in the directions indicated by the arrows. However, in most of the case, the error involved in not following the arrow is fairly small. Further, the WKB approximation breaks down when the turning points are close to the top of the barrier.

[3]There are other set of these relations. However, those also can be converted to the present one.

3. Criterion for validity of WKB approximation

While deriving the solution (10.14), we neglected the terms beyond S_1 in the expansion in equation (10.5). This approximation is possible only when

$$\frac{\hbar S_1}{S_0} \ll 1 \qquad \text{or} \qquad \hbar\frac{\mathrm{d}S_1/\mathrm{d}x}{\mathrm{d}S_0/\mathrm{d}x} \ll 1$$

After using the values of S_0 and S_1, this condition reduces to

$$\left|\frac{\mathrm{d}k/\mathrm{d}x}{2k^2}\right| \ll 1, \qquad \text{where} \qquad k^2 = \frac{2m}{\hbar^2}\,[E - V(x)]$$

Thus, the WKB approximation method is valid when k is large and $\mathrm{d}k/\mathrm{d}x$ is small. Hence, this approximation is valid for a system having large mass, high energy and slowly varying potential. For the de Broglie wavelength $\lambda = h/p = 2\pi/k$, this condition can also be expressed as

$$\left(\frac{\lambda}{2\pi}\right)\frac{1}{k}\left|\frac{\mathrm{d}k}{\mathrm{d}x}\right| \ll 1$$

This form of the condition says that the fractional change in k in the distance $\lambda/2\pi$ must be small as compared to unity.

4. Barrier penetration

One of the important applications of the WKB method is to discuss the problem of barrier penetration by particles. Figure 2 represents a barrier, with peak energy V_0, on which particles of energy $E < V_0$ are incident from left. Since energy E of each particle is smaller than the peak potential V_0 of the barrier, classically all the particles will be reflected back at $x = x_1$. However, because of wave nature of particles, some of them may penetrate through the barrier, resulting in a transmitted wave. In the region 3, only transmitted wave exists which can be represented by

$$\psi_3 = \frac{A}{k^{1/2}}\exp\left[i\left(\int_{x_2}^{x} k\,\mathrm{d}x - \frac{\pi}{4}\right)\right] \tag{10.22}$$

In order to apply the connection formulas, equation (10.22) can be rearranged as

$$\psi_3 = \frac{A}{k^{1/2}}\left[\cos\left(\int_{x_2}^{x} k\,\mathrm{d}x - \frac{\pi}{4}\right) + i\,\sin\left(\int_{x_2}^{x} k\,\mathrm{d}x - \frac{\pi}{4}\right)\right]$$

Using equations (10.20) and (10.21), the wave function for the region $x_1 < x < x_2$ is

$$\psi_2 = \frac{A}{2\gamma^{1/2}} \exp\Big(- \int_x^{x_2} \gamma \, dx\Big) - \frac{Ai}{\gamma^{1/2}} \exp\Big(- \int_x^{x_2} \gamma \, dx\Big) \qquad (10.23)$$

For getting the wave function in the region $x < x_1$, we have to apply the connection formulas (10.18) and (10.19) for which equation (10.23) can be rearranged to the following form.

$$\psi_2 = \frac{A}{2\gamma^{1/2}} \exp\Big(-\int_{x_1}^{x_2} \gamma \, dx + \int_{x_1}^{x} \gamma \, dx\Big) - \frac{Ai}{\gamma^{1/2}} \exp\Big(\int_{x_1}^{x_2} \gamma \, dx - \int_{x_1}^{x} \gamma \, dx\Big)$$

After applying the connections formulas (10.18) and (10.19), the wave function in region 1 can be written as

$$\psi_1 = -\frac{A}{2k^{1/2}\alpha} \sin\Big(\int_x^{x_1} k \, dx - \frac{\pi}{4}\Big) - \frac{2Ai}{k^{1/2}\alpha} \cos\Big(\int_x^{x_1} k \, dx - \frac{\pi}{4}\Big) \quad (10.24)$$

where

$$\alpha = \exp\Big(\int_{x_1}^{x_2} \gamma \, dx\Big)$$

On replacing the sine and cosine terms by exponentials, we get[4]

$$\begin{aligned} \psi_1 = & -\frac{Ai}{k^{\frac{1}{2}}}\Big(\alpha - \frac{1}{4\alpha}\Big)\exp\Big[i\Big(\int_x^{x_1} k \, dx - \frac{\pi}{4}\Big)\Big] \\ & -\frac{Ai}{k^{\frac{1}{2}}}\Big(\alpha + \frac{1}{4\alpha}\Big)\exp\Big[-i\Big(\int_x^{x_1} k \, dx - \frac{\pi}{4}\Big)\Big] \end{aligned} \qquad (10.25)$$

By using the rule

$$\exp\Big[i\Big(\int_x^{x_1} k \, dx - \frac{\pi}{4}\Big)\Big] = \exp\Big[-i\Big(\int_{x_1}^{x} k \, dx + \frac{\pi}{4}\Big)\Big]$$

---

[4] Let us consider

$$y = \int_x^{x_1} k \, dx - \frac{\pi}{4}$$

Thus, equation (10.24) becomes

$$\begin{aligned} \psi_1 &= -\frac{A}{2k^{\frac{1}{2}}\alpha}\Big[\frac{e^{iy} - e^{-iy}}{2i}\Big] - \frac{2Ai\alpha}{k^{\frac{1}{2}}}\Big[\frac{e^{iy} + e^{-iy}}{2}\Big] \\ &= e^{iy}\Big[\frac{Ai}{4k^{\frac{1}{2}}\alpha} - \frac{Ai\alpha}{k^{\frac{1}{2}}}\Big] + e^{-iy}\Big[-\frac{Ai}{4k^{\frac{1}{2}}\alpha} - \frac{Ai\alpha}{k^{\frac{1}{2}}}\Big] \\ &= -\frac{Ai}{k^{\frac{1}{2}}}\Big(\alpha - \frac{1}{4\alpha}\Big)e^{iy} - \frac{Ai}{k^{\frac{1}{2}}}\Big(\alpha + \frac{1}{4\alpha}\Big)e^{-iy} \end{aligned}$$

equation (10.25) can be rearranged as

$$\psi_1 = -\frac{Ai}{k^{\frac{1}{2}}}\Big(\alpha - \frac{1}{4\alpha}\Big)\exp\Big[-i\Big(\int_{x_1}^{x} k\ \mathrm{d}x + \frac{\pi}{4}\Big)\Big]$$

$$-\frac{Ai}{k^{\frac{1}{2}}}\Big(\alpha + \frac{1}{4\alpha}\Big)\exp\Big[i\Big(\int_{x_1}^{x} k\ \mathrm{d}x + \frac{\pi}{4}\Big)\Big] \qquad (10.26)$$

The first term on he right side of equation (10.26) represents a wave moving in the negative x-direction whereas the second term represents a wave moving in the positive x-direction. Obviously, the one traveling in the positive direction is the incident wave and the other in the negative direction is the reflected wave. Now, we can calculate the reflection coefficient R and the transmission coefficient T in the following way.

$$\text{Amplitude of the incident wave} = -\frac{Ai}{k^{\frac{1}{2}}}\Big(\alpha + \frac{1}{4\alpha}\Big)$$

$$\text{Amplitude of the reflected wave} = -\frac{Ai}{k^{\frac{1}{2}}}\Big(\alpha - \frac{1}{4\alpha}\Big)$$

$$\text{Amplitude of the transmitted wave} = \frac{A}{k^{\frac{1}{2}}}$$

Then, we have

$$T = \frac{|\text{Amplitude of transmitted wave}|^2}{|\text{Amplitude of incident wave}|^2} = \frac{1}{\Big(\alpha + \frac{1}{4\alpha}\Big)^2}$$

$$R = \frac{|\text{Amplitude of reflected wave}|^2}{|\text{Amplitude of incident wave}|^2} = \frac{\Big(\alpha - \frac{1}{4\alpha}\Big)^2}{\Big(\alpha + \frac{1}{4\alpha}\Big)^2}$$

Now,

$$R + T = \frac{\Big(\alpha - \frac{1}{4\alpha}\Big)^2}{\Big(\alpha + \frac{1}{4\alpha}\Big)^2} + \frac{1}{\Big(\alpha + \frac{1}{4\alpha}\Big)^2} = \frac{\Big(\alpha + \frac{1}{4\alpha}\Big)^2}{\Big(\alpha + \frac{1}{4\alpha}\Big)^2} = 1$$

For a broad, high barrier $\alpha >> 1$ and we have

$$T \cong \frac{1}{\alpha^2} = \exp\Big(-2\int_{x_1}^{x_2} \gamma\ \mathrm{d}x\Big) \qquad (10.27)$$

When the barrier has a square shape, *i.e.*, the potential V is constant, then γ is constant and equation (10.27) reduces to

$$T \cong \exp[-2\gamma(x_2 - x_1)]$$

This phenomenon of barrier tunneling is very important as it explains the emission of α-particles from a nucleus.

5. Alpha emission

Theoretical explanation of α-particle emission by a nucleus is an important example of barrier tunneling. Typical variation of potential versus the distance from the center of the nucleus is shown in Figure 3. Classically no alpha-particle can go out of the nucleus. However, quantum tunneling provides an explanation for the emission of alpha-particle from a nucleus. The alpha particle attraction r_1 is approximately equal to the nucleus radius r_0. Denoting the distance at which the alpha-particle energy E is equal to the Coulomb potential by r_2, we get for an α-particle with zero angular momentum:

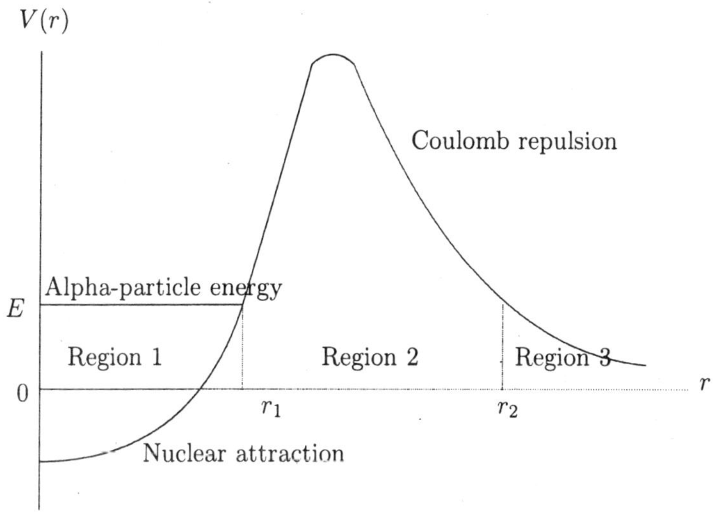

Figure 3: Typical variation of potential of alpha-particle as a function of its distance from the center of the nucleus.

$$r_2 = \frac{2Ze^2}{E}$$

where Z is the nuclear charge of the residual nucleus. For evaluation of the transmission coefficient, equation (10.27), we require the value of the integral $\int \gamma dr$, where $\gamma^2 = 2m(V(x) - E)/\hbar^2$, between the limits r_0 and r_2.

$$\int_{r_0}^{r_2} \gamma \, dr = \left(\frac{2m}{\hbar^2}\right)^{1/2} \int_{r_0}^{r_2} \left(\frac{2Ze^2}{r} - E\right)^{1/2} dr \qquad (10.28)$$

For convenience, we introduce a new variable ρ defined as[5]

$$\cos\rho = \left(\frac{rE}{2Ze^2}\right)^{\frac{1}{2}}$$

[5]We have

Thus, equation (10.28) becomes

$$\int_{r_0}^{r_2} \gamma dr = \left(\frac{2m}{\hbar^2}\right)^{1/2} \frac{2Ze^2}{E^{1/2}} \left[\cos^{-1}\left(\frac{r_0 E}{2Ze^2}\right)^{1/2} - \left(\frac{r_0 E}{2Ze^2}\right)^{\frac{1}{2}} \left(1 - \frac{r_0 E}{2Ze^2}\right)^{1/2}\right]$$

The transmission coefficient T is given by

$$T = \exp\left(-2\int_{r_0}^{r_2} \gamma \, dr\right)$$

6. Bound states in a potential well

Let us now apply WKB approximation to find the bound states of a particle moving in a one-dimensional potential well (Figure 4). For $E < 0$, according to the classical mechanics, the particle will oscillate back and forth between the turning points x_1 and x_2 where $E = V(x)$. However, the situation is different in quantum mechanics. Our requirement is that the wave function ψ must be finite and for that we should have only exponentially decreasing wave functions in the region $x < x_1$ (region 1) and $x > x_2$ (region 3). During the motion from the region 2 to region 1, the turning point is to the left of the classical region (barrier to the left of the turning point) and during the motion from region 2 to region

---

$$-\sin\rho \, d\rho = \sqrt{\frac{E}{2Ze^2}} \frac{1}{2r^{1/2}} \, dr$$

When $r = r_0$, we have $\rho = \cos^{-1}[\sqrt{r_0 E/2Ze^2}]$ and when $r = r_1$, we have $\rho = 0$. Thus,

$$\int_{r_0}^{r_2}\left(\frac{2Ze^2}{r} - E\right)^{1/2} dr = \int_{r_0}^{r_2}\sqrt{\frac{2Ze^2}{r}}\left[1 - \frac{rE}{2Ze^2}\right]^{1/2} dr$$

$$= \frac{4Ze^2}{E^{\frac{1}{2}}}\int_0^{\cos^{-1}\left[\sqrt{r_0 E/2Ze^2}\right]} (1-\cos^2\rho)^{1/2} \sin\rho \, d\rho = \frac{4Ze^2}{E^{1/2}}\int_0^{\cos^{-1}\left[\sqrt{r_0 E/2Ze^2}\right]} \sin^2\rho \, d\rho$$

$$= \frac{4Ze^2}{E^{\frac{1}{2}}}\int_0^{\cos^{-1}\left[\sqrt{r_0 E/2Ze^2}\right]} \frac{1-\cos 2\rho}{2} \, d\rho = \frac{4Ze^2}{E^{1/2}}\left[\frac{\rho}{2} - \frac{\sin 2\rho}{4}\right]_0^{\cos^{-1}\left[\sqrt{r_0 E/2Ze^2}\right]}$$

$$= \frac{2Ze^2}{E^{\frac{1}{2}}}\left[\rho - \cos\rho\sqrt{1-\cos^2\rho}\right]_0^{\cos^{-1}\left[\sqrt{r_0 E/2Ze^2}\right]}$$

$$= \frac{2Ze^2}{E^{1/2}}\left[\cos^{-1}\left[\sqrt{\frac{r_0 E}{2Ze^2}}\right] - \sqrt{\frac{r_0 E}{2Ze^2}}\sqrt{1 - \frac{r_0 E}{2Ze^2}}\right]$$

3, the turning point is to the right of the classical region (barrier to the right of the turning point).

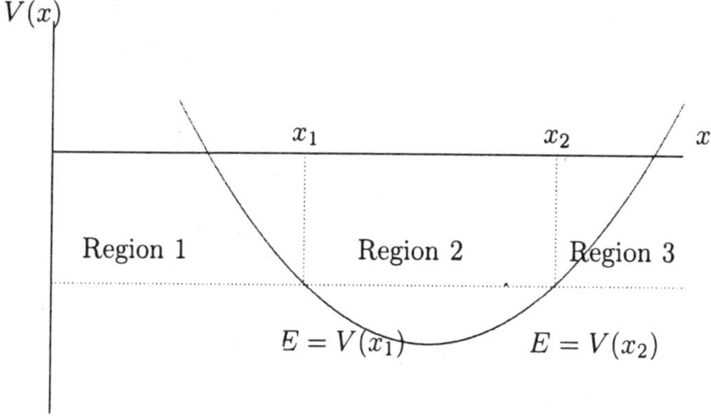

Figure 4: A potential well with turning points at x_1 and x_2.

Now, unnormalized exponentially decreasing wave function in region 1 can be written as

$$\psi_1 = \frac{1}{\gamma^{\frac{1}{2}}} \exp\Big(-\int_x^{x_1} \gamma \, dx\Big), \qquad x < x_1.$$

where

$$\gamma^2 = \frac{2m[V(x) - E]}{\hbar^2}$$

To obtain the wave function in region 2, we now apply the connections formulas (10.20) and (10.21) and obtain the wave function in region 2.

$$\psi_2 = \frac{2}{k^{1/2}} \cos\Big(\int_{x_1}^{x} k \, dx - \frac{\pi}{4}\Big), \qquad x_1 < x < x_2$$

where

$$k^2 = \frac{2m[E - V(x)]}{\hbar^2} \tag{10.29}$$

According to our requirement, the wave function in region 3 is also an exponentially decaying one. Following the connections formulas (10.18) and (10.19), the wave function which connects region 2 with the exponentially decaying wave function in region 3 is

$$\cos\Big(\int_{x_1}^{x_2} k \, dx - \frac{\pi}{4}\Big)$$

The wave function in region 2 can therefore be rearranged as

$$\psi_2 = \frac{2}{k^{1/2}} \cos\left(\int_{x_1}^{x_2} k\mathrm{d}x + \int_{x_2}^{x} k\ \mathrm{d}x - \frac{\pi}{4}\right) \tag{10.30}$$

Accounting for the properties $\cos(-\theta) = \cos\theta$ and $\sin(-\theta) = -\sin\theta$, equation (10.30) can be written as

$$\begin{aligned}\psi_2 &= \frac{2}{k^{1/2}} \cos\left(\int_{x_1}^{x_2} k\ \mathrm{d}x\right) \cos\left(\int_{x_1}^{x_2} k\ \mathrm{d}x + \frac{\pi}{4}\right) \\ &\quad + \frac{2}{k^{1/2}} \sin\left(\int_{x_1}^{x_2} k\ \mathrm{d}x\right) \sin\left(\int_{x}^{x_2} k\ \mathrm{d}x + \frac{\pi}{4}\right) \\ &= -\frac{2}{k^{1/2}} \cos\left(\int_{x_1}^{x_2} k\ \mathrm{d}x\right) \sin\left(\int_{x}^{x_2} k\ \mathrm{d}x - \frac{\pi}{4}\right) \\ &\quad + \frac{2}{k^{1/2}} \sin\left(\int_{x_1}^{x_2} k\ \mathrm{d}x\right) \cos\left(\int_{x}^{x_2} k\ \mathrm{d}x - \frac{\pi}{4}\right)\end{aligned} \tag{10.31}$$

From comparison of equation (10.31) with (10.18) and (10.19), it follows that the second term of equation (10.31) is the one that connects with the decreasing exponential in region 3 whereas the first term connects the increasing exponential. Since an increasing exponential in region 3 is not acceptable, the first term of equation (10.31) must vanish. This can be possible only when

$$\cos \int_{x_1}^{x_2} k\ \mathrm{d}x = 0 \qquad \text{or} \qquad \int_{x_1}^{x_2} k\ \mathrm{d}x = \left(n + \frac{1}{2}\right)\pi \tag{10.32}$$

where $n = 0, 1, 2, \ldots$. Substituting the value of k from equation (10.29) in (10.32), we have

$$\left(\frac{2m}{\hbar^2}\right)^{\frac{1}{2}} \int_{x_1}^{x_2} [E - V(x)]^{\frac{1}{2}}\ \mathrm{d}x = \left(n + \frac{1}{2}\right)\pi \tag{10.33}$$

where $n = 0, 1, 2, \ldots$. This equation gives the allowed energy values, and shows that it is possible to connect the wave functions in the regions 2 and 3 only for certain discrete values of energy.

Classical expression for linear momentum is $p = \sqrt{2m[E - V(x)]}$ and therefore equation (10.33) can be expressed as

$$2\int_{x_1}^{x_2} p\ \mathrm{d}x = \left(n + \frac{1}{2}\right)h \tag{10.34}$$

where $n = 0, 1, 2, \ldots$. This equation can be rearranges as

$$\int_{x_1}^{x_2} p\ \mathrm{d}x + \int_{x_2}^{x_1} -p\ \mathrm{d}x = \left(n + \frac{1}{2}\right)h \tag{10.35}$$

where $n = 0, 1, 2, \ldots$. Since the linear momentum is a vector quantity, the left side of this equation (10.35) is the integral of linear momentum over a complete cycle of motion (from x_1 to x_2 and back from x_2 to x_1) and therefore equation (10.34) can be expressed as

$$\oint p \, dx = \left(n + \frac{1}{2}\right)h$$

where $n = 0, 1, 2, \ldots$. It is simply the Sommerfeld's condition for quantization with half integer rather than integer quantum numbers.

Exercise 1. Obtain energy values of a one-dimensional harmonic oscillator of mass m and oscillating with a frequency ω with the help of the WKB approximation.

Solution: For a one-dimensional harmonic oscillator of mass m and oscillating with a frequency ω, the potential energy $V(x)$ is

$$V(x) = \frac{1}{2} m\omega^2 x^2$$

The classical turning points are those at which the energy is equal to the potential energy ($V(x) = E$). That is

$$\frac{1}{2} m\omega^2 x^2 = E$$

Thus, we have

$$x_1 = -\left(\frac{2E}{m\omega^2}\right)^{1/2} \qquad \text{and} \qquad x_2 = \left(\frac{2E}{m\omega^2}\right)^{1/2}$$

Energy of the oscillator is

$$E = \frac{p^2}{2m} + \frac{1}{2} m\omega^2 x^2$$

This equation gives the momentum

$$p = \left[2m\left(E - \frac{1}{2} m\omega^2 x^2\right)\right]^{1/2}$$

Now, the condition

$$2\int_{x_1}^{x_2} p \, dx = \left(n + \frac{1}{2}\right)h$$

gives

$$2\int_{x_1}^{x_2} \left[2m\left(E - \frac{1}{2} m\omega^2 x^2\right)\right]^{1/2} dx = \left(n + \frac{1}{2}\right)h$$

Introducing a new variable

$$\sin\theta = \left(\frac{m\omega^2}{2E}\right)^{1/2} x$$

we get

$$2\int_{-\pi/2}^{\pi/2} (2mE)^{1/2}\left(\frac{2E}{m\omega^2}\right)^{1/2} \cos^2\theta \; d\theta = \left(n+\frac{1}{2}\right)h$$

$$\frac{4E}{\omega}\int_{-\pi/2}^{\pi/2} \cos^2\theta \; d\theta = \left(n+\frac{1}{2}\right)h$$

$$\frac{4E}{\omega}\,\frac{\pi}{2} = \left(n+\frac{1}{2}\right)h \qquad \text{or} \qquad E = \left(n+\frac{1}{2}\right)\hbar\omega$$

7. Problems and questions

1. Discuss about the WKB approximation.
2. What do you mean by a classical turning point? How can one get the turning points for the known potential.
3. Write down the connection formulas and explain them.
4. Discuss about the validity criterion for the WKB approximation.
5. Explain why the WKB approximation is valid for a system in which the potential is slowly varying.
6. A particle is constrained to move between a pair of classical turning points in a potential well. How the energies are obtained in the WKB approximation.
7. Write notes on the following

    (i) WKB approximation

    (ii) Connection formulas

    (iii) Criterion for validity of WKB method

Time-dependent Perturbation

In a physical system, time dependence of Hamiltonian may be small, slowly varying or rapidly varying. These three situations correspond to the perturbation, adiabatic and sudden approximations, respectively. In the adiabatic case, one expects on physical grounds that solutions of Schrödinger equation can be approximated by means of stationary eigen functions of the instantaneous Hamiltonian.

We know that when Hamiltonian H of a system is independent of time, the time-dependent Schrödinger equation

$$i\hbar \frac{\partial \Psi}{\partial t} = \Big[-\frac{\hbar^2}{2m} \nabla^2 + V(\vec{r}) \Big] \Psi$$

could be separated into spatial and time coordinates, and the eigenfunctions of the equation were of the form

$$\Psi(\vec{r}, t) = \psi(\vec{r}) \exp\Big(-\frac{iEt}{\hbar} \Big)$$

where E is the energy of the system and $\psi(\vec{r})$ satisfies the time-independent Schrödinger equation

$$\Big[-\frac{\hbar^2}{2m} \nabla^2 + V(\vec{r}) \Big] \psi(\vec{r}) = E\psi(\vec{r})$$

The position probability density $|\Psi|^2$ is obviously independent of time, *i.e.*, $|\Psi|^2 = |\psi|^2$. Further, the expectation values of an observable does not depend on the time t, as the operator corresponding to the observable is independent of time. Hence, we can say that in spite of the time-dependent Schrödinger equation, every thing done so far can be put rightly under quantum statics(*i.e.*, under the quantum mechanics which does not depend on time). In some systems, the Hamiltonian may depend on time (this situation may be referred to as the quantum dynamics). In case the Hamiltonian can be written as a sum of two parts: (i) The main part being independent of time and (ii) rest part being dependent of time. Such situations can be dealt under the time-dependent perturbation theory.

1. Introduction

Let us now consider a system for which the Hamiltonian depends on time, showing the absence of stationary states. Further, let the Hamiltonian of the system can be expressed as

$$H(\vec{r},t) = H^0(\vec{r}) + H'(\vec{r},t) \tag{11.1}$$

where $H^0(\vec{r})$ is the time-independent part of the Hamiltonian for which the Schrödinger equation can be solved exactly, and $H'(\vec{r},t)$ the time-dependent part of Hamiltonian, such that $H'(\vec{r},t) << H^0(\vec{r})$. (Remember that in the preceding chapter, the peeturbation part H' was time independent.) Here, our aim is to solve the time-dependent Schrödinger equation

$$i\hbar \frac{\partial}{\partial t}\Psi(\vec{r},t) = H(\vec{r},t)\ \Psi(\vec{r},t) \tag{11.2}$$

Let $\psi_n^0(\vec{r})$, $n = 1, 2, \ldots$ be the stationary state eigenfunctions of $H^0(\vec{r})$.

$$H^o(\vec{r})\ \psi_n^0(\vec{r}) = E_n^0\ \psi_n^0(\vec{r})$$

Thus, we have

$$\Psi_n^0(\vec{r},t) = \psi_n^0(\vec{r})\ \exp\Big(-\frac{iE_n^0 t}{\hbar}\Big) \qquad n = 1, 2, \ldots \tag{11.3}$$

so that

$$i\hbar \frac{\partial}{\partial t}\Psi_n^0(\vec{r},t) = H^o(\vec{r})\ \Psi_n^0(\vec{r},t) \qquad n = 1, 2, \ldots \tag{11.4}$$

In order to include $H'(\vec{r},t)$ in the Hamiltonian, let the wave functions $\Psi(\vec{r},t)$ can be expressed as a linear combination of the wavefunctions $\Psi_n^0(\vec{r},t)$

$$\Psi(\vec{r},t) = \sum_n c_n(t)\ \Psi_n^0(\vec{r},t) = \sum_n c_n(t)\ \psi_n^0(\vec{r})\ \exp\Big(-\frac{iE_n^0 t}{\hbar}\Big) \tag{11.5}$$

where $c_n(t)$ are expansion coefficients. This way of expressing the wavefunction $\Psi(\vec{r},t)$ is known as the adiabatic approximation. Using equations (11.1) and (11.5) in (11.2), we have

$$i\hbar \frac{\partial}{\partial t}\Big[\sum_n c_n(t)\ \psi_n^0(\vec{r})\ \exp\Big(-\frac{iE_n^0 t}{\hbar}\Big)\Big]$$

$$= (H^0 + H')\sum_n c_n(t)\ \psi_n^0(\vec{r})\ \exp\Big(-\frac{iE_n^0 t}{\hbar}\Big)$$

$$i\hbar \sum_n \frac{\partial c_n}{\partial t}\, \psi_n^0(\vec{r}) \exp\Big(-\frac{iE_n^0 t}{\hbar}\Big) + \sum_n c_n(t)\, \psi_n^0(\vec{r}) \exp\Big(-\frac{iE_n^0 t}{\hbar}\Big)\, E_n^0$$

$$= (H^0 + H') \sum_n c_n(t)\, \psi_n^0(\vec{r}) \exp\Big(-\frac{iE_n^0 t}{\hbar}\Big) \qquad (11.6)$$

Using equations (11.3) and (11.4) in (11.6), we get

$$i\hbar \sum_n \frac{\partial c_n}{\partial t}\, \psi_n^0(\vec{r}) \exp\Big(-\frac{iE_n^0 t}{\hbar}\Big) = H' \sum_n c_n(t)\, \psi_n^0(\vec{r}) \exp\Big(-\frac{iE_n^0 t}{\hbar}\Big)$$

Multiplying this equation from left by $\langle k|$ and writing this equation in the bra and ket notation for convenience, we get

$$i\hbar \sum_n \frac{\partial c_n}{\partial t} \langle k|n\rangle \exp\Big(-\frac{iE_n^0 t}{\hbar}\Big) = \sum_n c_n(t) \langle k|H'|n\rangle \exp\Big(-\frac{iE_n^0 t}{\hbar}\Big)$$

$$i\hbar\, \frac{\partial c_k}{\partial t} \exp\Big(-\frac{iE_k^0 t}{\hbar}\Big) = \sum_n c_n(t)\, \langle k|H'|n\rangle \exp\Big(-\frac{iE_n^0 t}{\hbar}\Big)$$

$$i\hbar\, \frac{\partial c_k}{\partial t} = \sum_n c_n(t)\, \langle k|H'|n\rangle \exp\Big(\frac{i(E_k^0 - E_n^0)t}{\hbar}\Big)$$

$$= \sum_n c_n(t)\, H'_{kn} \exp(i\omega_{kn} t) \qquad (11.7)$$

where

$$H'_{kn} = \langle k|H'|n\rangle \qquad \text{and} \qquad \omega_{kn} = \frac{E_k^0 - E_n^0}{\hbar}$$

The summation symbol in equation (11.7) stands for the summation over the discrete states as well as for the integration over the continuum states. This equation is exact as we have not yet made any assumption.

2. First-order perturbation

As we did in case of time-independent perturbation theory, we replace H' by $\lambda H'$ and express the coefficient $c_n(t)$ as a power series in λ as

$$c_n(t) = c_n^{(0)}(t) + \lambda\, c_n^{(1)}(t) + \lambda^2\, c_n^{(2)}(t) + \ldots \qquad (11.8)$$

Using equation (11.8) and replacing H' by $\lambda H'$ in (11.7), we have

$$i\hbar\Big(\frac{\partial c_k^{(0)}}{\partial t} + \lambda\, \frac{\partial c_k^{(1)}}{\partial t} + \lambda^2\, \frac{\partial c_k^{(2)}}{\partial t} + \ldots\Big)$$

$$= \sum_n \Big(c_n^{(0)} + \lambda\, c_n^{(1)} + \lambda^2\, c_n^{(2)} + \ldots\Big) \langle k|\lambda H'|n\rangle \exp\Big(\frac{i(E_k^0 - E_n^0)t}{\hbar}\Big)$$

$$= \sum_n \left(c_n^{(0)} + \lambda\, c_n^{(1)} + \lambda^2\, c_n^{(2)} + \ldots\right)\lambda H'_{kn} \exp(i\omega_{kn}t) \tag{11.9}$$

Since λ is a free parametre, equation (11.9) can only be satisfied when coefficients of each power of λ is zero individually. Thus, the coefficients of λ^0, λ^1 and λ^2 give

$$i\hbar\frac{\partial c_k^{(0)}}{\partial t} = 0 \tag{11.10}$$

$$i\hbar\frac{\partial c_k^{(1)}}{\partial t} = \sum_n c_n^{(0)} H'_{kn}\exp(i\omega_{kn}t) \tag{11.11}$$

$$i\hbar\frac{\partial c_k^{(2)}}{\partial t} = \sum_n c_n^{(1)} H'_{kn}\exp(i\omega_{kn}t)$$

Equation (11.10) shows that $c_k^{(0)}$ is constant in time. It can be consistent with the fact that the zero-order Hamiltonian does not depend on time. Now, from equation (11.11), we have

$$c_k^{(1)}(t) = \frac{1}{i\hbar}\int \sum_n c_n^{(0)} H'_{kn}(\vec{r}, t)\exp(i\omega_{kn}t)\ \mathrm{dt}$$

Suppose the system initially ($t = 0$) is in the unperturbed state m, so that all $c_n^{(0)}$'s, except $c_m^{(0)}$, are zero and $c_m^{(0)}$ is unity. Thus, we have

$$c_k^{(1)}(t) = \frac{1}{i\hbar}\int_0^t H'_{km}(\vec{r}, t)\ \exp(i\omega_{km}t')\ \mathrm{d}t' \tag{11.12}$$

The parametre H' has introduced transitions to other states and after a time t, the probability of transition to a state k is given by $\left|c_k^{(1)}(t)\right|^2$.

Exercise 1. A constant perturbation $H'(r)$ is applied suddenly to a system in an unperturbed state n for a time t. Calculate the probability for transition of the system from the state n to k and show that the probability varies harmonically with an angular frequency $\omega = (E_k - E_n)/2\hbar$ and amplitude $4|H'_{kn}|^2/(E_k - E_n)^2$.

Solution: When a system goes suddenly from a state n to k under the action of a constant perturbation $H'(r)$, the coefficient of the first order perturbation is

$$c_k^{(1)}(t) = \frac{1}{i\hbar}\int_0^t H'_{kn}(r)\ \exp(i\omega_{kn}t')\ \mathrm{d}t'$$

where $H'_{kn}(r) = \langle k|H'(r)|n\rangle$ is the transition element. Since $H'(r)$ does not depend on time, we have

$$c_k^{(1)}(t) = \frac{H'_{kn}(r)}{i\hbar}\int_0^t \exp(i\omega_{kn}t')\,dt' = -\frac{H'_{kn}}{\hbar\omega_{kn}}\Big[\exp(i\omega_{kn}t) - 1\Big]$$

$$= -\frac{H'_{kn}}{\hbar\omega_{kn}}\exp\Big(i\omega_{kn}\frac{t}{2}\Big)\Big[\exp\Big(\frac{i\omega_{kn}t}{2}\Big) - \exp\Big(-\frac{i\omega_{kn}t}{2}\Big)\Big]$$

$$= -\frac{2iH'_{kn}}{\hbar\omega_{kn}}\exp\Big(\frac{i\omega_{kn}t}{2}\Big)\sin\Big(\frac{\omega_{kn}t}{2}\Big)$$

Thus, the probability for the transition from the state n to k is

$$|c_k^{(1)}(t)|^2 = \frac{4|H'_{kn}|^2}{\hbar^2\omega_{kn}^2}\sin^2\Big(\frac{\omega_{kn}t}{2}\Big)$$

It shows that the probability varies harmonically and has a

$$\text{angular frequency} = \frac{\omega_{kn}}{2} = \frac{E_k - E_n}{2\hbar}$$

and

$$\text{amplitude} = \frac{4|H'_{kn}|^2}{\hbar^2\omega_{kn}^2} = \frac{4|H'_{kn}|^2}{(E_k - E_n)^2}$$

3. Harmonic perturbation

Let us apply these general results to a system such as an atom or a molecule exposed to a harmonic perturbation (electromagnetic radiation). Further, we consider only the discrete states in the system. A harmonic perturbation with an angular frequency ω is

$$H'(\vec{r},t) = 2H_1(\vec{r})\cos(\omega t) = H_1[\exp(i\omega t) + \exp(-i\omega t)]$$

Thus, we have

$$\begin{aligned} H'_{km} &= \langle k|H_1(\vec{r})\ [\exp(i\omega\ t) + \exp(-i\omega\ t)]|m\rangle \\ &= \langle k|H_1|m\rangle\ [\exp(i\omega\ t) + \exp(-i\omega\ t)] \end{aligned} \tag{11.13}$$

Using equation (11.13) in (11.12), we have

$$c_k^{(1)}(t) = \frac{1}{i\hbar}\int_0^t \langle k|H_1|m\rangle\ [\exp(i\omega\ t) + \exp(-i\omega\ t)]\exp(i\omega_{km}t')\,dt'$$

$$= \frac{\langle k|H_1|m\rangle}{i\hbar}\int_0^t [\exp\{i(\omega_{km} + \omega)t\} + \exp\{i(\omega_{km} - \omega)t\}]\,dt'$$

$$= -\frac{\langle k|H_1|m\rangle}{\hbar}\left\{\frac{\exp[i(\omega_{km}+\omega)t]-1}{\omega_{km}+\omega}+\frac{\exp[i(\omega_{km}-\omega)t]-1}{\omega_{km}-\omega}\right\} \quad (11.14)$$

The first term on right side of equation (11.14) has a maximum value when

$$\omega_{km}+\omega=0 \qquad \frac{E_k^0-E_m^0}{\hbar}+\omega=0 \qquad E_k^0=E_m^0-\hbar\omega$$

The effect of harmonic perturbation with angular frequency ω is to get energy $\hbar\omega$ from the system. This corresponds to the stimulated emission or induced emission. Now, the second term on right side of equation (11.14) has a maximum value when

$$\omega_{km}-\omega=0 \qquad \frac{E_k^0-E_m^0}{\hbar}-\omega=0 \qquad E_k^0=E_m^0+\hbar\omega$$

Here, the effect of harmonic perturbation is to give energy $\hbar\omega$ to the system. This corresponds to the induced absorption or simply absorption.

Since only one term in equation (11.14) is important at one time, let us consider the second term. Now equation (11.14) is

$$c_k^{(1)}(t) = -\frac{\langle k|H_1|m\rangle}{\hbar}\,\frac{\exp[i(\omega_{km}-\omega)t]-1}{\omega_{km}-\omega}$$

$$= -\frac{\langle k|H_1|m\rangle}{\hbar}\,\frac{\exp[i(\omega_{km}-\omega)t/2]}{\omega_{km}-\omega}\left\{\exp\left[\frac{i(\omega_{km}-\omega)t}{2}\right] - \exp\left[\frac{i(\omega_{km}-\omega)t}{2}\right]\right\}$$

$$= -\frac{2i\langle k|H_1|m\rangle}{\hbar}\cdot\frac{\exp[i(\omega_{km}-\omega)t/2]}{\omega_{km}-\omega}\sin\frac{(\omega_{km}-\omega)t}{2}$$

When the system moves from the state n (at $t=0$) to a state k, the probability $P_{n\to k}(t)$ of finding the system in the discrete state k at a time t is $|c_k^{(1)}|^2$

$$P_{n\to k}(t)=|c_k^{(1)}(t)|^2=\frac{4|\langle k|H_1|m\rangle|^2}{\hbar^2}\,\frac{\sin^2[(\omega_{km}-\omega)t/2]}{(\omega_{km}-\omega)^2}$$

An interesting feature here that the transition probability oscillates sinusoidally as a function of time. Variation of the probability $P_{n\to k}(t)$ versus $(\omega_{km}-\omega)$ is shown in Figure 1. The central peak has a height of

$|\langle k|H_1|m\rangle|^2 t^2/\hbar^2|$ and a width of about $2\pi/t$ and therefore the area under the main peak is proportional to the time t. Thus, the probability of finding the system in the state k is proportional to t. Since the probability of transition is inversely proportional to $(\omega_{km} - \omega)^2$, the probability of finding the system in a state whose energy differs considerably from the initial state is smaller than the probability of finding the system in a neighbouring state.

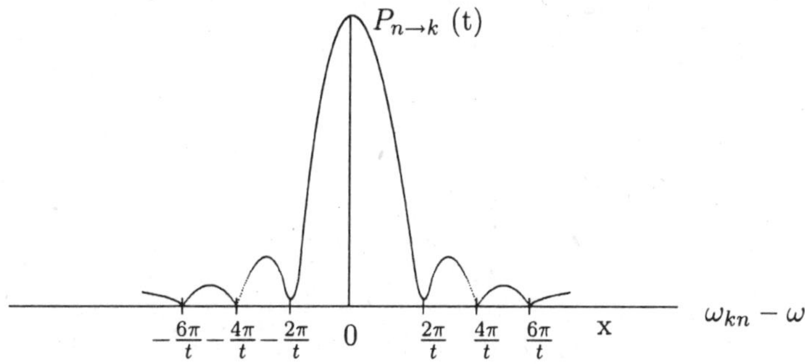

Figure 1: Variation of the probability $P_{n\to k}(t)$ versus $(\omega_{km} - \omega)$

4. Transitions to continuum states

So far we have accounted for the transitions between two discrete states. We now consider the transitions from a discrete state n to a continuum of states lying around the energy E_k, where the density of the states is $\rho(E_k)$. Let us denote these states by k. The number of states in the energy range from E_k to $(E_k + \mathrm{d}E_k)$ is therefore $\rho(E_k)\mathrm{d}E_k$ and the total probability of transition from the discrete state n to the continuum around E_k is

$$P(t) = \frac{4}{\hbar^2} \int_{\mathrm{d}E_k} |H'_{kn}|^2 \, \frac{\sin^2[(\omega_{kn} - \omega)t/2]}{(\omega_{kn} - \omega)^2} \, \rho(E_k) \, \mathrm{d}E_k \tag{11.15}$$

With the time t, the height of the main peak increases, but its width deceases. When t is large, the width of the main peak becomes small and only a limited number of final states contribute in this integral. Consequently, H'_{kn} and $\rho(E_k)$ may be treated as constant over the range of the integral.

For convenience, let us introduce a new variable x expressed as

$$x = \frac{(\omega_{kn} - \omega)t}{2} = \left(\frac{E_k - E_n}{\hbar} - \omega\right)\frac{t}{2}; \qquad \mathrm{d}x = \frac{t}{2\hbar}\,\mathrm{d}E_k$$

as the energy E_n is of the initial (fixed) state. Now, equation (11.15) gives

$$P(t) = \frac{2t}{\hbar}|H'_{kn}|^2 \ \rho(E_k) \int \frac{\sin^2 x}{x^2} \mathrm{d}x \tag{11.16}$$

The transition from the state E_n to a state E_K can be on either side of E_n. Though the integral is over a small range of $(\omega_{kn} - \omega)$, but since the integrand is small outside the actual range, the limits on x can be extended over the region from $-\infty$ to $+\infty$. Now, the integral is a standard one, for which we have

$$\int_{-\infty}^{\infty} \frac{\sin^2 x}{x^2} \ \mathrm{d}x = \pi \tag{11.17}$$

and equation (11.16) reduces to

$$P(t) = \frac{2\pi}{\hbar} t|H'_{kn}|^2 \ \rho(E_k)$$

Now, the number of transitions per unit time, called the transition probability, is usually denoted by the letter ω (This ω is different from the frequency ω of the harmonic peeturbation.)

$$\omega = \frac{2\pi}{\hbar}|H'_{kn}|^2 \rho(E_k) \tag{11.18}$$

This equation (11.18) is known as the 'Fermi's Golden Rule'. The transition probability is proportional to the square of the matrix element of the amplitude of the harmonic perturbing term between the states n and k and to the density of final states.

5. Absorption and emission of radiation

Description of absorption and emission of electromagnetic radiation by an atom or molecule requires the knowledge of quantization of electromagnetic radiation field. Without going for quantization of electromagnetic radiation field, here we shall discuss a semi-classical theory which would however provide a considerable insight into the phenomenon of absorption and emission.

5.1 Electromagnetic field

Let us consider two Maxwell equations

$$\nabla . \vec{B} = 0 \tag{11.19}$$

$$\nabla \times \vec{B} = \frac{1}{c}\frac{\partial \vec{E}}{\partial t} \tag{11.20}$$

From equation (11.19), using the identity that gradient of a curl of a vector quantity is zero, we have

$$\vec{B} = \nabla \times \vec{A} \tag{11.21}$$

where $\vec{A}$ is known as the vector potential. Using equation (11.21) in (11.20), we have

$$\nabla \times \vec{E} = -\frac{\partial}{\partial t}(\nabla \times \vec{A})$$

$$\nabla \times \left(\vec{E} + \frac{\partial \vec{A}}{\partial t} \right) = 0 \tag{11.22}$$

From equation (11.22), using the identity that curl of a gradient of a scalar is zero, we have

$$\vec{E} + \frac{\partial \vec{A}}{\partial t} = -\nabla\phi$$

where ϕ is known as the scalar potential. Thus, we have

$$\vec{E} - -\frac{\partial \vec{A}}{\partial t} - \nabla\phi \tag{11.23}$$

One may take $\phi = 0$ and $\vec{A}$ satisfies the wave equation

$$\nabla^2 \vec{A} - \frac{1}{c^2}\frac{\partial^2 \vec{A}}{\partial t^2} = 0 \tag{11.24}$$

Solution of equation (11.24) is

$$\vec{A}(\vec{r}, t) = \vec{A}_0(\omega) \exp[\pm i(\vec{k}.\vec{r} - \omega t)] \tag{11.25}$$

where $\vec{A}_0(\omega)$ does not depend on time, $\vec{k}$ is the propagation vector and ω the angular frequency. Substitution of equation (11.25) in (11.24) gives the dispersion relation $\omega = ck$. A general solution of equation (11.24) is

$$\vec{A}(\vec{r}, t) = \vec{A}_0(\omega)\left[\exp\{i(\vec{k}.\vec{r} - \omega t)\} + \exp\{-i(\vec{k}.\vec{r} - \omega t)\}\right]$$

For plane electromagnetic waves, we can select vector potential $\vec{A}$ and scalar potential ϕ such that

$$\nabla . \vec{A} = 0 \qquad \text{and} \qquad \phi = 0 \tag{11.26}$$

The electromagnetic field associated with such a plane wave is

$$\begin{aligned}\vec{E} &= -\frac{1}{c}\frac{\partial \vec{A}}{\partial t} \\ &= \vec{A}_0 \frac{\omega}{c} i \Big[\exp\{i(\vec{k} . \vec{r} - \omega t)\} - \exp\{-i(\vec{k} . \vec{r} - \omega t)\}\Big] \\ &= -2 \vec{A}_0 \frac{\omega}{c} \sin((\vec{k} . \vec{r} - \omega t)\end{aligned}$$

Thus, the average value of the electric field is

$$\bar{E}^2 = 4 \frac{\omega^2}{c^2} \mid \vec{A}_0 \mid^2 \frac{1}{T} \int_0^T \sin^2(\vec{k} . \vec{r} - \omega t)\, dt$$

where T is the time period and we have

$$\frac{1}{T} \int_0^T \sin^2(\vec{k} . \vec{r} - \omega t)\, dt = \frac{1}{2}$$

so that

$$\bar{E}^2 = 2 \frac{\omega^2}{c^2} \mid \vec{A}_0 \mid^2$$

The energy density $\rho(\omega)$ in an electromagnetic wave is

$$\rho(\omega) = \frac{\bar{E}^2}{2} = \frac{\omega^2}{c^2} \mid \vec{A}_0 \mid^2$$

An electromagnetic radiation is not strictly monochromatic, but has a small width in frequencies. Further, a radiation can be taken as the superposition of harmonic plane waves

$$\vec{A}(\vec{r}, t) = \int \vec{A}_0(\omega) \Big[\exp\{i(\vec{k} . \vec{r} - \omega t)\} + \exp\{-i(\vec{k} . \vec{r} - \omega t)\}\Big] d\omega \tag{11.27}$$

5.2 Charged particle in an electromagnetic field

The Hamiltonian H of a charged particle in an electromagnetic filed is

$$\begin{aligned}H &= \frac{1}{2m}\left(\vec{p} - \frac{e}{c}\vec{A}\right)^2 + e\phi \\ &= \frac{1}{2m}\left(\vec{p} . \vec{p} + \frac{e^2}{c^2} \mid \vec{A} \mid^2 - \frac{e}{c} \vec{p} . \vec{A} - \frac{e}{c} \vec{A} . \vec{p}\right) + e\phi\end{aligned}$$

where e is the charge on the particle. For the situation expressed by equation (11.26), the Hamiltonian operator is[1]

$$H = -\frac{\hbar^2}{2m}\nabla^2 + \frac{e^2}{2mc^2}|\vec{A}|^2 + \frac{ie\hbar}{mc}\vec{A}.\nabla$$

The second term on right side is very small in comparison to the other two terms, and we have

$$H = -\frac{\hbar^2}{2m}\nabla^2 + \frac{ie\hbar}{mc}\vec{A}.\nabla$$

After accounting for the potential energy $V(r)$, the Hamiltonian operator is

$$H = -\frac{\hbar^2}{2m}\nabla^2 + V(r) + \frac{ie\hbar}{mc}\vec{A}.\nabla = H^0 + H'$$

where the unperturbed Hamiltonian H^0 and the perturbation part H' are

$$H^0 = -\frac{\hbar^2}{2m}\nabla^2 + V(r) \qquad \text{and} \qquad H' = \frac{ie\hbar}{mc}\vec{A}.\nabla \qquad (11.28)$$

5.3 Electric dipole approximation

The wavelength of electromagnetic radiation lying in the visible as well as radio region is much larger than the size of the interacting system (atom, ion, molecule, etc., as their dimension is of the order Å). The smallest wavelength of radiation in the visible region is about $\lambda \approx 5 \times 10^{-7}$ m. Thus, we have $k = 2\pi/\lambda \approx 1.2 \times 10^7\ \text{m}^{-1}$. As the size of an atom (the smallest system) is $r \approx \times 10^{-10}$ m, we have $kr \approx 1.2 \times 10^{-3}$. For this value of kr, we have $\exp(\pm ik.r) \approx 1$ and equation (11.27) gives

$$\vec{A}(\vec{r},t) = \int \vec{A}_0(\omega)\Big[\exp(\omega t) + \exp(-i\omega t)\Big]d\omega \qquad (11.29)$$

[1]We have

$$(\vec{p}\,.\,\vec{A} + \vec{A}\,.\,\vec{p})\psi = -i\hbar\nabla.\,(\vec{A}\,\psi) - i\hbar\,\vec{A}\,.\nabla\psi$$
$$= -i\hbar\,\vec{A}\,.\,\nabla\psi - i\hbar\psi\,\nabla.\,\vec{A} - i\hbar\,\vec{A}\,.\,\nabla\psi$$

We have taken $\nabla.\,\vec{A} = 0$. Thus, we have

$$(\vec{p}\,.\,\vec{A} + \vec{A}\,.\,\vec{p})\psi = -2i\hbar\,\vec{A}\,.\,\nabla\psi$$

and therefore, we have

$$\vec{p}\,.\,\vec{A} + \vec{A}\,.\,\vec{p} = -2i\hbar\,\vec{A}\,.\,\nabla$$

Using equation (11.29) in the second of (11.28), we get

$$H' = \frac{ie\hbar}{mc}\int \left[\exp(i\omega t) + \exp(-i\omega t)\right] \vec{A}_0\,(\omega)\,.\nabla\, d\omega$$

$$= \frac{ie\hbar}{mc}\int \left[\exp(i\omega t) + \exp(-i\omega t)\right]\left|\,\vec{A}_0\,(\omega)\right|\,\nabla_A\, d\omega$$

where ∇_A is the component of ∇ along the polarization vector $\vec{A}_0\,(\omega)$.

We can now calculate the matrix element $H'_{kn}(t)$ of the perturbation H' between the sates k and n.

$$\begin{aligned} H'_{kn} &= \langle k|H'|n\rangle \\ &= \left\langle k\left|\frac{ie\hbar}{mc}\int [\exp(i\omega t) + \exp(-i\omega t)]\right|\vec{A}_0\,(\omega)\right|\nabla_A\, d\omega\left|n\right\rangle \\ &= \frac{ie\hbar}{mc}\langle k\,|\nabla_A|\,n\rangle\int[\exp(i\omega t) + \exp(-i\omega t)]A_0(\omega)\, d\omega \qquad (11.30)\end{aligned}$$

The matrix element $\langle k\,|\nabla_A|\,n\rangle$ can be transformed into the more familiar dipole moment matrix element in the following manner. The commutator of r with H^0 is

$$[r, H^0] = \left[r,\ \frac{p^2}{2m} + V\right] = \frac{1}{2m}\left[r, p^2\right] + \left[r, V\right]$$

Since $[r, V] = 0$, we have

$$[r, H^0] = \frac{1}{2m}\,[r, p.p] = \frac{p}{2m}\,[r, p] + \frac{1}{2m}\,[r, p]\,p$$

$$= \frac{i\hbar p}{m} = \frac{\hbar^2\nabla}{m} \qquad (11.31)$$

We have

$$\left\langle k\left|\left[r, H^0\right]\right|n\right\rangle = \left\langle k\left|rH^0\right|n\right\rangle - \left\langle k\left|H^0 r\right|n\right\rangle$$

Using equation (11.31) and the property that H^0 is Hermitian[2]

$$\frac{\hbar^2}{m}\langle k\,|\nabla|\,n\rangle = E_n^0\,\langle k\,|r|\,n\rangle - E_k^0\,\langle k\,|r|\,n\rangle$$

$$\langle k\,|\nabla|\,n\rangle = \frac{m}{\hbar^2}\left(E_n^0 - E_k^0\right)\langle k\,|r|\,n\rangle = -\frac{m}{\hbar}\,\omega_{kn}\,\langle k\,|r|\,n\rangle$$

$$\langle k\,|e\nabla_A|\,n\rangle = -\frac{m}{\hbar}\,\omega_{kn}\,\langle k\,|e\ r_A|\,n\rangle \qquad (11.32)$$

---

[2]For H^0 to be Hermitian, we have

$$\left\langle k\left|H^0 r\right|n\right\rangle = (\langle n\,|rH^0|\,k\rangle)^* = E_k(\langle n\,|r|\,k\rangle)^* = E_k\ \langle k\,|r|\,n\rangle$$

where r_A is the component of $\vec{r}$ in the direction of the polarization.

In the derivation of equation (11.32), we have assumed that $kr << 1$ and therefore have taken $\exp(ikr) \approx 1$. Since $e\,\vec{r}$ is an electric dipole moment, this approximation is equivalent to the replacement of the atom by an electric dipole. Therefore, this approximation is known as the 'electric dipole approximation'. The quantity $\langle k\,|er_A|\,n\rangle = \mu_{kn}$ is known as the transition dipole moment. It shows that when the wavelength λ of the electromagnetic radiation is quite large in comparison to the size of the interacting system, the radiation sees the system as a dipole. But when λ is small in comparison to the size of the system, the radiation sees the individual charges of the dipole and not the dipole as a whole. For that situation, the electric field at the individual charge will be different.

Substituting the value of $\langle k\,|\nabla_A|\,n\rangle$ in equation (11.30), we have

$$H'_{kn}(t) = -\frac{i\omega_{kn}}{c}\langle k\,|er_A|\,n\rangle \int [\exp(i\omega t) + \exp(-i\omega t)]\, A_0(\omega)\; d\omega \quad (11.33)$$

5.4 Transition probability

Since the factor $\langle k|er_A|n\rangle$ is the component of $\langle k|e\,\vec{r}\;|n\rangle$ along the polarization direction, we have

$$|\langle k\,|er_A|\,n\rangle|^2 = |\langle k\,|er|\,n\rangle|^2\;\cos^2\theta$$

where θ is the angle between the direction of polarization and $\langle k|e\,\vec{r}\;|n\rangle$. For isotropic behaviour of radiation, $\langle k|e\,\vec{r}\;|n\rangle$ is not restricted to a single plane as assumed. Considering all the directions, we have

$$|\langle k\,|er_A|\,n\rangle|^2 = \frac{1}{3}\left|\left\langle k\left|e\vec{r}\right|n\right\rangle\right|^2 = \frac{1}{3}|\langle k\,|\mu|\,n\rangle|^2 = \frac{1}{3}\,|\mu_{kn}|^2 \quad (11.34)$$

where

$$\mu_{kn} = \langle k\,|er|\,n\rangle$$

The probability that H' has induced a transition from the state n to k is given by $|c_k^{(1)}(t)|^2$. Using equation (11.33) in (11.12), we get

$$c_k^{(1)}(t) = -\frac{\omega_{kn}}{c\hbar}\int_0^t \langle k\,|er_A|\,n\rangle \int \left[e^{i\omega t'} + e^{-i\omega t'}\right] A_0(\omega)\; d\omega\; e^{i\omega_{kn}t'}\; dt'$$

The first term in the integral corresponds to the emission process whereas the second term corresponds to the absorption process.

(i) For the absorption process, we have

$$c_k^{(1)}(t) = -\frac{\omega_{kn}}{c\hbar} \langle k| er_A | n\rangle \int A_0(\omega) d\omega \int_0^t e^{i(\omega_{kn}-\omega)t'} dt'$$

$$= -\frac{\omega_{kn}}{ic\hbar} \langle k| er_A | n\rangle \int A_0(\omega) \frac{\exp[i(\omega_{kn}-\omega)t]-1}{\omega_{kn}-\omega} d\omega \quad (11.35)$$

The absorption has a large value when $\omega_{kn} - \omega = 0$ or $E_k = E_n + \hbar\omega$. In the absorption process, the electromagnetic field loses energy $\hbar\omega$ and the photon is lost. Squaring equation (11.35) and using equation (11.34) there, we get

$$|c_k^{(1)}(t)|^2 = \frac{4\omega_{kn}^2}{3c^2\hbar^2} |\mu_{kn}|^2 \int |A_0(\omega)|^2 \frac{\sin^2[(\omega_{kn}-\omega)t/2]}{(\omega_{kn}-\omega)^2} d\omega$$

The term $|A_0(\omega)|^2$ is fairly broad whereas the remaining factor in the integral is sharply peaked about $\omega = \omega_{kn}$. Thus, we can replace $|A_0(\omega)|^2$ by $|A_0(\omega_{kn})|^2$ and it can be taken outside the integral so that we have

$$|c_k^{(1)}(t)|^2 = \frac{4\omega_{kn}^2}{3c^2\hbar^2} |\mu_{kn}|^2 |A_0(\omega_{kn})|^2 \int \frac{\sin^2[(\omega_{kn}-\omega)t/2]}{(\omega_{kn}-\omega)^2} d\omega$$

In order to evaluate the integral, let introduce a new variable x defined as $x = (\omega - \omega_{kn})t/2$ so that

$$|c_k^{(1)}(t)|^2 = \frac{2t}{3\hbar^2} |\mu_{kn}|^2 \rho(\omega_{kn}) \int_{-\infty}^{\infty} \frac{\sin^2 x}{x^2} dx$$

where $\rho(\omega_{kn}) = \omega_{kn}^2 |A_0(\omega_{kn})|^2 / c^2$. The value of the integral is π as given in equation (11.17). Thus, we have

$$|c_k^{(1)}(t)|^2 = \frac{2\pi t}{3\hbar^2} |\mu_{kn}|^2 \rho(\omega_{kn})$$

The probability per unit time for the absorption is

$$\omega = \frac{2\pi}{3\hbar^2} |\mu_{kn}|^2 \rho(\omega_{kn}) \quad (11.36)$$

(ii) For the emission process, we have

$$c_k^{(1)}(t) = -\frac{\omega_{kn}}{c\hbar} \langle k| er_A | n\rangle \int A_0(\omega) d\omega \int_0^t e^{i(\omega_{kn}+\omega)t'} dt'$$

$$= \frac{\omega_{kn}}{ic\hbar} \langle k| er_A | n\rangle \int A_0(\omega) \frac{\exp[i(\omega_{kn}+\omega)t]-1}{\omega_{kn}+\omega} d\omega \quad (11.37)$$

The emission has a large value when $\omega_{kn}+\omega=0$ or $E_k=E_n-\hbar\omega$. In the emission process, the electromagnetic field gains energy $\hbar\omega$. Therefore, the output has two photons. Hence, there is amplification of the number of photons and it forms the basis for light amplification by stimulated emission of radiation (LASER). Squaring equation (11.37) and using equation (11.34) there, we get

$$|c_k^{(1)}(t)|^2 = \frac{4\omega_{kn}^2}{3c^2\hbar^2}\,|\mu_{kn}|^2 \int |A_0(\omega)|^2\, \frac{\sin^2[(\omega_{kn}+\omega)t/2]}{(\omega_{kn}+\omega)^2}\, d\omega$$

The term $|A_0(\omega)|^2$ is fairly broad whereas the remaining factor in the integral is sharply peaked about $\omega=-\omega_{kn}$. Thus, we can replace $|A_0(\omega)|^2$ by $|A_0(\omega_{kn})|^2$ and it can be taken outside the integral so that we have

$$|c_k^{(1)}(t)|^2 = \frac{4\omega_{kn}^2}{3c^2\hbar^2}\,|\mu_{kn}|^2\,|A_0(\omega_{kn})|^2 \int \frac{\sin^2[(\omega_{kn}+\omega)t/2]}{(\omega_{kn}+\omega)^2}\, d\omega$$

In order to evaluate the integral, let introduce a new variable x defined as $x=(\omega+\omega_{kn})t/2$ so that

$$|c_k^{(1)}(t)|^2 = \frac{2t}{3\hbar^2}\,|\mu_{kn}|^2\,\rho(\omega_{kn}) \int_{-\infty}^{\infty} \frac{\sin^2 x}{x^2}\, dx$$

where $\rho(\omega_{kn}) = \omega_{kn}^2\,|A_0(\omega_{kn})|^2/c^2$. The value of the integral is π as given in equation (11.17). Thus, we have

$$|c_k^{(1)}(t)|^2 = \frac{2\pi t}{3\hbar^2}\,|\mu_{kn}|^2\,\rho(\omega_{kn})$$

The probability per unit time for absorption is

$$\omega = \frac{2\pi}{3\hbar^2}\,|\mu_{kn}|^2\,\rho(\omega_{kn})$$

This expression is the same as equation (11.36). It shows that the probabilities of absorption and of emission are the same.

Exercise 2. Calculate square of the electric dipole moment $\langle 310|\mu|200\rangle$ for the hydrogen atom.

Solution: For hydrogen atom, we have

$$\psi_{200} = \left(\frac{1}{32\pi a_0^3}\right)^{1/2}\left(2-\frac{r}{a_0}\right)\exp\left(-\frac{r}{2a_0}\right)$$

and

$$\psi_{310} = \frac{4}{27(2\pi)^{1/2} a_0^{5/2}}\, r\left(1 - \frac{r}{6a_0}\right) \exp\left(-\frac{r}{3a_0}\right) \cos\theta$$

Substituting the expressions for ψ_{200} and ψ_{310}, we have

$$\langle 310|\mu_z|200\rangle = -e\langle 310|r\cos\theta|200\rangle$$

$$= -\frac{e}{54\pi a_0^4}\int_0^\infty \left(2r^4 - \frac{4r^5}{3a_0} + \frac{r^6}{6a_0^2}\right)\exp\left(-\frac{5r}{6a_0}\right)\, \mathrm{d}r \int_0^\pi \cos^2\theta\, \sin\theta\, \mathrm{d}\theta \int_0^{2\pi} \mathrm{d}\phi$$

We have

$$\int_0^{2\pi} \mathrm{d}\phi = 2\pi \qquad \text{and} \qquad \int_0^\pi \cos^2\theta\, \sin\theta\, \mathrm{d}\theta = \frac{2}{3}$$

$$\int_0^\infty r^4 \exp\left(-\frac{5r}{6a_0}\right)\, \mathrm{d}r = \frac{4!}{(5/6a_0)^5}$$

$$\int_0^\infty r^5 \exp\left(-\frac{5r}{6a_0}\right)\, \mathrm{d}r = \frac{5!}{(5/6a_0)^6}$$

$$\int_0^\infty r^6 \exp\left(-\frac{5r}{6a_0}\right)\, \mathrm{d}r = \frac{6!}{(5/6a_0)^7}$$

Using these integrals, we get

$$\langle 310|\mu_z|200\rangle = -\frac{e}{54\pi a_0^4}\frac{4\pi}{3}\left(\frac{4!\;2}{(5/6a_0)^5} - \frac{5!\;8}{6a_0(5/6a_0)^6} + \frac{6!}{6a_0^2(5/6a_0)^7}\right)$$

$$= -\frac{e}{54\pi a_0^4}\frac{4\pi}{3}\frac{144}{5}\left(\frac{6a_0}{5}\right)^5 = -\frac{32}{45}\left(\frac{6}{5}\right)^5 ea_0 = -1.77\; ea_0$$

Instead of calculating $\langle 310|\mu_x|200\rangle$ and $\langle 310|\mu_y|200\rangle$, for convenient we calculate

$$\langle 310|\mu_x \pm \mu_y|200\rangle = -e\langle 310|x \pm y|200\rangle = -e\langle 310|r\, \sin\theta\, \exp^{\pm i\phi}|200\rangle$$

Substituting the expressions for ψ_{200} and ψ_{310}, we have

$$\langle 310|\mu_x \pm \mu_y|200\rangle = -\frac{e}{54\pi a_0^4}\int_0^\infty \left(2r^4 - \frac{8r^5}{6a_0} + \frac{r^6}{6a_0^2}\right)\exp\left(-\frac{5r}{6a_0}\right)\, \mathrm{d}r$$

$$\times \int_0^\pi \cos\theta\, \sin^2\theta\, \mathrm{d}\theta \int_0^{2\pi} \mathrm{e}^{\pm i\phi}\mathrm{d}\phi$$

Since

$$\int_0^{2\pi} \mathrm{e}^{\pm i\phi}\mathrm{d}\phi = 0$$

we have

$$\langle 310|\mu_x \pm \mu_y|200\rangle = 0$$

and therefore, we have

$$\langle 310|\mu_x|200\rangle = 0 \qquad \text{and} \qquad \langle 310|\mu_y|200\rangle = 0$$

Finally, we have

$$|\langle 310|\mu|200\rangle|^2 = |\langle 310|\mu_x|200\rangle|^2 + |\langle 310|\mu_x|200\rangle|^2 + |\langle 310|\mu_z|200\rangle|^2$$

$$= 3.13\ e^2 a_0^2$$

Exercise 3. Show that there is no electric dipole transition between 1*s* and 2*s* states of the hydrogen atom.

Solution: For hydrogen atom, we have

$$\psi_{100} = \frac{1}{\sqrt{\pi a_0^3}} \exp\left(-\frac{r}{a_0}\right)$$

and

$$\psi_{200} = \left(\frac{1}{32\pi a_0^3}\right)^{1/2} \left(2 - \frac{r}{a_0}\right) \exp\left(-\frac{r}{2a_0}\right)$$

Substituting the expressions for ψ_{100} and ψ_{200}, we have

$$\langle 200|\mu_z|100\rangle = -e\langle 200|r\cos\theta|100\rangle$$

$$= -\frac{e}{4\sqrt{2}\ \pi a_0^3} \int_0^\infty \left(2r^3 - \frac{r^4}{a_0}\right) \exp\left(-\frac{3r}{2a_0}\right) \mathrm{d}r \int_0^\pi \cos\theta\ \sin\theta\ \mathrm{d}\theta \int_0^{2\pi} \mathrm{d}\phi$$

We have

$$\int_0^\pi \cos\theta\ \sin\theta\ \mathrm{d}\theta = 0$$

Therefore,

$$\langle 200|\mu_z|100\rangle = 0$$

Instead of calculating $\langle 200|\mu_x|100\rangle$ and $\langle 200|\mu_y|100\rangle$, for convenient we calculate

$$\langle 200|\mu_x \pm \mu_y|100\rangle = -e\langle 200|x \pm y|100\rangle = -e\langle 200|r\ \sin\theta\ \exp^{\pm i\phi}|100\rangle$$

Substituting the expressions for ψ_{100} and ψ_{200}, we have

$$\langle 200|\mu_x \pm \mu_y|100\rangle = -\frac{e}{4\sqrt{2}\,\pi a_0^3}\int_0^\infty \left(2r^3 - \frac{r^4}{a_0}\right)\exp\left(-\frac{3r}{2a_0}\right) dr$$

$$\times \int_0^\pi \sin^2\theta \, d\theta \int_0^{2\pi} e^{\pm i\phi} d\phi$$

Since

$$\int_0^{2\pi} e^{\pm i\phi} d\phi = 0$$

we have

$$\langle 310|\mu_x \pm \mu_y|200\rangle = 0$$

and therefore

$$\langle 310|\mu_x|200\rangle = 0 \qquad \text{and} \qquad \langle 310|\mu_y|200\rangle = 0$$

Finally, we have

$$|\langle 200|\mu|100\rangle|^2 = |\langle 200|\mu_x|100\rangle|^2 + |\langle 200|\mu_y|100\rangle|^2 + |\langle 200|\mu_z|100\rangle|^2 = 0$$

It shows that there is no dipole transition between $1s$ and $2s$ states of hydrogen atom.

6. Einstein A and B coefficients

In the last section, we discussed about the absorption and stimulated emission, induced by the external electromagnetic field. In order to account for all the contributions to the emission of radiation from an excited state, it was found necessary to account for one more emission process, called the spontaneous emission which could take place in absence of an external electromagnetic field. Einstein in 1917 interrelated these three processes, which allowed to calculate the transition probability for the spontaneous emission.

Einstein considered a system (an atom or molecule) with two energy states in thermal equilibrium with a radiation density $\rho(\nu)$. Let the number densities of the lower level n and upper level m are N_n and N_m, respectively. The system in the state n can go the state m after absorption of a photon. This transition probability is obviously proportional to the radiation density $\rho(\nu)$. Hence,

$$\text{The transition rate for absorption} = B_{nm}\rho(\nu) \tag{11.38}$$

where the proportionality constant B_{nm} is known as the Einstein B-coefficient for induced absorption.

$$\text{Number of transitions from state } n \text{ to state } m = N_n B_{nm} \rho(\nu)$$

Denoting Einstein coefficient for induced (stimulated) emission from the state m to n as B_{mn}

$$\text{Number of transitions from state } m \text{ to state } n = N_m B_{mn} \rho(\nu)$$

Einstein considered $B_{nm} = B_{mn}$ for non-degenerate states. In the thermal equilibrium, when there is no net transfer of energy between the system and the radiation field, we have

$$N_n B_{nm} \rho(\nu) = N_m B_{mn} \rho(\nu)$$

Since $B_{nm} = B_{mn}$, we have $N_n = N_m$, which violates the Boltzmann distribution

$$\frac{N_m}{N_n} = \exp\left(-\frac{h\nu_{mn}}{kT}\right) \tag{11.39}$$

where $\nu_{mn} = (E_m - E_n)/h$ is the frequency of photon corresponding to the transition between the states m and n, k the Boltzmann constant, T the kinetic temperature in Kelvin. In order to get an agreement with the Boltzmann distribution, Einstein proposed an additional contribution to the emission process which was independent of the radiation field. He simply added a term $N_m A_{mn}$, where A_{mn} is known as the Einstein's coefficient for the spontaneous emission. With this additional term, in the thermal equilibrium, we have

$$N_m A_{mn} + N_m B_{mn} \rho(\nu_{mn}) = N_n B_{nm} \rho(\nu_{mn}) \tag{11.40}$$

Assuming the Boltzmann distribution and using the value of N_n from equation (11.39) in (11.40), we get

$$A_{mn} = B_n \left[\exp\left(\frac{h\nu}{kT}\right) - 1\right] \rho(\nu_{mn}) \tag{11.41}$$

From the Planck's radiation law, we have

$$\rho(\nu_{mn}) = \frac{8\pi h \nu_{mn}^3}{c^3} \frac{1}{\exp(h\nu_{mn}/kT) - 1} \tag{11.42}$$

Using equation (11.42) in (11.41), we have

$$\frac{A_{mn}}{B_{mn}} = \frac{8\pi h \nu_{mn}^3}{c^3}$$

Since the transition rate for absorption is the same as the probability per unit time, equations (11.36) and (11.38) give

$$B_{nm}\ \rho(\nu_{nm}) = \frac{8\pi^3}{3h^2}\ |\mu_{mn}|^2\ \rho(\nu_{mn})$$

$$B_{mn} = \frac{8\pi^3}{3h^2}\ |\mu_{mn}|^2 = \frac{8\pi^3 e^2}{3h^2}\ |\langle m\,|r|\,n\rangle|^2$$

and therefore

$$A_{mn} = \frac{2h\nu_{mn}^3}{\pi^2 c^3}\ B_{mn}$$

$$= \frac{64\pi^4\nu_{mn}^3}{3hc^3}\ |\mu_{mn}|^2 = \frac{64\pi^4\nu_{mn}^3 e^2}{3hc^3}\ |\langle m\,|r|\,n\rangle|^2 \tag{11.43}$$

Equation (11.43) shows that the coefficient of spontaneous emission increases as the cube of the transition frequency. Using the fact that $B_{nm} = B_{mn}$ and equation (11.41), we get

$$\frac{\text{Spontaneous emission}}{\text{Stimulated emission}} = \frac{N_m}{N_m}\frac{A_{mn}}{B_{mn}\ \rho(\nu_{mn})} = \exp\left(\frac{h\nu}{kT}\right) - 1$$

Thus, at the thermal equilibrium, $h\nu << kT$ so that the stimulated emission exceeds the spontaneous emission. When $h\nu >> kT$, the spontaneous emission exceeds the stimulated emission. This is the reason why construction of a MASER (microwave amplification by stimulated emission of radiation) is easier than that of a LASER.

Exercise 4. Calculate Einstein B coefficient for transition between $2p$ and $1s$ states of hydrogen atom.

Solution: The Einstein B coefficient is given by

$$B_{mn} = \frac{8\pi^3 e^2}{3h^2}\ |\langle m\,|r|\,n\rangle|^2 = \frac{8\pi^3 e^2}{3h^2}\ |\langle 210\,|r|\,100\rangle|^2$$

For computation of the element $\langle 210|r|100\rangle$, we have to calculate $\langle 210|x|100\rangle$, $\langle 210|y|100\rangle$ and $\langle 210|z|100\rangle$. In spherical coordinates, we have $x = r\sin\theta\cos\phi$, $y = r\sin\theta\sin\phi$, $z = r\cos\theta$. Further, $x \pm y = r\sin\theta\ e^{\pm i\phi}$. The hydrogen wave functions are

$$\psi_{210} = \frac{1}{\sqrt{32\pi a_0^3}}\ \frac{r}{a_0}\ \exp\left(-\frac{r}{2a_0}\right)\cos\theta$$

and

$$\psi_{100} = \frac{1}{\sqrt{\pi a_0^3}} \exp\Big(-\frac{r}{a_0}\Big)$$

Substituting the expressions for ψ_{210} and ψ_{100}, we have

$$\langle 210|z|100\rangle = \langle 210|r\cos\theta|100\rangle$$

$$= \frac{1}{4\sqrt{2}\,\pi a_0^4}\int_0^\infty r^4 \exp\Big(-\frac{3r}{2a_0}\Big)\,\mathrm{d}r\int_0^\pi \cos^2\theta\,\sin\theta\,\mathrm{d}\theta\int_0^{2\pi}\mathrm{d}\phi$$

$$= \frac{1}{4\sqrt{2}\,\pi a_0^4}\frac{4!}{(3/2a_0)^5}\frac{2}{3}2\pi = 4\sqrt{2}\Big(\frac{2}{3}\Big)^5 a_0$$

For convenient we calculate

$$\langle 210|x \pm y|100\rangle = \langle 210|r\,\sin\theta\;\mathrm{e}^{\pm i\phi}|100\rangle$$

$$= \frac{1}{4\sqrt{2}\,\pi a_0^4}\int_0^\infty r^4 \exp\Big(-\frac{3r}{2a_0}\Big)\,\mathrm{d}r\int_0^\pi \cos\theta\,\sin^2\theta\,\mathrm{d}\theta\int_0^{2\pi}\mathrm{e}^{\pm i\phi}\,\mathrm{d}\phi$$

Since

$$\int_0^{2\pi}\mathrm{e}^{\pm i\phi}\mathrm{d}\phi = 0$$

we get

$$\langle 210|x \pm y|100\rangle = 0$$

Thus, we have

$$\langle 210|x|100\rangle = 0 \qquad \text{and} \qquad \langle 210|y|100\rangle = 0$$

Therefore,

$$|\langle 210|r|100\rangle|^2 = |\langle 210|x|100\rangle|^2 + |\langle 210|y|100\rangle|^2 + |\langle 210|z|100\rangle|^2$$

$$= 32\Big(\frac{2}{3}\Big)^{10} a_0^2 = 32\Big(\frac{2}{3}\Big)^{10}(5.295\times 10^{-11})^2 = 1.556\times 10^{-21}\ \mathrm{m}^2$$

In S.I. units, we have

$$e^2 = \frac{e'^2}{4\pi\epsilon_0} = \frac{(1.602\times 10^{-19})^2}{4\times 3.14159\times(8.854\times 10^{-12})} = 2.307\times 10^{-28}\ \mathrm{Nm}^2$$

Thus,

$$B = \frac{8\pi^3 e^2}{3h^2}\ |\langle 210\,|r|\,100\rangle|^2$$

$$= \frac{8\times(3.14159)^3\times(2.307\times 10^{-28})\times(1.556\times 10^{-21})}{3\times(6.626\times 10^{-34})^2}$$

$$= 6.76\times 10^{19}\ \mathrm{N}^{-1}\mathrm{m}^2\mathrm{s}^{-2}$$

7. Selection rules

In the calculation of transition probability for absorption, stimulated emission and spontaneous emission we evaluate the transition dipole moment

$$\mu_{kn} = \langle k|\mu|n\rangle$$

where $\mu = -e\vec{r}$ is the electric dipole moment. Radiative transitions between all the states are not allowed. The selection rules tell about the transitions which may take place on the basis of dipole approximation. The transitions for which the transition dipole moment is non-zero are knwon as the allowed transitions. On the other side, the transitions for which the transition dipole moment is zero are knwon as the forbidden transitions.

We already know that the parity of an atomic orbital with quantum number l is $(-1)^l$. Thus, the s and d orbitals have an even parity whereas the p and f orbitals have an odd parity. The parity of the integrand of the transition dipole moment μ_{kn} is

$$(-1)^{l_k}(-1)(-1)^{l_n} = (-1)^{l_k+l_n+1}$$

This is even when the two orbitals have opposite parity and odd when they have the same parity. Group analysis suggests that a transition dipole moment will be finite only when the integrand is symmetric under the symmetry operation of the system. Hence, the allowed electric dipole transition involves a change in parity. That is, a transition is allowed between the two levels whose parities are opposite to each other.

For hydrogenic atoms, the states are specified by the quantum numbers n, l, m. For a allowed electric dipole transition, the dipole moment matrix element $\langle|n'l'm'|\vec{r}|nlm\rangle$ must be finite. The matrix element is finite only when

$$l' - l = \Delta l = \pm 1$$

For getting the selection rule on the quantum number m, we may express the matrix element as

$$\langle n'l'm'|\vec{r}|nlm\rangle = \hat{i}\langle n'l'm'|x|nlm\rangle + \hat{j}\langle n'l'm'|y|nlm\rangle = \hat{k}\langle n'l'm'|z|nlm\rangle$$

When the radiation is plane polarized with the electric field along the z-axis, z-component of the dipole moment is the only relevant quantity and it can be expressed as $\langle n'l'm'|r\cos\theta|nlm\rangle$. The ϕ-part of this element

$$\int_0^{2\pi} \exp[i(m-m')\phi]\, d\phi$$

is finite only when $m' = m$ or $\Delta m = 0$.

When the radiation is polarized in the xy-plane, we need to calculate the matrix elements of x and y. It is however convenient to calculate the matrix elements of $x \pm iy$ and then to get the values corresponding to x and y with the help of the relations

$$x = \frac{1}{2}[(x+iy)+(x-iy)] \qquad\qquad y = \frac{1}{2i}[(x+iy)-(x-iy)]$$

In the spherical polar coordinates, we have

$$x \pm iy = r\sin\theta\cos\phi \pm ir\sin\theta\sin\phi = r\sin\theta\ \exp\ (\pm i\phi)$$

The matrix element of $x \pm iy$ is

$$\langle n'l'm'|r\sin\theta\ \exp(\pm i\phi)|nlm\rangle = f(r,\theta)\int_0^{2\pi}\exp\left[i(m-m'\pm 1)\phi\right]\mathrm{d}\phi$$

The integral on right side is non-vanishing only when $m' - m = \pm 1$ or $\Delta m = \pm 1$.

Thus, for arbitrary polarization, the general selection rule is

$$\Delta m = 0, \pm 1$$

The radial part of the element $\langle |n'l'm'|\ \vec{r}\ |nlm\rangle$ is always non-vanishing for all values of n', l', n, l. Thus, all values for Δn are permissible. Therefore, the selection rules for hydrogenic atoms are

$$\Delta l = \pm 1 \qquad \text{and} \qquad \Delta m = 0, \pm 1$$

The reason for the selection rule $\Delta l = \pm 1$ may be assigned to the spin angular momentum of photon which is equal to 1. Therefore, when a photon is emitted or absorbed, for conservation of total angular momentum, the orbital angular momentum of the electron in the atom changes by ± 1. Emission or absorption by atom of a photon may increase the orbital angular momentum of electron ($\Delta l = 1$) depending on the orientation of the electron's angular momentum with the angular momentum of the photon.

The selection rules discussed above are valid only for electric dipole transitions. The magnetic component of the electromagnetic may interact with the magnetic moment of the electron in the atom and may induce the magnetic dipole transitions. These transitions are 10^5 times weaker than those allowed under the electric dipole transitions. Since the selection rules for the two categories (electric dipole and magnetic dipole) of transitions are different, therefore the transitions which are allowed under one category may be forbidden under the other category.

Another kind of transitions is the electric quadrupole transitions in which the spatial variation of the electric field interacts with the electric quadrupole moment. Selection rule for these transitions are $\Delta l = 0, \pm 2$. These transitions are 10^8 times weaker than the electric dipole transitions.

Exercise 5. Which if the following transitions are electric dipole allowed: (i) $1s \rightarrow 2s$, (ii) $1s \rightarrow 2p$, (iii) $1s \rightarrow 3d$, (iv) $2p \rightarrow 3d$, (v) $1s \rightarrow 4p$.

Solution: (i) For $1s$, we have $l = 0$. For $2s$, we have $l = 0$. Since $\Delta l = 0$, the transition is not allowed.

(ii) For $1s$, we have $l = 0$ and parity is even. For $2p$, we have $l = 1$ and parity is odd. Since $\Delta l = 1$ and the parity changes, the transition is allowed.

(iii) For $1s$, we have $l = 0$. For $3d$, we have $l = 2$. Since $\Delta l = 2$, the transition is not allowed.

(iv) For $2p$, we have $l = 1$ and parity is odd. For $3d$, we have $l = 2$ and parity is even. Since $\Delta l = 1$ and the parity changes, the transition is allowed.

(v) For $1s$, we have $l = 0$ and parity is even. For $4p$, we have $l = 1$ and parity is odd. Since $\Delta l = 1$ and the parity changes, the transition is allowed.

8. Problems and questions

1. Discuss about the time-dependent first order perturbation. Describe the transitions to continuum states.
2. Discuss about the harmonic perturbation and derive the Fermi's Golden rule.
3. Derive expression for the energy density of electromagnetic field.
4. Distinguishes between the stimulated emission and spontaneous emission.
5. Explain the electric dipole approximation.
6. Show that there is no electric dipole transition between $1s$ and $2s$ states of hydrogen atom.
7. Obtain the electric dipole moment $\langle 310| \vec{\mu} |200\rangle$ for hydrogen atom.

8. Obtain the electric dipole moment $\langle 210|\ \vec{\mu}\ |100\rangle$ for hydrogen atom.

9. Obtain the electric dipole moment $\langle 310|\ \vec{\mu}\ |100\rangle$ for hydrogen atom.

10. Calculate Einstein B coefficient for transition between $3p$ and $2s$ states of hydrogen atom.

11. What do you understand by the selection rules? Under the electric dipole approximation and obtain selection selection rules for l, m and parity.

12. Describe the Einstein A and B coefficients. Obtain relations between them. Explain why construction of a MASER is easier than that of a LASER.

13. Derive expression for the transition probability.

14. Which of the following transitions are allowed under electric dipole approximation: (i) $1s \rightarrow 2d$, (ii) $1s \rightarrow 4p$, (iii) $1s \rightarrow 3s$, (iv) $2p \rightarrow 4d$, (v) $1p \rightarrow 4f$.

15. Write notes on the following

    (i) Fermi's golden rule

    (ii) Selection rules of l and m under electric dipole approximation

    (iii) Electric dipole approximation

    (iv) Einstein A and B coefficients and relation between them

    (v) Hamiltonian for a charged particle in an electromagnetic field

Many Electron Atoms

So far we have discussed about the atoms with one or two electrons. However, in practice most of the atoms have several electrons. As the atoms are building blocks of molecules, the information regarding the atoms plays key role in the understanding of structures of molecules and their properties. In the present chapter, we shall discuss about some fundamentals required for understanding the structures of atoms with many electrons. For the electrons being indistinguishable, we shall account for symmetry considerations for their wave functions. The required information would be applied to the energy levels of helium.

1. Identical particles

Two particles in a system are said to be identical (or indistinguishable) when they can be exchanged for each other without any change in the physical situation of the system. In an atom suppose we have n electrons, which are all indistinguishable. We may however specify their positions, but we cannot say which one of them is at the given position. Thus, for example, an exchange of coordinates of any two electrons would not change the Hamiltonian of the system. It is equivalence to say that the Hamiltonian of the system is symmetric with respect to the exchange of the coordinates of any pair of electrons. For example, for a pair of two electrons 1 and 2, we can write the Hamiltonian of the system as

$$H(1, 2, 3, \ldots, n) = H(2, 1, 3, \ldots, n) \tag{12.1}$$

The same is true for any number of exchanges of identical electrons in pairs.

1.1 Exchange operator

In order to understand about the particle exchange operator, let us consider a system of two particles. Suppose the Hamiltonian of the system is

$H(1,2)$. The energy eigenvalue equation with the wave function $\psi(1,2)$ may be expressed as

$$H(1,2)\psi(1,2) = E\psi(1,2) \tag{12.2}$$

where E is the energy of the system. On interchanging the electrons 1 and 2, the wave function becomes $\psi(2,1)$ and Hamiltonian is $H(2,1)$. Since the electrons are indistinguishable, the energy of the system must be the same and therefore we have

$$H(2,1)\psi(2,1) = E\psi(2,1)$$

In view of the equation (12.1), we have $H(2,1) = H(1,2)$ and this equation can be expressed as

$$H(1,2)\psi(2,1) = E\psi(2,1) \tag{12.3}$$

Let us define a particle exchange operator $P_{12}(\equiv P_{21})$ such that when it is applied on a wave function of the particles 1 and 2, the coordinates of the particles are interchanged. Thus, we have

$$P_{12}\psi(1,2) = \psi(2,1) \qquad \text{or} \qquad P_{12}\psi(2,1) = \psi(1,2) \tag{12.4}$$

Using equation (12.4) in (12.3), we have

$$H(1,2)P_{12}\psi(1,2) = EP_{12}\psi(1,2) = P_{12}E\psi(1,2)$$

Using equation (12.2) here, we have

$$H(1,2)P_{12}\psi(1,2) = P_{12}H(1,2)\psi(1,2)$$

It shows that

$$H(1,2)P_{12} = P_{12}H(1,2) \qquad \text{or} \qquad [P_{12}, H(1,2)] = 0$$

showing that the particle exchange operator P_{12} commutes with the Hamiltonian H of the system. Hence, the particle exchange operator is a constant of motion.

Exercise 1. Show that for two identical particles, an operator representing a physical property of the system must commute with the exchange operator of the particles.

Solution: Suppose $S(1,2)$ is a physical operator for a system of two identical particles. Let $\psi(1,2)$ is its eigen function and s be the corresponding eigen value. Then we have

$$S(1,2)\psi(1,2) = s\ \psi(1,2)$$

On interchanging the particles 1 and 2, the wave function becomes $\psi(2,1)$ and the operator becomes $S(2,1)$. Since the particles are identical, the eigen value of the system must be the same and therefore we have

$$S(2,1)\psi(2,1) = s\ \psi(2,1)$$

For the particle exchange operator $P_{12}(\equiv P_{21})$ we have

$$P_{12}\psi(1,2) = \psi(2,1) \qquad \text{or} \qquad P_{12}\psi(2,1) = \psi(1,2)$$

Considering that the physical operator $S(1,2) = S(2,1)$, we have

$$\begin{aligned} S(1,2)P_{12}\psi(1,2) &= S(1,2)\psi(2,1) = S(2,1)\psi(2,1) = s\ \psi(2,1) \\ &= s\ P_{12}\psi(1,2) = P_{12}\ s\ \psi(1,2) = P_{12}S(1,2)\psi(1,2) \end{aligned}$$

It shows that

$$S(1,2)P_{12} = P_{12}S(1,2) \qquad \text{or} \qquad [P_{12}, S(1,2)] = 0$$

showing that the particle exchange operator P_{12} commutes with the operator S of the system.

1.2 Eigen value of exchange operator

Let the function $\psi(1,2)$ be an eigenfunction of P_{12} with eigenvalue p. Then, we have

$$P_{12}\psi(1,2) = p\ \psi(1,2) \qquad \text{or} \qquad \psi(2,1) = p\ \psi(1,2) \qquad (12.5)$$

On operating both sides of equation (12.5) by P_{12}, we have

$$\begin{aligned} P_{12}\ \psi(2,1) &= P_{12}\ p\ \psi(1,2) \\ \psi(1,2) &= p\ P_{12}\ \psi(1,2) = p^2\ \psi(1,2) \end{aligned}$$

Thus, we have

$$p^2 = 1 \qquad \text{or} \qquad p = \pm 1$$

It shows that the eigenvalue of the particle exchange operator is either $+1$ or -1. It follows from equation (12.5) that

$$\psi(2,1) = \pm\psi(1,2) \qquad (12.6)$$

Hence, the wave function for an exchange of a pair of identical particles is either symmetric or antisymmetric.

1.3 Symmetric and antisymmetric wave functions

If interchange of a given pair of particles in a wave function leaves the wave function unchanged, the wave function is said to be symmetric for that pair of particles. For example, for the interchange of the particles 1 and 2, a symmetric wave function satisfies the condition: $\psi(1,2,3,\ldots,n) = \psi(2,1,3,\ldots,n)$. On the other side, when on interchange of a given pair of particles, the wave function changes by a sign, the wave function is said to be antisymmetric for that pair. For example, for the interchange of the particles 1 and 2, the antisymmetric wave function satisfies the condition: $\psi(1,2,3,\ldots,n) = -\psi(2,1,3,\ldots,n)$. If the wave function is symmetric for exchange of any pair of particles, the wave function is said to be symmetric, and when the wave function is antisymmetric for exchange of any pair of particles, the wave function is said to be antisymmetric.

Equation (12.6) therefore requires that for a system of two identical particles, an acceptable wave function is either symmetric or antisymmetric. Generalization of this requirement says that the wave function of the system of several identical particles must be either symmetric or antisymmetric with respect to the interchange of any pair of particles. This symmetric or antisymmetric nature of a wave function does not change with time.

Exercise 2. Show that symmetric or antisymmetric nature of a wave function does not change with time.

Solution: Let us first consider that the wave function $\Psi(1,2,3,\ldots,n,t)$ is symmetric at a particular time t. Then $H\Psi$ is symmetric since the Hamiltonian H is symmetric in its arguments. The time-dependent Schrödinger wave equation is

$$i\hbar\frac{\partial}{\partial t}\,\Psi(1,2,3,\ldots,n.t) = H(1,2,\ldots,t)\;\Psi(1,2,3,\ldots,n,t) \qquad (12.7)$$

Since right side $H\Psi$ is symmetric, $\partial\Psi/\partial t$ on the left side also must be symmetric. Consequently, the wave function at any infinitesimally latter time $t+\mathrm{d}t$ given by $\Psi + (\partial\Psi/\partial t)\,\mathrm{d}t$ is also symmetric. This process of changing time can be repeated in order to cover a large time interval. It shows that the symmetric nature of the wave function does not change with time.

In a similar fashion, it can be shown that an antisymmetric nature of a wave function does not change with time.

1.4 Construction of symmetric and antisymmetric wave functions

Solutions of a Schrödinger equation for a system of indistinguishable particles may neither be symmetric nor antisymmetric functions. The general solutions can however be used to construct symmetric and antisymmetric wave functions in the following manner. Solution of a Schrödinger equation (12.7) can be expressed as

$$\Psi(1,2,3,\ldots,n,t) = \psi(1,2,3,\ldots,n)\exp\left[-\frac{iEt}{\hbar}\right]$$

where E is the energy eigenvalue of the system. Suppose two particles are interchanged in equation (12.7), the equation remains unchanged due to the symmetry properties. Thus, for the exchange of any two particles, we have two solutions of the Schrödinger equation. Now, for exchange of n particles in pairs, there are $n!$ permutations, which give rise to $n!$ solutions for this equation (12.7), each corresponding to one of the $n!$ permutations of arguments. All these $n!$ solutions however correspond to the same energy E. Thus, a degeneracy, called exchange degeneracy, arises due to the indistinguishability of the particles. Any linear combination of these solutions is also a solution of the wave equation. The sum of all these functions is a symmetrical unnormalized wave function ψ_s. This is obvious as an interchange of any pair of particles changes one of the terms in ψ_s into the another one and the latter one into the former one. For example, in a system of two particles 1 and 2, the 2! solutions are $\psi(1,2)$ and $\psi(2,1)$. A sum of these wave functions is

$$\psi_s = \psi(1,2) + \psi(2,1)$$

An interchange of 1 and 2 changes $\psi(1,2)$ into $\psi(2,1)$ and vice versa.

In a system of three particles 1, 2 and 3, we have

$$\psi_s = \psi(1,2,3) + \psi(1,3,2) + \psi(3,2,1) + \psi(2,1,3) + \psi(2,3,1) + \psi(3,1,2)$$

These wave functions are symmetric for exchange of any pair of particles.

An antisymmetric unnormalized wave function ψ_{as} can be constructed by adding together all the permuted functions obtained from the original solution by an even number of interchange of pairs of particles and subtracting the sum of all the permuted functions obtained by an odd number of interchanges of pairs of particles. Thus, for a system of two particles 1 and 2, an antisymmetric wave function is

$$\psi_{as} = \psi(1,2) - \psi(2,1)$$

where we have started from $\psi(1, 2)$ and in $\psi(2, 1)$ there is a single interchange. For a system of three particles 1, 2 and 3, we have antisymmetric wave function

$$\psi_{as} = [\psi(1,2,3) + \psi(2,3,1) + \psi(3,1,2)] \\ -[\psi(2,1,3) + \psi(1,3,2) + \psi(3,2,1)]$$

Here, we have started from $\psi(1, 2, 3)$. The wave functions in the first pair of the square brackets are obtained after making even number of interchanges and those in the second pair of the square brackets is obtained after making odd number of interchanges.

This process of obtaining symmetric and antisymmetric wave functions can be applied for a system of any number of particles.

2. Pauli exclusion principle

Consider a system of n noninteracting identical particles. Since there is no interaction between the particles, the Hamiltonian of the system can be written as

$$H(1, 2, \ldots, n) = H(1) + H(2) + \ldots + H(n)$$

and the eigenfunction of the system will be a product of wave functions for individual particles. If $u_a(1), u_b(2), \ldots, u_n(n)$ are the n one particle eigenfunctions, then

$$\psi(1, 2, \ldots, n) = u_a(1)\ u_b(2) \ldots u_n(n)$$

Here, the particles 1, 2, ... n are in the states a, b, ... n, respectively. When the energies of the states a, b, ... n are E_a, E_b, ... E_n, respectively, the total energy E of the system is

$$E = E_a + E_b + E_c + \ldots + E_n$$

and $u_a(1)$, $u_b(2)$, ... and E_a, E_b, ... satisfy the following relations for individual particles.

$$H(1)u_a(1) = E_a u_a(1), \quad H(2)u_b(2) = E_b u_b(2), \\ \ldots, H(n)u_n(n) = E_n u_n(n)$$

This assumption that the state u_a is occupied by the particle 1, u_b by the particle 2 and so on is not correct since the particles are identical. We can only say is that one particle is in state u_a, a second is in u_b and so on. Thus, for a system consisting of two particles 1 and 2, the possible eigenfunctions are $u_a(1)u_b(2)$ and $u_a(2)u_b(1)$. The symmetric ψ_s and antisymmetric ψ_{as} combinations of these wave functions are:

$$\psi_s = u_a(1)u_b(2) + u_a(2)u_b(1)$$

and

$$\psi_{as} = u_a(1)u_b(2) - u_a(2)u_b(1) = \begin{vmatrix} u_a(1) & u_a(2) \\ u_b(1) & u_b(2) \end{vmatrix} \tag{12.8}$$

Symmetry or antisymmetry of a state function is an important law of nature. Experiments have shown that the symmetry or antisymmetry of a state function of a system is related to the spin of the identical particles in it. Pauli successfully proved as the following.

(1) A system of identical particles with half-integer spins (spin = $\frac{1}{2}, \frac{3}{2}, \ldots$) is described by an antisymmetric wave function. Such particles obey the Fermi-Dirac statistics and are known as the fermions.

(2) A system of identical particles with integer spins (spin = $0, 1, 2, \ldots$) is described by a symmetric wave function. Such particles obey the Bose-Einstein statistics and are known as the bosons.

In order to understand the Pauli exclusion principle, suppose we put two fermions in the same state, say u_a, then the antisymmetric wave function is

$$\psi_{as} = u_a(1)u_a(2) - u_a(2)u_a(1) = 0 \tag{12.9}$$

It shows that the antisymmetric wave function vanishes when two identical particles are in the same position. Thus, the two identical fermions cannot occupy the same state. This is one form of the Pauli's exclusion principle. As the spin of electron is 1/2, the Pauli's exclusion principle says that 'No two electrons can have the same set of quantum numbers'.

Exercise 3. Show that a system of fermions is expressed by antisymmetric wave functions.

Solution: Generalization of equation (12.8) for a system of n fermions the normalized antisymmetric combination can be written as

$$\psi_{as}(1,2,\ldots,n) = \frac{1}{\sqrt{n!}} \begin{vmatrix} u_a(1) & u_a(2) & \ldots & u_a(n) \\ u_b(1) & u_b(2) & \ldots & u_b(n) \\ \vdots & \vdots & & \vdots \\ u_n(1) & u_n(2) & \ldots & u_n(n) \end{vmatrix}$$

The factor $1/\sqrt{n!}$ is the normalization constant and this determinant is known as the Slater determinant. This function $\psi_{as}(1,2,\ldots,n)$ is antisymmetric, as the interchange of two electrons is equivalent to the interchange of two columns of the determinant, which amounts to multiplying its value by a factor of -1.

3. Inclusion of spin

So far we have not talked explicitly about the spin of identical particles though each of the numbers, 1, 2, 3, ..., n, in the Hamiltonian $H(1,2,\ldots,n)$ includes both position as well as spin coordinates. The spin of identical particles can be included in the formulation when the Hamiltonian of the system does not contain space-spin interaction terms. For such situation, an eigenfunction of a particle including spin can be written as the product of space and spin parts:

$$\psi(\vec{r}, m_s) = \phi(\vec{r})\ \chi(m_s)$$

Here, $\phi(\vec{r})$ is the space function and the spin function $\chi(m_s)$ describes only the orientation of particle's spin. Each of the space coordinates can take a large number of values whereas a particle having spin s can have only $2s+1$ orientations, ranging from $-s$ to s with the step of unity.

A convenient set of spin eigenfunctions are column vectors of $(2s+1)$ rows with zeros in all the positions except one. We have already seen the two-spin eigenfunctions of S_z. For example, for a system of spin $s = 1/2$, we have

$$\chi\left(\tfrac{1}{2}\right) \equiv \alpha = \begin{bmatrix} 1 \\ 0 \end{bmatrix}, \qquad \chi\left(-\tfrac{1}{2}\right) \equiv \beta = \begin{bmatrix} 0 \\ 1 \end{bmatrix}$$

Conventionally, α represents the spin-up position whereas β represents the spin-down position. The three-spin eigenfunctions for S_z for $s = 1$ are

$$\chi(1) = \begin{bmatrix} 1 \\ 0 \\ 0 \end{bmatrix}, \qquad \chi(0) = \begin{bmatrix} 0 \\ 1 \\ 0 \end{bmatrix}, \qquad \chi(-1) = \begin{bmatrix} 0 \\ 0 \\ 1 \end{bmatrix}$$

For $s = 3/2$, we have

$$\chi\left(\frac{3}{2}\right) = \begin{bmatrix} 1 \\ 0 \\ 0 \\ 0 \end{bmatrix}, \quad \chi\left(\frac{1}{2}\right) = \begin{bmatrix} 0 \\ 1 \\ 0 \\ 0 \end{bmatrix}, \quad \chi\left(-\frac{1}{2}\right) = \begin{bmatrix} 0 \\ 0 \\ 1 \\ 0 \end{bmatrix}, \quad \chi\left(-\frac{3}{2}\right) = \begin{bmatrix} 0 \\ 0 \\ 0 \\ 1 \end{bmatrix}$$

Including spin functions, the symmetric and antisymmetric states can be expressed in the following manner. For a boson, the total wave function must be symmetric and it can be possible with the following combinations:

$$\psi_s = \begin{cases} \phi_s(\text{spatial}) & \chi_s(\text{spin}) \\ \phi_{as}(\text{spatial}) & \chi_{as}(\text{spin}) \end{cases}$$

That is, both the *phi* and χ must be either symmetric or antisymmetric. For a boson, the total wave function must be antisymmetric and it can be possible with the following combinations:

$$\psi_{as} = \begin{cases} \phi_s(\text{spatial}) & \chi_{as}(\text{spin}) \\ \phi_{as}(\text{spatial}) & \chi_s(\text{spin}) \end{cases}$$

That is, when *phi* is symmetric, χ must be antisymmetric, and vice versa.

3.1 System of two-electrons

For a single electron with spin $s = 1/2$, the wave functions are α (spin-up) and β (spin-down). In a system of two identical electrons, the possible spin product functions are $\alpha\alpha$, $\alpha\beta$, $\beta\alpha$ and $\beta\beta$. These product functions are orthonormal as the one electron spin functions are orthonormal. Of these, $\alpha\alpha$ and $\beta\beta$ are symmetric and other two $\alpha\beta$ and $\beta\alpha$ are neither symmetric nor antisymmetric. Symmetric and antisymmetric combinations of $\alpha\beta$ and $\beta\alpha$ can however be formed. Spin product functions of a system of two electrons are given in the following table.

Spin product functions*	Nature
$\alpha\alpha$	Symmetric
$(\alpha\beta + \beta\alpha)/\sqrt{2}$	Symmetric
$\beta\beta$	Symmetric
$(\alpha\beta - \beta\alpha)/\sqrt{2}$	Antisymmetric

*$(1/\sqrt{2})$ is the normalization factor

Exercise 4. N noninteracting bosons are in an infinite potential well given by $V(x) = 0$ for $0 < x < a$; $V(x) = \infty$ for $x < 0$ and for $x > a$. Find the ground state energy of the system. What would be the ground state energy in case of the fermions.

Solution: As discussed earlier, the energy eigenvalues of a particle in the infinite square well are

$$E_n = \frac{\pi^2\hbar^2 n^2}{2ma^2} \qquad n = 1, 2, \ldots$$

Here the particles are bosons, there is no Pauli's exclusion principle and therefore all the N particles will be in the $n = 1$ state. Hence, the total energy

$$E = \frac{N\pi^2\hbar^2}{2ma^2}$$

In case of are fermions, the Pauli's exclusion principle is applicable and therefore a state can have only two electrons, one with spin-up and the other with spin-down. Thus, the lowest $N/2$ states will be filled. The total ground energy is (when N is an even number)

$$E = 2\frac{\pi^2\hbar^2}{2ma^2}\left[1^2 + 2^2 + \ldots + \left(\frac{N}{2}\right)^2\right]$$

$$= \frac{\pi^2\hbar^2}{ma^2}\frac{1}{6}\left[\frac{N}{2}\left(\frac{N}{2}+1\right)\left(N+1\right)\right] = \frac{\pi^2\hbar^2}{24ma^2}N(N+1)(N+2)$$

When N is an odd number, the total energy is

$$E = 2\frac{\pi^2\hbar^2}{2ma^2}\left[1^2 + 2^2 + \ldots + \left(\frac{N}{2}-1\right)^2\right] + \frac{\pi^2\hbar^2}{2ma^2}\left(\frac{N}{2}\right)^2$$

$$= \frac{\pi^2\hbar^2}{ma^2}\frac{1}{6}\left[\left(\frac{N}{2}-1\right)\left(\frac{N}{2}\right)\left(N-1\right)\right] + \frac{\pi^2\hbar^2 N^2}{8ma^2} = \frac{\pi^2\hbar^2}{24ma^2}(N^3 + 2N)$$

Exercise 5. A system of two noninteracting electrons is described by the Hamiltonian

$$H = \frac{p_1^2}{2m} + \frac{p_2^2}{2m} + V(x_1) + V(x_2)$$

where $V(x) = 0$ for $0 < x < a$; $V(x) = \infty$ for $x < 0$ and for $x > a$. When both the electrons are in the same spin state what is the lowest energy and eigenfunction of the system.

Solution: Since the Hamiltonian can be written as a sum of Hamiltonian for individual electrons, the wave function of the system $\psi(1,2)$ can be written as

$$\psi(1,2) = \psi(1)\ \psi(2)$$

Now, the Schrödinger equation for the system reduces into the following equations, each for one electron:

$$-\frac{\hbar^2}{2m}\frac{d^2}{dx_1^2}\psi(1) + V(x_1)\ \psi(1) = E^{(1)}\ \psi(1)$$

and

$$-\frac{\hbar^2}{2m}\frac{d^2}{dx_2^2}\psi(2) + V(x_2)\,\psi(2) = E^{(2)}\,\psi(2)$$

where $E^{(1)} + E^{(2)} = E$ is the total energy of the system. As discussed earlier, the energies and eigenfunctions for a single particle in the given potential is

$$E_{n_1}^{(1)} = \frac{\pi^2\hbar^2 n_1^2}{2ma^2} \qquad \psi_{n_1}(1) = \sqrt{\frac{2}{a}}\,\sin\left(\frac{n_1\pi x_1}{a}\right) \qquad n_1 = 1, 2, \ldots$$

and

$$E_{n_2}^{(2)} = \frac{\pi^2\hbar^2 n_2^2}{2ma^2} \qquad \psi_{n_2}(2) = \sqrt{\frac{2}{a}}\,\sin\left(\frac{n_2\pi x_2}{a}\right) \qquad n_2 = 1, 2, \ldots$$

Since both the electrons are in the same spin state, the possible combination of spin functions are $\alpha\alpha$ or $\beta\beta$, which are symmetric. For the total wave function to be antisymmetric, the space function must be antisymmetric. Thus, the product $\psi_{n_1}(1)\psi_{n_2}(2)$ must be antisymmetric. This can only be possible when n_1 is even, n_2 must be odd and vice versa. Let us take $n_1 = 1, n_2 = 2$. Thus, we have

$$\text{Total energy} = \frac{\pi^2\hbar^2}{2ma^2} + \frac{4\pi^2\hbar^2}{2ma^2} = \frac{5\pi^2\hbar^2}{2ma^2}$$

and

$$\psi(1,2) = \frac{2}{a}\sin\left(\frac{\pi x_1}{a}\right)\,\sin\left(\frac{2\pi x_2}{a}\right)$$

For $n_1 = 2$, $n_2 = 1$, we have

$$\text{Total energy} = \frac{4\pi^2\hbar^2}{2ma^2} + \frac{\pi^2\hbar^2}{2ma^2} = \frac{5\pi^2\hbar^2}{2ma^2}$$

and

$$\psi(2,1) = \frac{2}{a}\sin\left(\frac{2\pi x_1}{a}\right)\,\sin\left(\frac{\pi x_2}{a}\right)$$

Since both the states have the same energy, the space wave function of the system is a linear combination of the two functions. The antisymmetric combination is $\psi(1,2) - \psi(2,1)$. To get the complete energy eigenfunction, this space part is multiplied by $\alpha\alpha$ or $\beta\beta$.

3.2 System of three-electrons

A system with three identical electrons can be considered as a system of $1+2$ electrons. We can then combine the α and β of the single electron ($s = 1/2$) with the spin functions for three-electrons. Spin product functions of two-electrons are given in the following table.

Spin product functions
$\alpha\alpha\alpha$
$(\alpha\alpha\beta + \alpha\beta\alpha + \beta\alpha\alpha)/\sqrt{3}$
$(\alpha\beta\beta + \beta\alpha\beta + \beta\beta\alpha)/\sqrt{3}$
$\beta\beta\beta$
$(2\beta\alpha\alpha - \alpha\alpha\beta - \alpha\beta\alpha)/\sqrt{6}$.
$(\beta\alpha\beta + \beta\beta\alpha - 2\alpha\beta\beta)/\sqrt{6}$
$(\alpha\alpha\beta - \alpha\beta\alpha)/\sqrt{2}$
$(\beta\alpha\beta - \beta\beta\alpha)/\sqrt{2}$

Here the first four functions are symmetric for the exchange of any pair of electrons, the next two are symmetric for the exchange of the electrons 2 and 3, while the last two are antisymmetric for the exchange of the electrons 2 and 3.

Exercise 6. Show that we cannot construct a completely antisymmetric spin function for three electrons.

Solution: Let a, b, c stand for three functions and 1, 2, 3 for three identical particles. In the function $a(1)$ $b(2)$ $c(3)$, particle 1 is in a, particle 2 in b and particle 3 in c. When we do not specify that these functions correspond to space or spin functions. The Slater determinant for this case is

$$\frac{1}{\sqrt{6}}\begin{vmatrix} a(1) & a(2) & a(3) \\ b(1) & b(2) & b(3) \\ c(1) & c(2) & c(3) \end{vmatrix}$$

This is completely antisymmetric as the interchange of two spins amounts to the interchange of two columns of the determinant, which multiplies it by -1. Suppose, we specify the functions a, b, c as due to the electron spins. Let $a = \alpha$, $b = \beta$ and $c = \beta$ in the above determinant. Then we have

$$\frac{1}{\sqrt{6}}\begin{vmatrix} \alpha(1) & \alpha(2) & \alpha(3) \\ \beta(1) & \beta(2) & \beta(3) \\ \beta(1) & \beta(2) & \beta(3) \end{vmatrix}$$

Here the second and third rows of the determinant are identical and therefore, the value of the determinant is zero. In whatever way we select a, b, c for electron spins, two rows of the determinant will be equal and the value of the determinant is zero. It shows that we cannot construct a completely antisymmetric three-electron spin functions.

4. Helium atom

Let us consider a helium atom which has two electrons. For discussion of its ground state, we applied perturbation and variation methods in the earlier chapters, where we did not account for the symmetry effects of the spin of electrons. We shall now calculate the energy of the ground and the first excited states by considering the fact that electrons are identical fermions and therefore the total wave function must be antisymmetric. For convenience, here we shall use the perturbation technique for calculation.

All the terms except the last one in the Hamiltonian for the helium atom

$$H = \Big(- \frac{\hbar^2}{2m} \nabla_1^2 - \frac{Ze^2}{r_1} \Big) + \Big(- \frac{\hbar^2}{2m} \nabla_2^2 - \frac{Ze^2}{r_2} \Big) + \frac{e^2}{r_{12}}$$

can be used as the unperturbed part. The unperturbed eigenfunction is the product of hydrogen wave function ($Z = 2$)

$$\psi^0 = \psi_1^0 \, \psi_2^0 = \frac{4}{\pi a_0^3} \exp\Big[- \frac{2(r_1 + r_2)}{a_0} \Big]$$

and the unperturbed energy is

$$E^0 = -2Z^2 W_H = -8 \times 13.6\text{eV} = -108.8\text{eV} \tag{12.10}$$

4.1 Ground state

The ground state of helium is $1s^2$. Here, both the electrons are in the ψ_{100} state, the wave function for the ground state is $\psi^0 = \psi_{100}(r_1)\,\psi_{100}(r_2)$. This space wave function is symmetric and therefore, the spin part that multiplies it must be antisymmetric. Using the antisymmetric spin product function for two electrons, the zeroth-order helium wave function for the ground state of helium is

$$\psi^0 = \psi_{100}(r_1)\,\psi_{100}(r_2)\,\frac{1}{\sqrt{2}}[\alpha(1)\beta(2) - \beta(1)\alpha(2)]$$

As the Hamiltonian does not contain spin-dependent terms, the energy here remains unchanged after inclusion of the spin part. Therefore, the energy of the ground state is the same as the given in equation (12.10).

4.2 First excited state

In the first excited state of the helium atom, one electron remains in the hydrogenic ground state $1s$ while the other is in an excited state. The wave function $\psi(\vec{r})$ is then

$$\psi(\vec{r}) = \psi_{100}\ \psi_{nlm}$$

For the first excited state, the possible electronic configurations are $1s2s$ and $1s2p$. Except for the electron exchange, the first configuration is non-degenerate whereas the second is a three-fold degenerate as the electron in the p-state has three possible values for m, *i.e.*, $m = 0, \pm 1$. In the present discussion, we shall consider the $1s2s$ state.

In the perturbation method, the term e^2/r_{12} in the Hamiltonian is treated as perturbation on the unperturbed part. Taking exchange degeneracy into account, the two unperturbed states are

$$\psi_{100}(r_1)\ \psi_{200}(r_2) \qquad \text{and} \qquad \psi_{100}(r_2)\ \psi_{200}(r_1)$$

Here also the spin part of the wave function does not contribute to the energy and as discussed earlier, the perturbation for these two-degenerate states can be calculated. The perturbation matrix has the form

$$\begin{bmatrix} J & K \\ K & J \end{bmatrix}$$

where Coulomb energy J and exchange energy are

$$J = \iint \psi^*_{100}(r_1)\ \psi^*_{200}(r_2)\ \frac{e^2}{r_{12}}\ \psi_{100}(r_1)\ \psi_{200}(r_2)\ d\tau_1\ d\tau_2$$

and

$$K = \iint \psi^*_{100}(r_1)\ \psi^*_{200}(r_2)\ \frac{e^2}{r_{12}}\ \psi_{100}(r_2)\ \psi_{200}(r_1)\ d\tau_1\ d\tau_2$$

Both J and K are obviously positive. The eigenvalues for the perturbation matrix are obtained by solving the determinant

$$\begin{vmatrix} J - E^{(1)} & K \\ K & J - E^{(1)} \end{vmatrix} = 0$$

where $E^{(1)}$ is the first-order correction to energy. From the determinant, we get

$$E^{(1)} = J + K \qquad \text{and} \qquad E^{(1)} = J - K$$

These energies correspond to the normalized eigenfunctions

$$\psi_s = \frac{1}{\sqrt{2}}\Big[\psi_{100}(r_1)\ \psi_{200}(r_2) + \psi_{100}(r_2)\ \psi_{200}(r_1)\Big]$$

and

$$\psi_{as} = \frac{1}{\sqrt{2}}\Big[\psi_{100}(r_1)\ \psi_{200}(r_2) - \psi_{100}(r_2)\ \psi_{200}(r_1)\Big]$$

respectively. The first function ψ_s is symmetric for the interchange of r_1 and r_2 whereas the second ψ_{as} is antisymmetric. Combining these wave functions with the spin functions for a two-electron system with the condition that the total wave function is antisymmetric, we get

$$\psi(s) = \psi_s\ \frac{1}{\sqrt{2}}[\alpha\beta - \beta\alpha]$$

$$\psi(t) = \psi_{as}\begin{cases}\alpha\alpha\\(\alpha\beta + \beta\alpha)/\sqrt{2}\\\beta\beta\end{cases}$$

where $\psi(s)$ and $\psi(t)$ refer to the singlet and triplet states, respectively. the energy of the singlet (E_s) and triplet (E_t) states corrected to first order are

$$E_s = -8W_H + J + K$$

and

$$E_t = -8W_H + J - K$$

Exercise 7. Write the zeroth-order wave functions for helium atom (i) in the ground state $(1s^2)$ (ii) in the excited state $1s2s$. Also, express them in the form of the Slater determinants.

Solution: (i) The zeroth-order wave function for helium atom in the $1s^2$ state is

$$1s(1)1s(2)\frac{1}{\sqrt{2}}[\alpha(1)\beta(2) - \beta(1)\alpha(2)]$$

In terms of Slater determinant, it can be expressed as

$$\frac{1}{\sqrt{2}}\begin{vmatrix}1s(1)\alpha(1) & 1s(2)\alpha(2)\\1s(1)\beta(1) & 1s(2)\beta(2)\end{vmatrix}$$

(ii) The zeroth wave functions in the $1s2s$ state are

$$\psi_1 = \frac{1}{2}\Big[1s(1)2s(2) + 1s(2)2s(1)\Big]\Big[\alpha(1)\beta(2) - \beta(1)\alpha(2)\Big]$$

$$\psi_2 = \frac{1}{\sqrt{2}}\left[1s(1)2s(2) - 1s(2)2s(1)\right]\alpha(1)\alpha(2)$$

$$\psi_3 = \frac{1}{2}\left[1s(1)2s(2) - 1s(2)2s(1)\right]\left[\alpha(1)\beta(2) + \beta(1)\alpha(1)\right]$$

$$\psi_4 = \frac{1}{\sqrt{2}}\left[1s(1)2s(2) - 1s(2)2s(1)\right]\ \beta(1)\beta(2)$$

For 1s2s configuration, we have the following spin-orbital combinations: $1s\alpha$, $1s\beta$, $2s\alpha$ and $2s\beta$ leading to four Slater determinants (the normalization factor $1/\sqrt{2}$ is not included here).

$$D_1 = \begin{vmatrix} 1s(1)\alpha(1) & 1s(2)\alpha(2) \\ 2s(1)\alpha(1) & 2s(2)\alpha(2) \end{vmatrix}$$

$$D_2 = \begin{vmatrix} 1s(1)\alpha(1) & 1s(2)\alpha(2) \\ 2s(1)\beta(1) & 2s(2)\beta(2) \end{vmatrix}$$

$$D_3 = \begin{vmatrix} 1s(1)\beta(1) & 1s(2)\beta(2) \\ 2s(1)\alpha(1) & 2s(2)\alpha(2) \end{vmatrix}$$

$$D_4 = \begin{vmatrix} 1s(1)\beta(1) & 1s(2)\beta(2) \\ 2s(1)\beta(1) & 2s(2)\beta(2) \end{vmatrix}$$

A comparison of the above wave functions with these determinants shows that $\psi_1 = (D_2 - D_3)/2$, $\psi_2 = D_1/\sqrt{2}$, $\psi_3 = (D_2 + D_3)/2$ and $\psi_4 = D_4/\sqrt{2}$.

5. Problems and questions

1. Show that the particle exchange operator commutes with the Hamiltonian of the system.

2. Define particle exchange operator for identical electrons and find out its eigen values.

3. Show that the wave function for two identical electrons are either symmetric or antisymmetric with respect to the exchange of electrons.

4. Show that the wave function for an exchange of a pair of electrons is either symmetric or antisymmetric.

5. State Pauli exclusion principle. Show that a system of fermions is expressed by antisymmetric wave functions.

6. Suppose $\psi(1, 2, \ldots, n)$ is eigenfunction of a symmetric Hamiltonian that correspond to a non-degenerate eigenvalue. Show that it is either symmetric or antisymmetric.

7. Sixteen noninteracting particles are confined in a potential $V(x) = \infty$ for $x < 0$ and $x > a$; $V(x) = 0, 0 < x < a$. (*i*) What is the ground state energy when the particles are bosons? (*ii*) What is the ground state energy when the particles are electrons?

8. What is the ground state energy and wave function for two identical particles in the potential defined, in a potential $V(x) = \infty$ for $x < 0$ and $x > a$; $V(x) = 0 for 0 < x < a$, if the two particles are (*i*) bosons and (*ii*) fermions.

9. Consider two identical particles described by the Hamiltonian

$$H = \frac{p_1^2}{2m} + \frac{p_2^2}{2m} + \frac{1}{2}\, m\omega^2 x_1^2 + \frac{1}{2}\, m\omega^2 x_2^2$$

   Obtain the energy spectrum of this system. Discuss its degeneracy.

10. Obtain the zeroth-order wave function for helium atom in the excited state $1s2p$.

11. Obtain the zeroth-order wave function for the state $1s2s$ of helium atom.

12. Write notes on the following

    (i) Particle exchange operator

    (ii) Symmetric and antisymmetric wave functions

    (iii) Pauli exclusion principle

Scattering

When a A beam of particles after passing through a potential field deviates from its original direction, this process of deviation is known as the scattering. Thus, scattering is a process in which an incident beam deviates from its direction after interaction with the scattering field. In this process, there may or may not be exchange of energy between the incident particle and the target (scatterer). When there is no exchange of energy, the scattering is known as the elastic scattering. On the other side, in the inelastic scattering, exchange of energy between the incident particle and the target takes place. Each system has bound and discrete energy sates. In the elastic scattering, the incident particle as well as the target remain in their original states. But in the inelastic scattering, the states of the target and/or the incident particle are changed. The process of scattering in one of the best techniques for studying the properties of atoms, nuclei and interactions of elementary particles among themselves as well as with the atoms and nuclei.

In the scattering process, we are mainly interested in angular distribution of the scattered particles.

1. Scattering cross-section

Consider a beam of particles, each of mass m, traveling along the z-axis with a velocity v. Let the beam is scattered by a potential V (target) centered at the origin of the coordinate system (Figure 1). The incident particles experience a force only when they are within a sphere of radius r_0 whose center is at the origin. Radius of this sphere is known as the range of the scattering potential. The incident particles are scattered in all directions, symmetrical about the z-axis, due to their interaction with the potential. After leaving the range of the scattering potential, the particles again move in the straight lines in various directions. The angle between the initial and final directions of motion of a particle is known as the scattering angle θ.

Results of an scattering experiment are generally expressed in terms of the differential scattering cross section or total scattering cross section. Let N be the number of incident particles crossing per unit area normal to the direction of the incident beam per unit time and n be the number of particles scattered into a solid angle $\mathrm{d}\Omega$ in the direction (θ, ϕ) in a unit time. For the scattered particles, we shall use the spherical polar coordinates (r, θ, ϕ) with origin at the target. The differential scattering cross section $\sigma(\theta, \phi)$ is defined as

$$\sigma(\theta, \phi) = \frac{n/\mathrm{d}\Omega}{N} \tag{13.1}$$

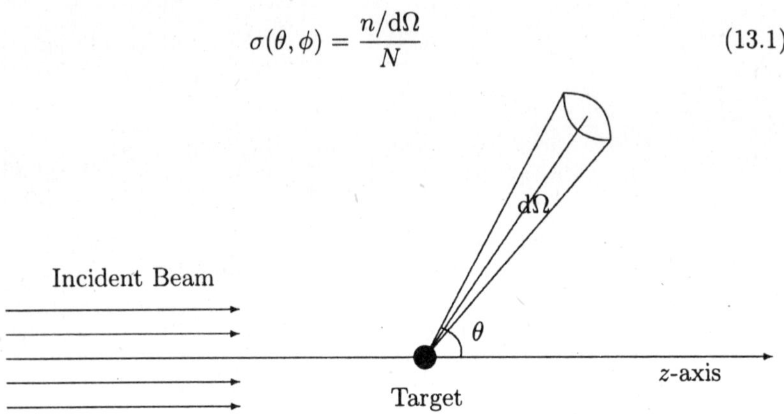

Figure 1: Schematic diagram for scattering. $\mathrm{d}\Omega$ is the solid angle in the direction $(\theta,\ \phi)$. θ is the scattering angle and the scattering is symmetrical about the z-axis.

The solid angle $\mathrm{d}\Omega$ in the direction (θ, ϕ) is

$$\mathrm{d}\Omega = \frac{(r\ \mathrm{d}\theta)(r \sin\theta\ \mathrm{d}\phi)}{r^2} = \sin\theta\ \mathrm{d}\theta\ \mathrm{d}\phi$$

The total scattering cross section σ is

$$\sigma = \int \sigma(\theta, \phi)\ \mathrm{d}\Omega = \int_0^{\pi}\int_0^{2\pi} \sigma(\theta, \phi) \sin\theta\ \mathrm{d}\theta\ \mathrm{d}\phi$$

Both $\sigma(\theta, \phi)$ and σ have dimensions of area and therefore are named cross-section. For spherically symmetric potential, $\sigma(\theta, \phi)$ is independent of ϕ and we have

$$\sigma = 2\pi \int_0^{\pi} \sigma(\theta) \sin\theta\ \mathrm{d}\theta$$

2. Scattering amplitude

In the scattering process, when the potential V depends only on the relative distance r between the particle and the target (spherically symmetric potential), the Schrödinger equation for the motion of a particle relative to the target is

$$-\frac{\hbar^2}{2\mu}\nabla^2\psi + V(r)\,\psi = E\,\psi \qquad \mu = \frac{mM}{m+M} \tag{13.2}$$

where m and M are the masses of the incident particle and the target, respectively. At the large distances from the target, the interaction between the particles and target is negligible ($V = 0$) and the particles move along the straight lines and their wave functions are represented by a plane wave

$$\psi = A\,\exp(i\vec{k}.\vec{r})$$

where $k^2 = 2\mu E/\hbar^2$. As the incident particle is moving along z-axis (incident direction), we have

$$\psi_i = A\,e^{ikr\cos\theta} = A\,e^{ikz} \tag{13.3}$$

A detector used for detection of the scattered particles is generally kept far away from the scatterer (target) and therefore we would like to have asymptotic form of the scattered wave. As the intensity of scattered wave depends on the angles θ and ϕ, the spherically diverging scattered wave can be expressed as

$$\psi_s \underset{r\to\infty}{\longrightarrow} A\,f(\theta,\phi)\,\frac{e^{ikr}}{r}$$

where $f(\theta,\phi)$ is known as the scattering amplitude. For spherically symmetric potential, the scattering amplitude does not depend on ϕ and we have

$$\psi_s \underset{r\to\infty}{\longrightarrow} A\,f(\theta)\,\frac{e^{ikr}}{r} \tag{13.4}$$

The most general asymptotic solution can be written as

$$\psi \underset{r\to\infty}{\longrightarrow} A\left[e^{ikz} + f(\theta)\,\frac{e^{ikr}}{r}\right]$$

2.1 Relation between $f(\theta)$ and $\sigma(\theta)$

A relation between the scattering amplitude $f(\theta)$ and the differential scattering cross section $\sigma(\theta)$ can be obtained in the following manner. For the incident wave, the probability current density is[1]

$$\vec{j}_i\,(\vec{r},t) = \frac{i\hbar}{2\mu}(\Psi_i \nabla \Psi_i^* - \Psi_i^* \nabla \Psi_i)$$

Using equation (13.3) here, we get

$$\begin{aligned} j_i &= \frac{i\hbar}{2\mu}\Big[A e^{ikz}\frac{\partial}{\partial z}\Big(A^* e^{-ikz}\Big) - A^* e^{-ikz}\frac{\partial}{\partial z}\Big(A e^{ikz}\Big)\Big] \\ &= \frac{i\hbar}{2\mu}(-2ikAA^*) = \frac{\hbar k|A|^2}{\mu} \end{aligned}$$

For the scattered wave, the probability current density is

$$\begin{aligned} \vec{j}_s\,(\vec{r},t) &= \frac{i\hbar}{2\mu}(\Psi_s \nabla \Psi_s^* - \Psi_s^* \nabla \Psi_s) \\ &= \frac{i\hbar}{2\mu}\Big[A f(\theta)\,\frac{e^{ikr}}{r}\Big(\frac{\partial}{\partial r}\hat{r} + \frac{1}{r}\frac{\partial}{\partial \theta}\hat{\theta}\Big)\Big(A^* f^*(\theta)\,\frac{e^{-ikr}}{r}\Big) \\ &\quad -A^* f^*(\theta)\,\frac{e^{-ikr}}{r}\,\Big(\frac{\partial}{\partial r}\hat{r} + \frac{1}{r}\frac{\partial}{\partial \theta}\hat{\theta}\Big)\Big(A f(\theta)\,\frac{e^{ikr}}{r}\Big)\Big] \\ &= \frac{i\hbar}{2\mu}\Big[-\frac{|A|^2|f|^2}{r^3}(ikr+1)\hat{r} + \frac{|A|^2}{r^2}f\frac{\partial f^*}{\partial \theta}\hat{\theta} \\ &\quad -\frac{|A|^2|f|^2}{r^3}(ikr-1)\hat{r} - \frac{|A|^2}{r^2}f^*\frac{\partial f}{\partial \theta}\hat{\theta}\Big] \end{aligned}$$

Considering $f(\theta)$ as real, *i.e.*, $f^*(\theta) = f(\theta)$, we have

$$j_s = \frac{i\hbar}{2\mu}\Big[-2ik\frac{|A|^2|f|^2}{r^2}\Big] = \frac{\hbar k}{\mu}\,\frac{|A|^2|f|^2}{r^2}$$

where $1/r^2$ is the solid angle subtended by the unit area of the detector at the target. Thus,

$$\sigma(\theta) = \frac{\text{Probability density of scattered wave per unit solid angle}}{\text{Probability density of incident wave}}$$

---

[1] In Cartesian coordinates, we have

$$\nabla \equiv \frac{\partial}{\partial x}\hat{i} + \frac{\partial}{\partial y}\hat{j} + \frac{\partial}{\partial z}\hat{k}$$

In the spherical polar coordinates, we have

$$\nabla \equiv \frac{\partial}{\partial r}\hat{r} + \frac{1}{r}\,\frac{\partial}{\partial \theta}\hat{\theta} + \frac{1}{r\sin\theta}\,\frac{\partial}{\partial \phi}\hat{\phi}$$

$$= \frac{\hbar k|A|^2|f(\theta)|^2/\mu}{\hbar k|A|^2/\mu} = |f(\theta)|^2$$

The scattering amplitude $f(\theta)$ is thus related to the differential scattering cross section $\sigma(\theta)$ which is an observable quantity. Since $\sigma(\theta)$ has dimension of (length)^2, the $f(\theta)$ has the dimension of length.

3. Partial waves

A plane wave e^{ikz} can be expanded as a linear combination of spherical waves[2]

---

[2]In the spherical polar coordinates (r, θ, ϕ) solution of Schrödinger equation for free particle was obtained as

$$\psi_k(r,\theta,\phi) = \sum_{l=0}^{\infty} A_l Y_{lm}(\theta,\phi)\, j_l(kr)$$

For the motion of the particles along the z-axis, ϕ-dependence disappears so that $m = 0$ and the free-particle wave function is

$$e^{ikr\cos\theta} = \sum_{l=0}^{\infty} A_l P_l(\cos\theta)\, j_l(kr)$$

Multiplying both sides by $P_m(\cos\theta)$ and integrating with respect to $\cos\theta$ over the limits from -1 to 1, we have

$$\int_{-1}^{1} e^{ikr\cos\theta} P_m(\cos\theta)\, d(\cos\theta) = \int_{-1}^{1}\sum_{l=0}^{\infty} A_l P_l(\cos\theta)\, P_m(\cos\theta)\, j_l(kr)\, d(\cos\theta)$$

$$\left[\frac{P_m(\cos\theta)\, e^{ikr\cos\theta}}{ikr}\right]_{-1}^{1} - \int_{-1}^{1}\frac{e^{ikr\cos\theta}}{ikr} P'_m(\cos\theta)\, d(\cos\theta)$$

$$= \sum_{l=0}^{\infty} A_l\, j_l(kr) \int_{-1}^{1} P_l(\cos\theta)\, P_m(\cos\theta)\, d(\cos\theta)$$

We have $P_m(1) = 1$ and $P_m(-1) = (-1)^m P_m(1) = e^{im\pi}$. Further integration of the second term gives $1/r^2$, $1/r^3$, ... and therefore vanish as r tends to infinity. Now, we have

$$\frac{1}{ikr}\left(e^{ikr} - e^{-ikr}e^{im\pi}\right) = A_m\, j_m(kr)\, \frac{2}{2m+1}$$

$$\frac{e^{im\pi/2}}{ikr}\left[\exp\left\{i\left(kr - \frac{m\pi}{2}\right)\right\} - \exp\left\{-i\left(kr - \frac{m\pi}{2}\right)\right\}\right] = \frac{2A_m}{kr(2m+1)}\sin\left(kr - \frac{m\pi}{2}\right)$$

$$A_m = (2m+1)e^{im\pi/2} = (2m+1)\, i^m$$

Thus,

$$e^{ikz} = \sum_{l=0}^{\infty} i^l\,(2l+1)\, j_l(kr)\, P_l(\cos\theta)$$

$$e^{ikz} = \sum_{l=0}^{\infty} i^l (2l+1) j_l(kr) P_l(\cos\theta) \tag{13.5}$$

where $j_l(kr)$ is the spherical Bessel function of order l given by

$$j_l(kr) \underset{r \to \infty}{\longrightarrow} \frac{1}{kr} \sin\left(kr - \frac{l\pi}{2}\right) \tag{13.6}$$

The $P_l(\cos\theta)$ is the Legendre polynomial.[3] Each term on the right side of equation (13.5) represents a spherical wave. Hence, a plane wave is equivalent to a supervision of an infinite number of spherical waves. These spherical waves are also known as the partial waves. The partial waves with $l = 0, 1, 2, \ldots$ are, respectively, known as the s-wave, p-wave, d-wave and so on. Using equation (13.6) in (13.5) and expressing $\sin(kr - l\pi/2)$ as the exponentials, we get

$$e^{ikz} = \sum_{l=0}^{\infty} \frac{i^l(2l+1)}{2ikr} P_l(\cos\,\theta) \left[\exp\left(ikr - i\frac{l\pi}{2}\right) - \exp\left(-ikr + i\frac{l\pi}{2}\right)\right] \tag{13.7}$$

This equation shows that each partial wave can be represented as a sum of an incoming spherical wave and an outgoing spherical wave.

In the investigation of scattering, the first few spherical waves play important role. Since $P_0(\cos\theta) = 1$, the s-partial wave is independent of the angle θ

4. Scattering by a central potential: partial wave analysis

We shall use the method of partial wave analysis for scattering by a central potential.

4.1 Asymptotic solution

For scattering, the Schrödinger equation is

$$\nabla^2\psi + \frac{2\mu}{\hbar^2}\left[E - V(r)\right]\psi = 0 \tag{13.8}$$

[3] We have

$$P_0(x) = 1 \qquad P_1(x) = x \qquad P_2(x) = \frac{1}{2}(3x^2 - 1) \qquad P_3(x) = \frac{1}{2}(5x^3 - 3x)$$

Since the incident particles are moving along the z-axis, the wave function ψ would be independent of ϕ. After removing the ϕ dependence, equation (13.8) in the spherical coordinates is

$$\frac{1}{r^2}\frac{\partial}{\partial r}\left(r^2\frac{\partial\psi}{\partial r}\right)+\frac{1}{r^2\,\sin\theta}\frac{\partial}{\partial\theta}\left(\sin\theta\frac{\partial\psi}{\partial\theta}\right)+\frac{2\mu}{\hbar^2}\left(E-V\right)\psi=0 \qquad (13.9)$$

Since the potential V, being spherically symmetric, depends on r only, this equation can be solved by the method of the separation of variables. In this technique of the separation of variables, the wave function is written as a product of distinct functions, each depending on one variable, as the following.

$$\psi(r,\theta)=R(r)P(\theta)\,. \qquad (13.10)$$

Using equation (13.10) in (13.9) and multiplying by r^2/RP, we get

$$\frac{1}{R}\frac{\mathrm{d}}{\mathrm{d}r}\left(r^2\frac{\mathrm{d}R}{\mathrm{d}r}\right)+\frac{1}{P\sin\theta}\frac{\mathrm{d}}{\mathrm{d}\theta}\left(\sin\theta\frac{\mathrm{d}P}{\mathrm{d}\theta}\right)+\frac{2\mu}{\hbar^2}\left(E-V(r)\right)r^2=0$$

$$\frac{1}{R}\frac{\mathrm{d}}{\mathrm{d}r}\left(r^2\frac{\mathrm{d}R}{\mathrm{d}r}\right)+\frac{2\mu}{\hbar^2}\left(E-V(r)\right)r^2=-\frac{1}{P\sin\theta}\frac{\mathrm{d}}{\mathrm{d}\theta}\left(\sin\theta\frac{\mathrm{d}P}{\mathrm{d}\theta}\right) \qquad (13.11)$$

In equation (13.11), the left side depends only on r whereas the right side depends only on θ. This can be possible only when each of the sides is equal to some constant, say $l(l+1)$. Then we have

$$\frac{1}{\sin\theta}\,\frac{\mathrm{d}}{\mathrm{d}\theta}\left(\sin\theta\,\frac{\mathrm{d}P}{\mathrm{d}\theta}\right)+l(l+1)P=0 \qquad (13.12)$$

and

$$\frac{1}{r^2}\frac{\mathrm{d}}{\mathrm{d}r}\left(r^2\frac{\mathrm{d}R}{\mathrm{d}r}\right)+\frac{2\mu}{\hbar^2}\left(E-V\right)R-\frac{l(l+1)}{r^2}R=0 \qquad (13.13)$$

Equation (13.12) can be solved conveniently by introducing a new variable $u=\cos\,\theta$. On differentiating, we get

$$\mathrm{d}u=-\sin\;\theta\;\mathrm{d}\theta$$

We have

$$\frac{\mathrm{d}P}{\mathrm{d}\theta}=\frac{\mathrm{d}u}{\mathrm{d}\theta}\,\frac{\mathrm{d}P}{\mathrm{d}u}=-\sin\theta\,\frac{\mathrm{d}P}{\mathrm{d}u}=-\left(1-u^2\right)^{1/2}\frac{\mathrm{d}P}{\mathrm{d}u}$$

On using these relations, equation (13.12) reduces to

$$\frac{\mathrm{d}}{\mathrm{d}u}\left[\left(1-u^2\right)\frac{\mathrm{d}P(u)}{\mathrm{d}u}\right]+l(l+1)P(u)=0 \qquad (13.14)$$

This is a Legendre equation and its solution $P_l(u)$ is the Legendre polynomial of degree l. The orthonormality condition for the Legendre polynomial is

$$\int_{-1}^{1} P_l(u)P_k(u)\, du = \frac{2}{2l+1}\delta_{lk}$$

Equation (13.13) can be expressed as

$$\frac{1}{r^2}\frac{d}{dr}\left(r^2\frac{dR_l}{dr}\right) + \left[\frac{2\mu E}{\hbar^2} - \frac{2\mu V}{\hbar^2} - \frac{l(l+1)}{r^2}\right]R_l = 0 \qquad (13.15)$$

Outside range of the central potential, we have $V = 0$ and equation (13.15) reduces to the free-particle equation

$$\frac{d^2 R_l}{dr^2} + \frac{2}{r}\frac{dR_l}{dr} + \left[k^2 - \frac{l(l+1)}{r^2}\right]R_l = 0 \qquad (13.16)$$

where $k^2 = 2\mu E/\hbar^2$. Using $\rho = kr$, equation (13.16) can be expressed as

$$\frac{d^2 R_l(\rho)}{d\rho^2} + \frac{2}{\rho}\frac{dR_l}{d\rho} + \left[1 - \frac{l(l+1)}{\rho^2}\right]R_l(\rho) = 0 \qquad (13.17)$$

Using $R(\rho) = Z(\rho)\rho^{-1/2}$ in equation (13.17) we get

$$\frac{d^2 Z(\rho)}{d\rho^2} + \frac{1}{\rho}\frac{dZ(\rho)}{d\rho} + \left[1 - \left(\frac{l+(\frac{1}{2})}{\rho^2}\right)^2\right]Z(\rho) = 0$$

This is a Bessel equation with general solution

$$Z(\rho) = A' J_{l+1/2}(\rho) + B'\, J_{-(l+1/2)}(\rho)$$

Thus,

$$R(\rho) = \frac{A}{\rho^{1/2}} J_{l+1/2}(\rho) + \frac{B}{\rho^{1/2}} J_{-(l+1/2)}(\rho) = A'\, j_l(\rho) - B'\, \eta_l(\rho)$$

where A' and B' are constants, $j_l(\rho)$ and $\eta_l(\rho)$ are, respectively, the spherical Bessel and the spherical Neumann functions, defined as

$$j_l(\rho) = \left(\frac{\pi}{2\rho}\right)^{1/2} J_{l+1/2}(\rho) \qquad (13.18)$$

and

$$\eta_l(\rho) = (-1)^{l+1}\left(\frac{\pi}{2\rho}\right)^{1/2} J_{-(l+1/2)}(\rho)$$

Though $\eta_l(\rho)$ is not finite at the origin, it is retained as we want to have asymptotic solution.

For small and large values of ρ, the functions j_l and η_l vary as

$$R_l(kr) \underset{r\to\infty}{=} \frac{A'}{kr}\sin\left(kr - \frac{l\pi}{2}\right) + \frac{B'}{kr}\cos\left(kr - \frac{l\pi}{2}\right)$$

Now we consider the case of $V \neq 0$, then we have to solve equation (13.15) inside the scattering region. Now we solve equation (13.15) with $k'^2 = 2\mu(E-V)/\hbar^2$. Writing new constants A_l and δ_l in terms of A' and B' as $A' = A_l \cos\delta_l$ and $B' = A_l \sin\delta_l$, we have asymptotic solution of (13.15) as

$$\begin{aligned} R_l(kr) &= \frac{A_l\cos\delta_l}{kr}\sin\left(kr - \frac{l\pi}{2}\right) + \frac{A_l\sin\delta_l}{kr}\cos\left(kr - \frac{l\pi}{2}\right) \\ &= \frac{A_l}{kr}\sin\left(kr - \frac{l\pi}{2} + \delta_l\right) \end{aligned} \tag{13.19}$$

where $l = 0, 1, 2, \ldots$ and δ_l is known as the phase shift corresponding to the angular momentum number l. It measures the amount by which the phase of the radial function differs from the corresponding case of $V = 0$. The asymptotic solution is then

$$\psi(r,\theta) = \sum_{l=0}^{\infty}\frac{A_l}{kr}\sin\left(kr - \frac{l\pi}{2} + \delta_l\right)P_l(\cos\theta) \tag{13.20}$$

where A_l is the asymptotic amplitude.

4.2 Scattering amplitude

Equating this asymptotic form of the wave function with that given by equation (13.4), we have

$$e^{ikz} + f(\theta)\frac{e^{ikr}}{r} = \sum_{l=0}^{\infty}\frac{A_l}{kr}\sin\left(kr - \frac{l\pi}{2} + \delta_l\right)P_l(\cos\theta) \tag{13.21}$$

Using equations (13.5) and (13.6) in (13.21), we get

$$\begin{aligned} &\sum_{l=0}^{\infty}\frac{i^l(2l+1)}{kr}\sin\left(kr - \frac{l\pi}{2}\right)P_l(\cos\theta) + f(\theta)\frac{e^{ikr}}{r} \\ &\qquad = \sum_{l=0}^{\infty}\frac{A_l}{kr}\sin\left(kr - \frac{l\pi}{2} + \delta_l\right)P_l(\cos\theta) \end{aligned}$$

$$\sum_{l=0}^{\infty}\frac{i^l(2l+1)}{2ikr}\left[\exp\left(ikr - i\frac{l\pi}{2}\right) - \exp\left(-ikr + i\frac{l\pi}{2}\right)\right]P_l(\cos\theta) + f(\theta)\frac{e^{ikr}}{r}$$

$$=\sum_{l=0}^{\infty}\frac{A_l}{2ikr}\left[\exp\left(ikr-i\frac{l\pi}{2}+i\delta_l\right)-\exp\left(-ikr+i\frac{l\pi}{2}-i\delta_l\right)\right]P_l(\cos\theta) \quad (13.22)$$

Equating the coefficients of e^{-ikr} on the two sides of equation (13.22), we have

$$\sum_{l=0}^{\infty}\frac{i^l(2l+1)}{2ikr}\exp\left(\frac{il\pi}{2}\right)P_l(\cos\theta)=\sum_{l=0}^{\infty}\frac{A_l}{2ikr}\exp\left(i\frac{l\pi}{2}-i\delta_l\right)P_l(\cos\theta)$$

Since this equation is valid for all values of θ, we have

$$A_l=i^l(2l+1)\exp(i\delta_l) \quad (13.23)$$

Equating the coefficients of e^{ikr} on the two sides of equation (13.22), we have

$$\sum_{l=0}^{\infty}\frac{i^l(2l+1)}{2ikr}\exp\left(-\frac{il\pi}{2}\right)P_l(\cos\theta)+\frac{f(\theta)}{r}$$

$$=\sum_{l=0}^{\infty}\frac{A_l}{2ikr}\exp\left(-\frac{il\pi}{2}+i\delta_l\right)P_l(\cos\theta) \quad (13.24)$$

Using equation (13.23) in (13.24) and rearranging, we get

$$f(\theta)=\frac{1}{2ik}\sum_{l=0}^{\infty}i^l(2l+1)e^{-il\pi/2}[\exp(2i\delta_l)-1]P_l(\cos\theta)$$

We have $i^l=\exp(il\pi/2)$ and therefore

$$[\exp(2i\delta_l)-1]=\exp(i\delta_l)\left[\exp(i\delta_l)-\exp(-i\delta_l)\right]=2i\exp(i\delta_l)\sin\delta_l$$

Thus, we have

$$f(\theta)=\frac{1}{k}\sum_{l=0}^{\infty}(2l+1)\exp(i\delta_l)\sin\delta_l P_l(\cos\theta) \quad (13.25)$$

Using equation (13.23) in (13.20), we get

$$\psi(r,\theta)=\sum_{l=0}^{\infty}\frac{i^l(2l+1)\exp(i\delta_l)}{2ikr}\left[\exp\left(ikr-i\frac{l\pi}{2}+i\delta_l\right)\right.$$

$$\left.-\exp\left(-ikr+i\frac{l\pi}{2}-i\delta_l\right)\right]P_l(\cos\theta)$$

$$=\sum_{l=0}^{\infty}\frac{i^l(2l+1)}{2ikr}\left[\exp\left(ikr-i\frac{l\pi}{2}+i2\delta_l\right)\right.$$

$$\left.-\exp\left(-ikr+i\frac{l\pi}{2}\right)\right]P_l(\cos\theta) \quad (13.26)$$

Comparison of equations (13.7) and (13.26) shows that the effect of the scattering potential is to shift the phase of the outgoing waves as compared to that of the incoming ones

4.3 Scattering cross section

The differential cross section is

$$\sigma(\theta) = |f(\theta)|^2 = \frac{1}{k^2}\left|\sum_{l=0}^{\infty}(2l+1)\exp(i\delta_l)\sin\delta_l\; P_l(\cos\theta)\right|^2 \qquad (13.27)$$

The total cross section is

$$\begin{aligned}\sigma &= 2\pi\int_0^{\pi}\sigma(\theta)\sin\theta\; d\theta \\ &= 2\pi\int_0^{\pi}\frac{1}{k^2}\left|\sum_{l=0}^{\infty}(2l+1)\exp(i\delta_l)\sin\delta_l\; P_l(\cos\theta)\right|^2\sin\theta\; d\theta \\ &= 2\pi\int_0^{\pi}\frac{1}{k^2}\left[\sum_{l=0}^{\infty}(2l+1)\exp(i\delta_l)\sin\delta_l\; P_l(\cos\theta)\right] \\ &\qquad\left[\sum_{m=0}^{\infty}(2m+1)\exp(-i\delta_m)\sin\delta_m\; P_m(\cos\theta)\right]\sin\theta\; d\theta \\ &= \frac{2\pi}{k^2}\sum_{l=0}^{\infty}(2l+1)^2\frac{2}{2l+1}\sin^2\delta_l = \frac{4\pi}{k^2}\sum_{l=0}^{\infty}(2l+1)\sin^2\delta_l \qquad (13.28)\end{aligned}$$

where we used orthonormality condition for the Legendre polynomials

$$\int_0^{\pi} P_n(\cos\theta)\; P_m(\cos\theta)\; \sin\theta\; d\theta = \frac{2}{2n+1}\delta_{mn}$$

In this way, the differential as well as the total cross sections are expressed in terms of the phase shifts δ_l of various partial waves. Hence, in order to understand the scattering problem, we should know the phase shifts δ_l of the waves corresponding to $l = 0, 1, 2, \ldots$.

For the s-wave scattering, the differential cross section $\sigma_0(\theta)$ and total cross section σ_0 are

$$\sigma_0(\theta) = \frac{1}{k^2}|e^{i\delta_0}\sin\delta_0|^2 = \frac{\sin^2\delta_0}{k^2} \qquad \text{and} \qquad \sigma_0 = \frac{4\pi}{k^2}\sin^2\delta_0 \qquad (13.29)$$

It is interesting to note that for the s-wave, both the cross sections are independent of the scattering angle θ. In the most experiments, the contribution of the s-wave is found much larger than that of the others. Equation (13.27) shows that the differential cross section $\sigma(\theta)$ contains the terms representing the interaction between various partial waves whereas equation (13.28) shows that the total cross section does not have any terms representing such interaction. Suppose all the phase shifts, except δ_0 and δ_1, are negligibly small. Then we have

$$\sigma(\theta) = \frac{1}{k^2}\left|e^{i\delta_0}\sin\delta_0 + 3e^{i\delta_1}\sin\delta_1\cos\theta\right|^2$$

$$= \frac{1}{k^2}\left[e^{i\delta_0}\sin\delta_0 + 3e^{i\delta_1}\sin\delta_1\cos\theta\right]\left[e^{-i\delta_0}\sin\delta_0 + 3e^{-i\delta_1}\sin\delta_1\cos\theta\right]$$

$$= \frac{1}{k^2}\Big[\sin^2\delta_0 + 3e^{i(\delta_0-\delta_1)}\cos\theta\ \sin\delta_0\ \sin\delta_1$$

$$+3e^{-i(\delta_0-\delta_1)}\cos\theta\ \sin\delta_0\ \sin\delta_1 + 9\cos^2\theta\ \sin^2\delta_1\Big]$$

$$= \frac{1}{k^2}\Big[\sin^2\delta_0 + 6\cos\theta\ \sin\delta_0\ \sin\delta_1\cos(\delta_0-\delta_1) + 9\cos^2\theta\sin^2\delta_1\Big] \qquad (13.30)$$

and

$$\sigma = \frac{4\pi}{k^2}\left[\sin^2\delta_0 + 3\sin^2\delta_1\right] \qquad (13.31)$$

The first term on the right side of equation (13.30) does not depend on θ. The second term has a factor $\cos^2\theta$ which has the same value at θ and $\pi - \theta$ and therefore it is symmetric between the backward and the forward directions. The presence of $\cos\theta$ in the third term produces a forward-backward asymmetry and may become large even if $\sin\delta_0 >> \sin\delta_1$. In the total cross section, though the contribution of the p-wave is negligible, it however contributes in the differential cross section through the interaction terms with the lower partial waves.

4.4 Optical theorem

Substituting $\theta = 0$ in equation (13.25), we get

$$f(0) = \frac{1}{k}\sum_{l=0}^{\infty}(2l+1)\ \exp(i\delta_l)\ \sin\delta_l$$

Imaginary part of this scattering amplitude is

$$\text{Im}f(0) = \frac{1}{k}\sum_{l=0}^{\infty}(2l+1)\sin^2\delta_l \qquad (13.32)$$

Comparison of equation (13.28) with (13.32) gives

$$\sigma = \frac{4\pi}{k}\ \text{Im}f(0)$$

This relation is known as the 'optical theorem'.

5. Significant number of partial waves

At sufficiently low energies in the scattering, the partial wave analysis says that only s-wave ($l = 0$) scattering is dominant. With the increase of energy, higher l waves start to contribute in the scattering. Now, a question arises to know about the partial waves up to what value of l be accounted for a given energy. An answer to this question can be given in the following manner.

We know that effect of a spherically symmetric potential is experienced within a sphere of radius r_0 around the scattering center. Classically, for a particle with the impact parameter b, linear momentum $p = k\hbar$ and the angular momentum $l\hbar$, we have

$$\vec{l} = \vec{r} \times \vec{p} \qquad l\hbar = bk\hbar \qquad l = bk$$

When the impact parameter $b > r_0$, the particle does not have any scattering. Hence, the particle does not get scattered when $l > kr_0$ and therefore, we need to account for the partial waves up to $l = kr_0$. At sufficiently low energies $k \approx 0$, the scattering is mainly due to the s-wave only. When $kr_0 \approx 1$, the contribution due to p-wave also comes in the picture. At the high energies, contributions of the waves corresponding to the values of l from zero to kr_0 are to be accounted for. However, with the inclusion of the higher partial waves, the treatment becomes quite complicated. For higher energies, use of the Born approximation is applicable.

6. Scattering by an attractive square-well potential

Let us consider a spherically symmetric potential $V(r)$ expressed as

$$V(r) = \begin{cases} -V_0 & 0 < r < a \\ 0 & r > a \end{cases}$$

For such potential, the radial part of the Schrödinger equation is

$$\frac{1}{r^2}\frac{\mathrm{d}}{\mathrm{d}r}\left(r^2\frac{\mathrm{d}R}{\mathrm{d}r}\right) + \frac{2\mu}{\hbar^2}\left(E + V_0\right)R - \frac{l(l+1)}{r^2}R = 0$$

Writing $R = u/r$, this equation gives[4]

$$\frac{d^2u}{dr^2} + \frac{2\mu}{\hbar^2}\left(E + V_0\right)u - \frac{l(l+1)}{r^2}u = 0 \tag{13.33}$$

For s-wave, we have

$$\frac{d^2u}{dr^2} + k_1^2 u = 0 \qquad \text{where} \quad k_1^2 = \frac{2\mu}{\hbar^2}\left(E + V_0\right) \qquad r < a \tag{13.34}$$

and

$$\frac{d^2u}{dr^2} + k^2 u = 0 \qquad \text{where} \quad k^2 = \frac{2\mu E}{\hbar^2} \qquad r > a \tag{13.35}$$

Solution of equation (13.34) is

$$u = A\ \sin(k_1 r) + B'\ \cos(k_1 r)$$

Since u must be zero at $r = 0$, we have $B' = 0$. Solution of equation (13.35) is

$$u = C\ \sin(kr) + D\ \cos(kr) \tag{13.36}$$

Considering $C = B\ \cos\ \delta_0$ and $D = B\ \sin\ \delta_0$ in equation (13.36), we have

$$u = B\ \sin(kr + \delta_0)$$

After accounting for the continuity of the wave function and its derivative at $r = a$, we have

$$A\ \sin(k_1 a) = B\ \sin(ka + \delta_0) \tag{13.37}$$

and

$$A\ k_1\ \cos(k_1 a) = B\ k\ \cos(ka + \delta_0) \tag{13.38}$$

Dividing equation (13.37) by (13.38), we have

$$\tan(ka + \delta_0) = \frac{k}{k_1}\ \tan(k_1 a) \tag{13.39}$$

or

$$\delta_0 = \tan^{-1}\left[\frac{k}{k_1}\ \tan(k_1 a)\right] - ka$$

Use of δ_0 in equation (13.20) provides an estimate for the cross-section for the s-wave scattering. Next section deals with some important features for this low energy scattering.

---

[4]We have

$$\frac{dR}{dr} = \frac{1}{r}\frac{du}{dr} - \frac{u}{r^2} \qquad\qquad r^2\frac{dR}{dr} = r\frac{du}{dr} - u$$

$$\frac{d}{dr}\left(r^2\frac{dR}{dr}\right) = r\frac{d^2u}{dr^2} \qquad\qquad \frac{1}{r^2}\frac{d}{dr}\left(r^2\frac{dR}{dr}\right) = \frac{1}{r}\frac{d^2u}{dr^2}$$

7. Breit-Wigner formula

In the preceding section, we dealt with the solution of the radial part of Schrödinger equation that describes s-wave scattering by an attractive square-well potential. In this section we shall discuss the case of very low energy scattering by such a potential. From equation (13.39), we have

$$\frac{\tan ka + \tan \delta_0}{1 - \tan ka \, \tan \delta_0} = \frac{k}{k_1} \tan k_1 a$$

$$\tan ka + \tan \delta_0 = \frac{k}{k_1} \tan k_1 a - \frac{k}{k_1} \tan ka \, \tan \delta_0 \, \tan k_1 a$$

$$\tan \delta_0 \Big[1 + \frac{k}{k_1} \tan ka \, \tan k_1 a\Big] = \frac{k}{k_1} \tan k_1 a - \tan ka \qquad (13.40)$$

When the potential is shallow, for very low energy scattering ($ka << 1$ and have $\tan ka \approx ka$), k/k_1 is very small and therefore the factor in square bracket on the left side of equation (13.40) becomes 1. Hence, equation (13.40) reduces to

$$\tan \delta_0 = \frac{k}{k_1} \tan k_1 a - ka$$

$$= ka\Big[\frac{\tan k_1 a}{k_1 a} - 1\Big]$$

On increasing the depth of the potential gradually, at certain stage, the value of $k_1 a$ will become $\pi/2$. When $k_1 a$ passes through $\pi/2$, the value of $\tan\delta_0$ becomes infinite. This shows that the phase shift also passes through $\pi/2$. At $\delta_0 = \pi/2$, the total cross section (equation 13.31) is

$$\sigma = \frac{4\pi}{k^2}$$

It is the maximum value of σ. It shows that if there is a bound state as energy tends to zero and the phase shift δ_0 passes through $\pi/2$, the total cross section attains the maximum value. This situation is known as the 'resonance'.

The behaviour of the cross section near the resonance can be investigated in the following manner. Suppose that the resonance energy is E_τ. The phase shift δ_0 will be a function of E_τ. Since the phase shift at the resonance is $\pi/2$, we have $\sin[\delta_0(E_\tau)] = 1$ and $\cos[\delta_0(E_\tau)] = 0$. On expanding $\sin \delta_0$ and $\cos \delta_0$ near $E = E_\tau$ by Taylor series, we get

$$\sin[\delta_0(E)] = \sin[\delta_0(E_\tau)] + \Big[\frac{\partial \sin \delta_0}{\partial E}\Big]_{E=E_\tau} (E - E_\tau)$$

$$= \sin[\delta_0(E_\tau)] + \left[\cos\delta_0 \frac{\partial \delta_0}{\partial E}\right]_{E=E_\tau}(E - E_\tau)$$

$$= \sin(\pi/2) + \left[\cos\,(\pi/2)\frac{\partial \delta_0}{\partial E}\right]_{E=E_\tau}(E - E_\tau) \approx 1 \quad (13.41)$$

and

$$\cos[\delta_0(E)] = \cos[\delta_0(E_\tau)] + \left[\frac{\partial \cos\delta_0}{\partial E}\right]_{E=E_\tau}(E - E_\tau)$$

$$= \cos[\delta_0(E_\tau)] - \left[\sin\,\delta_0\frac{\partial \delta_0}{\partial E}\right]_{E=E_\tau}(E - E_\tau)$$

$$= \cos(\pi/2) - \left[\sin\,(\pi/2)\frac{\partial \delta_0}{\partial E}\right]_{E=E_\tau}(E - E_\tau)$$

$$= -\frac{\partial \delta_0(E_\tau)}{\partial E}(E - E_\tau) = -\frac{2}{\Gamma}(E - E_\tau) \quad (13.42)$$

where we have

$$\frac{2}{\Gamma} = \frac{\partial \delta_0(E_\tau)}{\partial E}$$

The parameter Γ will be interpreted as the width of the resonant curve. The scattering amplitude $f_0(\theta)$, equation (13.25), can be written as

$$f_0(\theta, E) = \frac{1}{k}\exp[i\delta_0(E)]\ \sin[\delta_0(E)]$$

Dependence of the scattering amplitude on k is very weak. The strong dependence comes from the phase shift δ_0. Expressing the contribution due to the phase shift part by $f_0(\delta)$, we have

$$f_0(\delta) = \exp[i\delta_0(E)]\sin[\delta_0(E)] = \frac{\sin[\delta_0(E)]}{\exp[-i\delta_0(E)]} = \frac{\sin\delta_0}{\cos\delta_0 - i\sin\delta_0}$$

Using the values of sin δ_0 and cos δ_0 from equations (13.41) and (13.42) here, we get

$$f_0(\delta) = \frac{1}{[-2(E - E_\tau)/\Gamma] - i} = -\frac{\Gamma/2}{(E - E_\tau) + i\Gamma/2}$$

Now, the total cross section is

$$\sigma = \frac{4\pi}{k^2}\,|f_0(\delta)|^2 = \frac{4\pi}{k^2}\,\frac{\Gamma^2/4}{(E - E_\tau)^2 + \Gamma^2/4}$$

This is the Breit-Wigner formula for resonant cross-sections. The parameter Γ obviously represent the width of the resonant curve.

8. Scattering length

When energy E of incident particle or the value of kr_0, where r_0 is the range of the potential, is very low so that only s-wave contribute in the scattering, we have

$$f_0(\theta) = \frac{1}{k} \exp(i\delta_0) \sin\delta_0$$

Now, as $E \to 0$, the value of $-f(\theta)$ is defined as the scattering length, usually denoted by a:

$$a = \lim_{E \to 0} [-f(\theta)] = -\frac{1}{k} \exp(i\delta_0) \sin\delta_0 \tag{13.43}$$

We have

$$a^2 = |a|^2 = \frac{1}{k^2} \sin^2\delta_0 \tag{13.44}$$

From equations (13.29) and (13.44), we have

$$\sigma_0 = 4\pi a^2$$

When the potential $V(r)$ is weak, δ_0 is very small and thus, $\exp(i\delta_0) \approx 1$ and $\sin\delta_0 \approx \delta_0$. Then equations (13.43) reduces to

$$a = \lim_{E \to 0} \left(-\frac{\delta_0}{k}\right)$$

Thus, in the zero energy limit, we have $\delta_0 = -ka$.

9. Expressions for the phase shifts

For deriving an expression for phase shift, let us consider the radial part of Schrödinger equation describing the scattering

$$\frac{d^2u_l}{dr^2} + \left[\frac{2\mu E}{\hbar^2} - \frac{2\mu V}{\hbar^2} - \frac{l(l+1)}{r^2}\right]u_l = 0$$

where $u_l = r\ R_l(r)$. On the incident side which is away from the scattering region we have $V = 0$ and therefore

$$\frac{d^2u_l}{dr^2} + \left[k^2 - \frac{l(l+1)}{r^2}\right]u_l = 0 \tag{13.45}$$

where $k^2 = 2\mu E/\hbar^2$. Solution of this equation (13.45) is

$$u_l(kr) = kr\ j_l(kr)$$

Asymptotic expression for this solution is

$$u_l(kr) \underset{r \to \infty}{\longrightarrow} \sin\left(kr - \frac{l\pi}{2}\right) \tag{13.46}$$

Similarly the asymptotic solution of

$$\frac{d^2 w_l}{dr^2} + \left[k^2 - \frac{2\mu V}{\hbar^2} - \frac{l(l+1)}{r^2}\right] w_l = 0 \tag{13.47}$$

is

$$w_l(kr) \underset{r \to \infty}{\longrightarrow} \sin\left(kr - \frac{l\pi}{2} + \delta_l\right) \tag{13.48}$$

Multiplying equation (13.45) by w_l, equation (13.47) by u_l and subtracting the latter from the former, we get

$$w_l \frac{d^2 u_l}{dr^2} - u_l \frac{d^2 w_l}{dr^2} = -\frac{2\mu V}{\hbar^2} u_l w_l$$

This equation can be rearranged as

$$\frac{d}{dr}\left(w_l\ \frac{du_l}{dr} - u_l \frac{dw_l}{dr}\right) = -\frac{2\mu V}{\hbar^2} u_l w_l$$

Integrating this equation with respect to r over the limit from zero to r and remembering that $u_l(0) = 0$ and $w_l(0) = 0$, we get

$$w_l\ \frac{du_l}{dr} - u_l \frac{dw_l}{dr} = -\frac{2\mu}{\hbar^2} \int_0^r V(r') u_l(r')\ w_l(r')\ dr'$$

For r tending to infinity and using the values of $u_l(r)$ and $w_l(r)$ from equations (13.46) and (13.48), we get

$$k\ \sin\left(kr - \frac{l\pi}{2} + \delta_l\right)\ \cos\left(kr - \frac{l\pi}{2}\right) - k\ \sin\left(kr - \frac{l\pi}{2}\right)\ \cos\left(kr - \frac{l\pi}{2} + \delta_l\right)$$

$$= -\frac{2\mu}{\hbar^2} \int_0^\infty V(r) u_l(kr) w_l(kr)\ dr$$

or

$$k\ \sin\ \delta_l = -\frac{2\mu}{\hbar^2} \int_0^\infty V(r)\ u_l(kr)\ w_l(kr) dr$$

This expression for $\sin\delta_l$ is exact. At high energies for weak potential (Born approximation), the phase shifts are small and we have

$$u_l(kr) \equiv kr\ j_l(kr) \qquad \text{and} \qquad w_l(kr) \equiv kr\ j_l(kr)$$

Relation between spherical Bessel function and ordinary Bessel function is as the following.

$$j_l(kr) = \left(\frac{\pi}{2kr}\right)^{1/2} J_{l+1/2}(kr)$$

Now, we have

$$\begin{aligned}\sin\delta_l \equiv \delta_l &= -\frac{2\mu k}{\hbar^2}\int_0^\infty V(r)\ j_l^2(kr)\ r^2 \mathrm{d}r \\ &= -\frac{\pi\mu}{\hbar^2}\int_0^\infty V(r)\ [J_{l+1/2}(kr)]^2\ r^2 \mathrm{d}r \qquad (13.49)\end{aligned}$$

This expression is the Born approximation for the phase shifts. It is valid only for weak potentials. Equation (13.49) shows that an attractive potential ($V \le 0$) leads to positive phase shifts whereas a repulsive potential ($V \ge 0$) leads to negative phase shifts.

10. Integral equation

The partial wave analysis of scattering where we expanded a plane wave in terms of the spherical waves is more suitable for spherically symmetric potentials. Further, it is applicable at relatively low energies. For a general analysis, we need to express the wave function in the form of an integral equation, discussed in the present section.

The Schrödinger equation for the scattering of two-particles system, equation (13.2) can be expressed in the convenient form as

$$(\nabla^2 + k^2)\Psi(\vec{r}) = U(r)\ \Psi(\vec{r}) \qquad (13.50)$$

where

$$k^2 = \frac{2\mu E}{\hbar^2} \qquad U(r) = \frac{2\mu V(r)}{\hbar^2}$$

Here, the requirement is to obtain a solution of equation (13.50) whose asymptotic form is

$$\Psi(\vec{r}) = \mathrm{e}^{i\vec{k}\cdot\vec{r}} + \psi_s \qquad (13.51)$$

Using equation (13.51) in (13.50), we get

$$(\nabla^2 + k^2)(\mathrm{e}^{i\vec{k}\cdot\vec{r}} + \psi_s) = U(r)\ (\mathrm{e}^{i\vec{k}\cdot\vec{r}} + \psi_s)$$

For a free-particle, we have

$$(\nabla^2 + k^2)\, e^{i\vec{k}.\vec{r}} = 0$$

and therefore, we have

$$(\nabla^2 + k^2)\psi_s = U(r)\ \Psi(\vec{r}) = \rho(\vec{r}) \tag{13.52}$$

Equation (13.52) is an inhomogeneous equation and the inhomogeneous term depends on $\Psi(\vec{r})$. Solution of this inhomogeneous equation can be obtained by the Green's function method. Let the Green's function $G(\vec{r}, \vec{r}')$ is the solution of inhomogeneous equation with a delta function source term. That is

$$(\nabla^2 + k^2)G(\vec{r}, \vec{r}') = \delta(\vec{r} - \vec{r}')$$

The solution of the scattering equation for the density function $\rho(\vec{r})$ cane be expressed as

$$\psi_s = \int G(\vec{r}, \vec{r}')\ \rho(\vec{r}')\ d\tau'$$

After applying Green's function technique and contour integration, we can have

$$G(\vec{r}, \vec{r}') = -\frac{\exp(ik|\ \vec{r} - \vec{r}'\ |)}{4\pi|\ \vec{r} - \vec{r}'\ |}$$

so that

$$\psi_s(\vec{r}) = -\frac{1}{4\pi}\int \frac{\exp(ik|\ \vec{r} - \vec{r}'\ |)}{|\ \vec{r} - \vec{r}'\ |}\ U(r')\Psi(\vec{r}')\ d\tau' \tag{13.53}$$

where $\vec{r}$ is the position of the scattered particle after scattering in the region $\vec{r}' \approx 0$. The scattered wave ψ_s at the point $\vec{r}$ has the form of the superposition of spherical waves originating from all points $\vec{r}'$ with amplitudes $U(r')\Psi(\vec{r}')$. Geometry of the vectors $\vec{r}$ and $\vec{r}'$ is shown in Figure 2. Suppose $\hat{n}$ be the unit vector along the direction of the vector $\vec{r}$. Since $\vec{r}$ is very large, Figure 2 shows that

$$|\ \vec{r} - \vec{r}'\ | \approx r - \hat{n}.\ \vec{r}'$$

In the center-of-mass system, for elastic scattering the magnitude of momentum $\vec{k}'$ of the scattered particle is equal to the magnitude of momentum $\vec{k}$ of the incident particle, *i.e.*, $k' = k$. So that we can write $\vec{k}' = k\,\hat{n}$. Therefore, we have

$$k \mid \vec{r} - \vec{r}' \mid = k(r - \hat{n}.\,\vec{r}') = kr - \vec{k}'.\,\vec{r}'$$

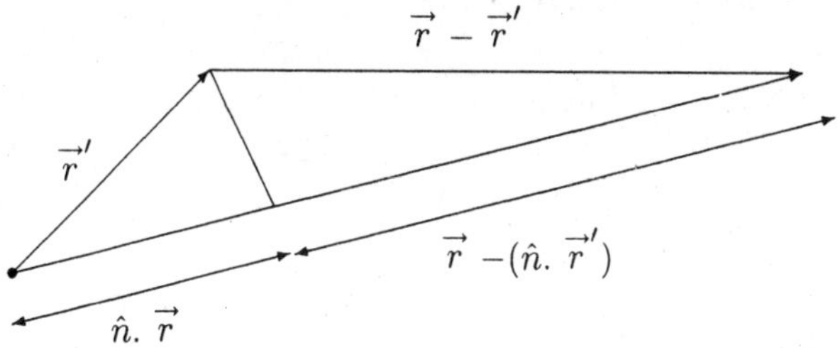

Figure 2: The vectors $\vec{r}$ and $\vec{r}'$ in a scattering problem.

Replacing $\mid \vec{r} - \vec{r}' \mid$ by r in the denominator of equation (13.53), the wave function $\Psi(\vec{r})$ is

$$\Psi(\vec{r}) \underset{r \to \infty}{\longrightarrow} \exp(i\vec{k}.\vec{r}) - \frac{1}{4\pi}\int \frac{\exp[i(kr - \vec{k}'.\vec{r}')]}{r}\, U(r')\Psi(\vec{r}')\, d\tau' \quad (13.54)$$

This is the integral equation for the wave function as $\Psi(\vec{r})$ appears in the argument of the integral on the right side. Comparison of equation (13.54) with (13.4) gives

$$f(\theta) = -\frac{1}{4\pi}\int \exp(-i\vec{k}'.\vec{r}')U(r')\Psi(\vec{r}')\, d\tau' \qquad (13.55)$$

which can be used in calculating the differential scattering cross-section $\sigma(\theta)$.

11. Born approximation

Equation (13.55) shows that the wave function $\Psi(\vec{r}')$ is required for computation of $f(\theta)$ which is a part of the same wave function (equation 13.54). This creates non-linearity in the picture. Born made an approximation that in the wave function $\Psi(\vec{r}')$, we can neglect the scattering part in comparison to the incident wave. That is, we can replace $\Psi(\vec{r}')$ by $\exp(i\,\vec{k}\,.\,\vec{r}')$ under the Born approximation. This calculated value of $f(\theta)$ is used in equation (13.54) for obtaining new value of $\Psi(\vec{r}')$ which is in turn used in equation (13.55) for obtaining a new value of

$f(\theta)$. This is known as the second order Born approximation. This iterative method goes on and we go on calculating the values in the higher order Born approximations, till the value of $\Psi(\vec{r}')$ is reproduced within the required accuracy. We shall however consider here the first order Born approximation so that we get

$$f(\theta) = -\frac{1}{4\pi}\int \exp(-i\vec{k}'.\vec{r}')U(r')\exp(i\vec{k}.\vec{r}')\,d\tau'$$

$$= -\frac{1}{4\pi}\int \exp[i(\vec{k} - \vec{k}').\,\vec{r}']U(r')\,d\tau' \tag{13.56}$$

where $\vec{k}$ and $\vec{k}'$ are the wave vectors in the incident and scattered directions (Figure 3). For the elastic scattering, magnitudes of them are equal. From Figure 3 we have $\vec{q}=\vec{k} - \vec{k}'$ and

$$\sin(\theta/2) = \frac{q/2}{k} \qquad\qquad q = 2k\ \sin(\theta/2) \tag{13.57}$$

Using the spherical polar coordinates, equation (13.56) reduces to

$$f(\theta) = -\frac{1}{4\pi}\int \exp(i\vec{q}.\vec{r}']U(r')\,d\tau'$$

$$= -\frac{1}{4\pi}\int_0^\infty\int_0^\pi\int_0^{2\pi} \exp(iqr'\cos\theta')U(r')r'^2\sin\theta'\,dr'\,d\theta'\,d\phi'$$

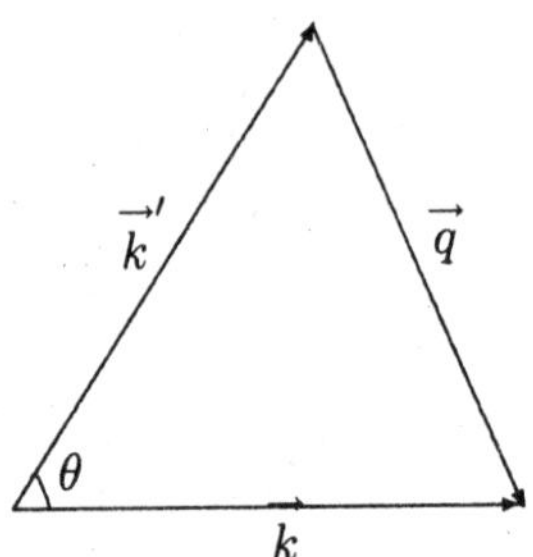

Figure 3: Representation of vectors $\vec{k}$, $\vec{k}'$ and $\vec{q}$. The vector $\vec{k}$ is in the incident direction and the vector $\vec{k}'$ is in the scattered direction.

Integration over ϕ gives 2π. For the integration over θ, let

$$u = \cos\theta' \qquad\qquad du = -\sin\theta'\,d\theta'$$

and thus,

$$\int_0^{\pi} \exp(iqr'\cos\theta')\sin\theta'\, d\theta' = \int_{-1}^{1} \exp(iqr'u)\, du$$

$$= \frac{\exp(iqr') - \exp(-iqr')}{iqr'} = \frac{2\sin(qr')}{qr'}$$

Thus, we have

$$f(\theta) = -\frac{2\mu}{\hbar^2}\int_0^{\infty} \frac{\sin(qr')}{qr'} V(r')\, r'^2\, dr' \tag{13.58}$$

This can be used for the calculation of the cross section $\sigma(\theta)$. Equation (13.58) shows that the only variable in $f(\theta)$ is the magnitude of the momentum transfer $q\hbar$, where q is expressed by equation (13.57). Hence, the scattering cross-section depends on the momentum of the incident particle $k\hbar$ and the scattering angle θ through the combination $q = 2k\sin(\theta/2)$.

12. Scattering by a screened coulomb potential

As an example of the first Born approximation, let us consider scattering of a particle having charge $Z'e$ by an atomic nucleus Ze. The interaction between the particle and the nucleus is obviously screened by electrons revolving around the nucleus. The potential representing interaction may be written as

$$V(r) = -\frac{ZZ'e^2}{r}\, e^{-\alpha r} \tag{13.59}$$

where α is a measure of screening by atomic electrons. Using equation (13.59) in (13.58), we get[5]

$$f(\theta) = \frac{2\mu ZZ'e^2}{\hbar^2 q}\int_0^{\infty} \frac{\sin(qr')}{qr'} e^{-\alpha r}\, dr'$$

$$= \frac{2\mu ZZ'e^2}{\hbar^2 q}\,\frac{q}{q^2+\alpha^2} = \frac{2\mu ZZ'e^2}{\hbar^2(q^2+\alpha^2)}$$

[5]We have

$$\int_0^{\infty} e^{-ax}\sin(bx)dx = \frac{e^{-ax}}{a^2+b^2}\Big[-a\sin(bx) - b\cos(bx)\Big]_0^{\infty} = \frac{b}{a^2+b^2}$$

Thus, the differential cross section is

$$\sigma(\theta) = |f(\theta)|^2 = \left(\frac{2\mu Z Z' e^2}{\hbar^2}\right)^2 \frac{1}{(q^2+\alpha^2)^2}$$

Special case: When α is negligibly small, we have coulomb potential

$$V(r) = -\frac{ZZ'e^2}{r} \qquad \text{and} \qquad f(\theta) = \frac{2\mu ZZ'e^2}{\hbar^2 q^2}$$

Using $q = 2k\sin(\theta/2)$, we get

$$f(\theta) = \frac{\mu ZZ'e^2}{2\hbar^2 k^2 \sin^2(\theta/2)}$$

and

$$\sigma(\theta) = \frac{\mu^2 Z^2 Z'^2 e^4}{4\hbar^4 k^4 \sin^4(\theta/2)}$$

It is the Rutherford's scattering formula.

13. Validity of the Born approximation

In the Born approximation, $\Psi(\vec{r}\,')$ is replaced by $\exp(i\vec{k}\,.\,\vec{r}\,')$, which is only possible when ψ_s is small in comparison to the incident wave.

$$|\psi_s| << 1$$

$$\left|\frac{1}{4\pi}\int \frac{\exp(-ikr')}{r'} U(r')\exp(i\vec{k}\,.\,\vec{r}\,')\ d\tau'\right| << 1$$

$$\left|\frac{1}{4\pi}\int_0^\infty\int_0^\pi\int_0^{2\pi} \frac{\exp(-ikr')}{r'} U(r')\exp(ikr'\cos\theta')r'^2\sin\theta'\ dr'\ d\theta'\ d\phi'\right| << 1$$

Integration over ϕ gives 2π. For integration over θ, let

$$u = \cos\theta' \qquad du = -\sin\theta'\ d\theta'$$

and thus, we have

$$\int_0^\pi \exp(ikr'\cos\theta')\sin\theta'\ d\theta' = \int_{-1}^1 \exp(ikr'u)\ du$$

$$= \frac{\exp(ikr') - \exp(-ikr')}{ikr'} = \frac{2\sin(kr')}{kr'}$$

Thus, we have

$$\left|\frac{2\mu}{k\hbar^2}\int_0^\infty \exp(-ikr')\sin(kr')V(r')\ dr'\right| << 1$$

This is the condition for validity of the Born approximation.

14. Laboratory and center-of-mass coordinate systems

In the study of scattering, two coordinate systems: (i) the laboratory coordinate system (L) and (ii) the center-of-mass coordinate system (CM), are used. In the laboratory system, the target T (scattering center) of mass M remains at rest and the incident particle P of mass m moves. In the other system, the target as well as the incident particle move in such a manner that the center of the masses of T and P remains at stationary.

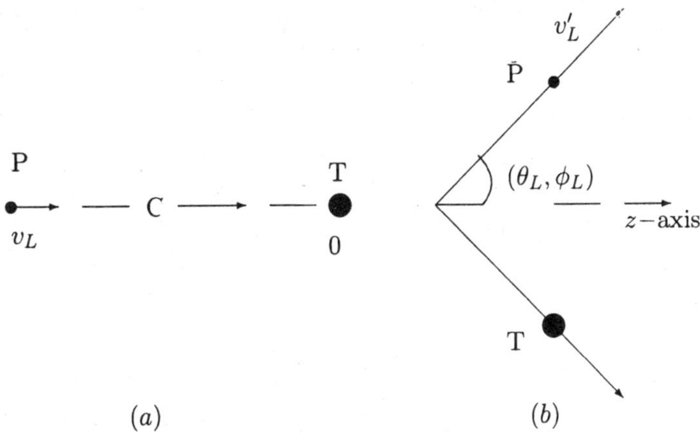

Figure 4: Motion of a particle in the laboratory system: (a) before collision, the target T is at rest and the incident particle P is moving with velocity v_L along $z-$axis, and (b) after collision, both the T and P move in different directions with respect to $z-$axis. Here, C denotes the center of the masses of T and P. The ratio of the distances of T and P from C is m/M.

Experimental observations are taken in the laboratory system whereas the center-of-mass system makes theoretical treatment convenient, as the motion of two particles can be expressed like the motion of two separate particles.

In the laboratory system, the target of mass M is at rest and the incident particle of mass m moves with velocity v_L. Here, the center-of-mass will move with a velocity v_{CM} towards the target T. This velocity v_{CM} can be calculated by considering the conservation of linear momentum as the following.

$$m\, v_L + 0 = v_{CM}(m + M)$$
$$v_{CM} = \frac{m\, v_L}{m + M}$$

After scattering, P is moving with a velocity v'_L in a direction (θ_L, ϕ_L).

Let us examine how this experiment is seen by an observer sitting at the center of mass. The target T looks as moving towards the observer with a velocity $-mv_L/(m+M)$ and the particle P looks as moving towards the observer with a velocity

$$v_c = v_L - v_{CM} = v_L - \frac{mv_L}{m+M} = \frac{Mv_L}{m+M}$$

For the elastic scattering, in the center-of-mass system, in order to keep C as stationary, the two particles must be scattered in the opposite directions with their initial speeds, as there is no exchange of energy between T and P. The scattering process in the center-of-mass system is shown in Figure 5. After the scattering, P is moving in a direction (θ_c, ϕ_c).

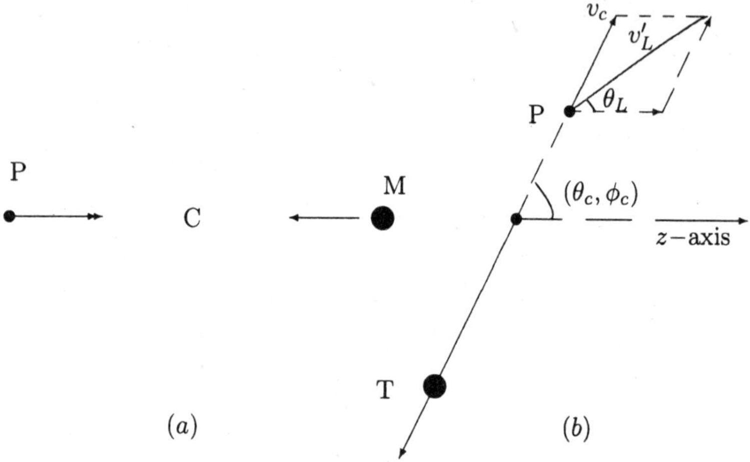

Figure 5: Motion of a particle in the center-of-mass system: (a) before collision and (b) after collision. In this system, the center of the masses C remains stationary. Hence, before the collision, P and T move with the velocities $Mv_L/(m+M)$ and $-mv_L/(m+M)$, respectively, relative to C. After the scattering, the particles are considered to move in such a manner that still the center-of-mass C remains stationary.

While the center of the masses in the laboratory system moves with a velocity v_{CM}, it is at rest in the center-of-mass system. From Figure 5, we have

$$v'_L \cos\theta_L - v_{CM} = v_c \cos\theta_c \tag{13.60}$$

$$v'_L \sin\theta_L = v_c \sin\theta_c \tag{13.61}$$

The angle ϕ obviously does not depend on the coordinate system and thus, $\phi_L = \phi_c$. Dividing equation (13.61) by (13.60), we get

$$\tan\theta_L = \frac{v_c \sin\theta_c}{v_c \cos\theta_c + v_{CM}} = \frac{\sin\theta_c}{\cos\theta_c + \alpha} \tag{13.62}$$

$\alpha = v_{CM}/v_c = m/M$. When the mass of the target T is infinitely larger than that of the particle P, we have $\alpha \approx 0$ and $\theta_L = \theta_c$.

The relation between the differential scattering cross-sections in the two systems can be obtained by considering the fact that the number of particles scattered in a direction are the same in the two systems (equation 13.1).

$$\begin{aligned}
&n = N\sigma_c(\theta_c, \phi_c)\ d\Omega_c = N\sigma_L(\vartheta_L, \phi_L)\ d\Omega_L \\
&\sigma_c(\theta_c, \phi_c) \sin\theta_c\ d\theta_c\ d\phi_c = \sigma_L(\theta_L, \phi_L) \sin\theta_L\ d\theta_L\ d\phi_L \\
&\sigma_c(\theta_c, \phi_c) \sin\theta_c\ d\theta_c = \sigma_L(\theta_L, \phi_L) \sin\theta_L\ d\theta_L
\end{aligned} \tag{13.63}$$

Squaring equation (13.62), we have

$$\frac{\sin^2\theta_L}{\cos^2\theta_L} = \frac{\sin^2\theta_c}{(\cos\theta_c + \alpha)^2}$$

$$\frac{1 - \cos^2\theta_L}{\cos^2\theta_L} = \frac{\sin^2\theta_c}{(\cos\theta_c + \alpha)^2}$$

$$\frac{1}{\cos^2\theta_L} = \frac{\sin^2\theta_c}{(\cos\theta_c + \alpha)^2} + 1$$

$$\frac{1}{\cos^2\theta_L} = \frac{\sin^2\theta_c + \cos^2\theta_c + \alpha^2 + 2\alpha\ \cos\theta_c}{(\cos\theta_c + \alpha)^2} = \frac{1 + \alpha^2 + 2\alpha\ \cos\theta_c}{(\cos\theta_c + \alpha)^2}$$

$$\cos\theta_L = \frac{|\alpha + \cos\theta_c|}{(1 + \alpha^2 + 2\alpha\ \cos\theta_c)^{1/2}}$$

On differentiating, we get

$$\sin\theta_L\ d\theta_L = \frac{|1 + \alpha\ \cos\theta_c|}{(1 + \alpha^2 + 2\alpha\ \cos\theta_c)^{3/2}} \sin\theta_c\ d\theta_c \tag{13.64}$$

Using equation (13.64) in (13.63), we have

$$\sigma_c(\theta_c, \phi_c)\ \sin\theta_c\ d\theta_c = \sigma_L(\theta_L, \phi_L) \frac{|1 + \alpha\ \cos\theta_c|}{(1 + \alpha^2 + 2\alpha\ \cos\theta_c)^{3/2}} \sin\theta_c\ d\theta_c$$

$$\sigma_L(\theta_L, \phi_L) = \frac{(1 + \alpha^2 + 2\alpha\ \cos\theta_c)^{3/2}}{|1 + \alpha\ \cos\theta_c|} \sigma_c(\theta_c, \phi_c)$$

When $M >> m$, we have $\alpha \approx 0$ and $\sigma_L(\theta_L, \phi_L) = \sigma_c(\theta_c, \phi_c)$. When $M = m$, we have $\alpha = 1$ and

$$\sigma_L(\theta_L, \phi_L) = \frac{(2 + 2\cos\theta_c)^{3/2}}{|1 + \cos\theta_c|}\,\sigma_c(\theta_c, \phi_c)$$

$$= 2\sqrt{2}(1 + \cos\theta_c)^{1/2}\,\sigma_c(\theta_c, \phi_c)$$

$$= 4\cos^2(\theta_c/2)\,\sigma_c(\theta_c, \phi_c) \qquad (13.65)$$

Using $\alpha = 1$ in equation (13.62), we have

$$\tan\theta_L = \frac{\sin\theta_c}{\cos\theta_c + 1} = \tan(\theta_c/2)$$

Thus, $\theta_L = \theta_c/2$, and equation (13.65) gives

$$\sigma_L(\theta_L, \phi_L) = 4\cos^2\theta_L\,\sigma_c(2\theta_L, \phi_c)$$

Since θ_L varies from 0 to π, in this case θ_c varies from 0 to $\pi/2$. Hence, this is the case of no back-scattering.

15. Problems and questions

1. Define differential scattering cross section and total scattering cross section.

2. Define scattering amplitude $f(\theta)$ for spherically symmetric potential and obtain relation between the differential scattering cross section $\sigma(\theta)$ and the scattering amplitude.

3. Show that a plane wave can be expressed as a linear combination of partial waves. Further, show that each partial wave can be represented as a sum of an incoming wave and an outgoing wave.

4. Obtain an expressions for the scattering amplitude as well as for the scattering cross section in the partial wave analysis of scattering by a central potential.

5. Which method is used for low energy scattering. Why this method is not applied for high energy scattering. Which method is used for high energy scattering.

6. Obtain expression for a phase-shift of s-wave in the scattering by an attractive square-well potential. Discuss the phenomenon of resonance in the low energy scattering.

7. Derive an expression for phase-shifts.

8. Derive integral equation for general wave function in the scattering. What is the first Born approximation? Discuss about the validity of the first Born approximation.

9. In the first Born approximation, for a pure potential $-ZZ'e^2/r$, show that the differential cross section $\sigma(\theta)$ is

$$\sigma(\theta) = \frac{\mu^2 Z^2 Z'^2 e^4}{4\hbar^4 k^4 \sin^4(\theta/2)}$$

where θ is the scattering angle for the elastic scattering.

10. Describe the laboratory and the center-of-mass coordinate systems. Obtain relationship between the differential cross sections in the two systems.

11. Write notes on the following

    (i) Partial waves

    (ii) Relation between the scattering amplitude $f(\theta)$ and the differential scattering cross section $\sigma(\theta)$

    (iii) Optical theorem

    (iv) Significant number of partial waves in the low energy scattering

    (v) Scattering length

    (vi) Laboratory and center of mass coordinate systems in scattering

Relativistic Wave Mechanics

We have yet considered nonrelativistic Hamiltonian, therefore the quantum mechanics developed so far can be applied to the nonrelativistic systems only. Further, the nonrelativistic wave equation is of first order in time and of second order in space coordinates. Thus, it does not satisfy the condition of space-time symmetry, required for the relativity. For a relativistic motion, we obviously need a relativistic wave equation. Two ways for obtaining the relativistic wave equations have been suggested by the scientists. In one approach, the nonrelativistic Hamiltonian for a free particle is replaced by relativistic one. In the second approach, a first order wave equation both in space as well as in time is obtained. Klein and Gordon used the first idea and derived an equation, known as the 'Klein-Gordon equation', in 1926. This equation however does not say anything about the spin of the particles. Therefore, it is used to describe the particles with zero spin, such as π-meson, K-meson, etc. Based on the second idea, Dirac developed another relativistic wave equation, where the concept of spin evolved automatically and this equation is applicable to a system of particles with one-half spin. In this chapter, we shall discuss about these two relativistic wave equations.

1. Klein-Gordon equation

In the preceding chapters, the nonrelativistic Schrödinger equation was obtained by replacing the momentum $\vec{p}$ and energy E by the corresponding operators $-i\hbar\nabla$ and $i\hbar\frac{\partial}{\partial t}$, respectively. In the nonrelativistic case, the expression for energy E of a free particle is $E = p^2/2m$. Here, p is momentum and m the mass of the particle. In the non-relativistic case, the mass of the particle does not vary with velocity and therefore it is taken equal to the rest mas of the particle. In this relation between the energy and momentum, each physical quantity was replaced by the corresponding operator, and after getting the operator equation, it was allowed to operate on a wave function. Similar approach can be followed here for the relativistic case. For the relativistic motion, the energy E and the momentum $\vec{p}$ for a free particle are related through the relation

$$E^2 = c^2p^2 + m_0^2c^4 \tag{14.1}$$

where m_0 is the rest mass of the particle and c is the speed of light. (Mass of a relativistic particle varies with velocity.) This energy is the kinetic energy of the particle. This energy is taken as the total energy, as the potential energy of the particle is zero. Here, also E and $\vec{p}$ are replaced by the same operators $-i\hbar\nabla$ and $i\hbar\frac{\partial}{\partial t}$, respectively, and we get the operator equation

$$-\hbar^2\frac{\partial^2}{\partial t^2} = -c^2\hbar^2\nabla^2 + m_0^2c^4$$

On operating this operator equation on a wave function $\Psi(\vec{r}, t)$, we get

$$-\hbar^2\frac{\partial^2}{\partial t^2}\Psi(\vec{r}, t) = -\hbar^2c^2\nabla^2\Psi(\vec{r}, t) + m_0^2c^4\Psi(\vec{r}, t) \tag{14.2}$$

This is the 'Klein-Gordon equation' or Schrödinger's relativistic equation. Equation (14.2) can be rearranged as

$$\left[\nabla^2 - \frac{1}{c^2}\frac{\partial^2}{\partial t^2}\right]\Psi(\vec{r}, t) = \frac{m_0^2c^2}{\hbar^2}\Psi(\vec{r}, t) \tag{14.3}$$

or

$$\Box^2\Psi(\vec{r}, t) = \frac{m_0^2c^2}{\hbar^2}\Psi(\vec{r}, t) \tag{14.4}$$

Here, the symbol $\Box^2$ defined as

$$\Box^2 \equiv \left[\nabla^2 - \frac{1}{c^2}\frac{\partial^2}{\partial t^2}\right]$$

is known as the de Alembertian operator which is invariant under Lorentz transformations. Therefore, equation (14.4) is relativistically invariant if $\Psi(\vec{r}, t)$ transforms like a scalar.

Exercise 1. Show that the operator $\nabla^2 - \frac{1}{c^2}\frac{\partial^2}{\partial t^2}$ in invariant under Lorentz transformations.

Solution: The Lorentz transformations are

$$x' = \gamma(x - Vt); \qquad y' = y; \qquad z' = z; \qquad t' = \gamma(t - Vx/c^2)$$

where $\gamma = 1/\sqrt{1-\beta^2}$ and $\beta = V/c$. It gives

$$\frac{\partial}{\partial x'} = \frac{\partial}{\partial x}\frac{\partial x}{\partial x'} + \frac{\partial}{\partial t}\frac{\partial t}{\partial x'} \qquad \frac{\partial}{\partial y'} = \frac{\partial}{\partial y}\frac{\partial y}{\partial y'} \tag{14.5}$$

$$\frac{\partial}{\partial z'} = \frac{\partial}{\partial z}\frac{\partial z}{\partial z'} \qquad \frac{\partial}{\partial t'} = \frac{\partial}{\partial x}\frac{\partial x}{\partial t'} + \frac{\partial}{\partial t}\frac{\partial t}{\partial t'} \tag{14.6}$$

The inverse Lorentz transformations are

$$x = \gamma(x' + Vt'); \qquad y = y'; \qquad z = z'; \qquad t = \gamma(t' + Vx'/c^2)$$

It gives

$$\frac{\partial x}{\partial x'} = \gamma \qquad \frac{\partial x}{\partial t'} = \gamma V \qquad \frac{\partial y}{\partial y'} = 1 \tag{14.7}$$

$$\frac{\partial z}{\partial z'} = 1 \qquad \frac{\partial t}{\partial t'} = \gamma \qquad \frac{\partial t}{\partial x'} = \frac{\gamma V}{c^2} \tag{14.8}$$

From equations (14.5), (14.6), (14.7) and (14.8), we have

$$\frac{\partial}{\partial x'} = \gamma\left(\frac{\partial}{\partial x} + \frac{V}{c^2}\frac{\partial}{\partial t}\right) \qquad \frac{\partial}{\partial y'} = \frac{\partial}{\partial y} \tag{14.9}$$

$$\frac{\partial}{\partial z'} = \frac{\partial}{\partial z} \qquad \frac{\partial}{\partial t'} = \gamma\left(V\frac{\partial}{\partial x} + \frac{\partial}{\partial t}\right) \tag{14.10}$$

Now, we have

$$\begin{aligned}
\nabla'^2 - \frac{1}{c^2}\frac{\partial^2}{\partial t'^2} &= \frac{\partial^2}{\partial x'^2} + \frac{\partial^2}{\partial y'^2} + \frac{\partial^2}{\partial z'^2} - \frac{1}{c^2}\frac{\partial^2}{\partial t'^2} \\
&= \frac{\partial}{\partial x'}\left(\frac{\partial}{\partial x'}\right) + \frac{\partial}{\partial y'}\left(\frac{\partial}{\partial y'}\right) + \frac{\partial}{\partial z'}\left(\frac{\partial}{\partial z'}\right) - \frac{1}{c^2}\frac{\partial}{\partial t'}\left(\frac{\partial}{\partial t'}\right)
\end{aligned}$$

Using equations (14.9) and (14.10) here we have

$$\begin{aligned}
\nabla'^2 - \frac{1}{c^2}\frac{\partial^2}{\partial t'^2} &= \gamma^2\left(\frac{\partial}{\partial x} + \frac{V}{c^2}\frac{\partial}{\partial t}\right)\left(\frac{\partial}{\partial x} + \frac{V}{c^2}\frac{\partial}{\partial t}\right) + \frac{\partial}{\partial y}\left(\frac{\partial}{\partial y}\right) + \frac{\partial}{\partial z}\left(\frac{\partial}{\partial z}\right) \\
&\quad - \frac{\gamma^2}{c^2}\left(V\frac{\partial}{\partial x} + \frac{\partial}{\partial t}\right)\left(V\frac{\partial}{\partial x} + \frac{\partial}{\partial t}\right) \\
&= \gamma^2\left(\frac{\partial^2}{\partial x^2} + \frac{2V}{c^2}\frac{\partial^2}{\partial x\partial t} + \frac{V^2}{c^4}\frac{\partial^2}{\partial t^2}\right) + \frac{\partial^2}{\partial y^2} + \frac{\partial^2}{\partial z^2} \\
&\quad - \frac{\gamma^2}{c^2}\left(V^2\frac{\partial^2}{\partial x^2} + 2V\frac{\partial^2}{\partial x\partial t} + \frac{\partial^2}{\partial t^2}\right)
\end{aligned}$$

$$= \gamma^2\left(1 - \frac{V^2}{c^2}\right)\frac{\partial^2}{\partial x^2} - \frac{\gamma^2}{c^2}\left(1 - \frac{V^2}{c^2}\right)\frac{\partial^2}{\partial t^2} + \frac{\partial^2}{\partial y^2} + \frac{\partial^2}{\partial z^2}$$

$$= \frac{\partial^2}{\partial x^2} + \frac{\partial^2}{\partial y^2} + \frac{\partial^2}{\partial z^2} - \frac{1}{c^2}\frac{\partial^2}{\partial t^2} = \nabla^2 - \frac{1}{c^2}\frac{\partial^2}{\partial t^2}$$

1.1 Plane-wave solution

We have seen earlier that

$$\Psi(\vec{r},t) = \exp[i(\vec{k}\,.\,\vec{r} - \omega t)] \tag{14.11}$$

represents a plane wave and it is an eigenfunction of both energy and momentum operators with the eigenvalues $\hbar\omega$ and $k\hbar$, respectively. Let us assume that for the relativistic equation also this expression (14.11) represents a plane wave. Substitution of equation (14.11) in (14.2) gives

$$(\hbar\omega)^2 = c^2\hbar^2k^2 + m_0^2c^4$$

$$\hbar\omega = \pm(c^2\hbar^2k^2 + m_0^2c^4)^{1/2}$$

This shows that the energy eigenvalue can have both positive as well as negative values. Klein and Gordon however could not provide any acceptable explanation for this negative energy.

2. Interpretation of Klein-Gordon equation

In the nonrelativistic case, the position probability density $P(\vec{r},t)$ is defined as $|\vec{\Psi}(\vec{r},t)|^2$ and the probability current density $\vec{j}(\vec{r},t)$ is expressed as

$$\vec{j}(\vec{r},t) = \frac{i\hbar}{2m_0}(\Psi\nabla\Psi^* - \Psi^*\nabla\Psi)$$

where the mass m_0 of the particle does not vary with velocity. The quantities $P(\vec{r},t)$ and $\vec{j}(\vec{r},t)$ satisfy the equation

$$\frac{\partial}{\partial t}P(\vec{r},t) + \nabla.\,\vec{j}(\vec{r},t) = 0$$

This equation is the equation of continuity for the non-relativistic case. Since equation (14.4) is invariant under the Lorentz transformation, we can therefore expect the same continuity relation for the relativistic case also. Complex conjugate of equation (14.2) is

$$-\hbar^2 \frac{\partial^2}{\partial t^2} \Psi^*(\vec{r},t) = -\hbar^2 c^2 \nabla^2 \Psi^*(\vec{r},t) + m_0^2 c^4 \Psi^*(\vec{r},t) \qquad (14.12)$$

Multiplying equation (14.2) on the left by Ψ^*, and equation (14.12) on the left by Ψ and then subtracting the latter from the former, we get

$$\Psi^* \frac{\partial^2 \Psi}{\partial t^2} - \Psi . \frac{\partial^2 \Psi^*}{\partial t^2} = c^2(\Psi^* \nabla^2 \Psi - \Psi \nabla^2 \Psi^*)$$

This equation can be rearranged as

$$\frac{\partial}{\partial t}\left[\Psi^* \frac{\partial \Psi}{\partial t} - \Psi \frac{\partial \Psi^*}{\partial t}\right] = c^2 \nabla.(\Psi^* \nabla \Psi - \Psi \nabla \Psi^*) \qquad (14.13)$$

Here, we have used the identity[1] $\nabla.(\phi\nabla\theta) = \phi\nabla^2\theta + \nabla\phi.\nabla\theta$. If we define

$$P(\vec{r},t) = \frac{i\hbar}{2m_0c^2}\left[\Psi^* \frac{\partial \Psi}{\partial t} - \Psi \frac{\partial \Psi^*}{\partial t}\right] \qquad (14.14)$$

and

$$\vec{j}(\vec{r},t) = \frac{i\hbar}{2m_0}(\Psi \nabla \Psi^* - \Psi^* \nabla \Psi) \qquad (14.15)$$

so that equation (14.13) is

$$\frac{\partial}{\partial t} P(\vec{r},t) + \nabla . \vec{j}(\vec{r},t) = 0$$

This is the equation of continuity for the relativistic case. This expression for the equation of continuity is the same as for the non-relativistic case. It shows that when the position probability density is expressed by

---

[1] We have

$$\begin{aligned}
\nabla.(\phi \vec{A}) &= \frac{\partial}{\partial x}(\phi A_x) + \frac{\partial}{\partial y}(\phi A_y) + \frac{\partial}{\partial z}(\phi A_z) \\
&= \phi \frac{\partial A_x}{\partial x} + A_x \frac{\partial \phi}{\partial x} + \phi \frac{\partial A_y}{\partial y} + A_y \frac{\partial \phi}{\partial y} + \phi \frac{\partial A_z}{\partial z} + A_z \frac{\partial \phi}{\partial z} \\
&= \phi\left[\frac{\partial A_x}{\partial x} + \frac{\partial A_y}{\partial y} + \frac{\partial A_z}{\partial z}\right] + A_x \frac{\partial \phi}{\partial x} + A_y \frac{\partial \phi}{\partial y} + A_z \frac{\partial \phi}{\partial z} \\
&= \phi \nabla . \vec{A} + \vec{A} . \nabla \phi
\end{aligned}$$

For $\vec{A} = \nabla\theta$, we have

$$\nabla.(\phi\nabla\theta) = \phi \nabla.\nabla\theta + \nabla\theta.\nabla\phi = \phi\nabla^2\theta + \nabla\theta.\nabla\phi$$

equation (14.14) and the probability current density by equation (14.15), the equation of continuity is obtained. It can be noted that this expression for $\vec{j}(\vec{r},t)$ is the same as that for the nonrelativistic case. However, the expression for $P(\vec{r},t)$ is completely different from that for the non-relativistic case. Further, for a real value of Ψ, the position probability density $P(\vec{r},t)$ vanishes and we just have $\nabla.\vec{j}(\vec{r},t) = 0$, which is physically unacceptable. Therefore, for a physically acceptable situation, Ψ must be a complex quantity.

Equation (14.14) can be written as

$$\begin{aligned} P(\vec{r},t) &= \frac{1}{2m_0c^2}\left[\Psi^* i\hbar\frac{\partial\Psi}{\partial t} + \Psi\Big(-i\hbar\frac{\partial}{\partial t}\Big)\Psi^*\right] \\ &= \frac{1}{2m_0c^2}[\Psi^* E\Psi + \Psi(E\Psi)^*] = \frac{E|\Psi|^2}{m_0c^2} \end{aligned} \tag{14.16}$$

Here, we have used the fact that the energy E is a real quantity. This equation (14.16) shows that the probability density $P(\vec{r},t)$ takes both positive as well as negative values depending on the value of E. Since the position probability density cannot be a negative quantity, the $P(\vec{r},t)$ given by (14.14) should not represent the position probability density.

This had been a reason not to accept the Klein-Gordon equation for some time. This problem was resolved by Pauli and Wesskopf in 1934 by interpreting eP as the electrical charge density and $e\vec{j}$ as the corresponding electric current density. This explanation is acceptable as a charge can have positive as well as negative values. If the system has a single particle of a given charge, the $P(\vec{r},t)$ cannot have different signs at different points. It shows that this theory is useful only for a system of particles having both positive as well as negative charges. This interpretation says that the relativistic equation is for a system of arbitrary number of particles and their antiparticles.

3. Inclusion of potential field

As we included potential field in the non-relativistic case, here also it can be included. After accounting for the potential field $V(r)$, equation (14.1) can be expressed as

$$(E-V)^2 = c^2p^2 + m_0^2c^4 \tag{14.17}$$

Now, E represents the total energy of the particle, so that the kinetic energy is $(E-V)$. For a Coulomb field between the two charges Ze and $-e$, we have $V = -Ze^2/r$. Now we replace E and $\vec{p}$ by their operators

$-i\hbar\nabla$ and $i\hbar\frac{\partial}{\partial t}$, respectively, and finally operate this operator equation on the wave function $\Psi(\vec{r},t)$ to get

$$\left[i\hbar\frac{\partial}{\partial t}+\frac{Ze^2}{r}\right]\Psi(\vec{r},t)=\left(-c^2\hbar^2\nabla^2+m_0^2c^4\right)\Psi(\vec{r},t) \tag{14.18}$$

Since the potential V does not depend on time, the wave function $\Psi(\vec{r},t)$ can be expressed as a product of two parts, one depending on $\vec{r}$ only and the other on t only, by writing

$$\Psi(\vec{r},t)=\psi(\vec{r})\exp\left(-\frac{iEt}{\hbar}\right) \tag{14.19}$$

Using equation (14.19) in (14.18) we get

$$\left[E+\frac{Ze^2}{r}\right]\psi(\vec{r})=\left(-c^2\hbar^2\nabla^2+m_0^2c^4\right)\psi(\vec{r})$$

This equation can be rearranged as

$$\nabla^2\psi(\vec{r})=\frac{1}{c^2\hbar^2}\left[m_0^2c^4-\left(E+\frac{Ze^2}{r}\right)^2\right]\psi(\vec{r})$$

In the spherical polar coordinates, this equation is expressed as

$$\frac{1}{r^2}\frac{\partial}{\partial r}\left(r^2\frac{\partial\psi}{\partial r}\right)+\frac{1}{r^2\sin\theta}\frac{\partial}{\partial\theta}\left(\sin\theta\frac{\partial\psi}{\partial\theta}\right)+\frac{1}{r^2\sin^2\theta}\frac{\partial^2\psi}{\partial\phi^2}$$
$$=\frac{1}{c^2\hbar^2}\left[m_0^2c^4-\left(E+\frac{Ze^2}{r}\right)^2\right]\psi(\vec{r}) \tag{14.20}$$

Equation (14.20) can be separated into radial and angular parts (as we did earlier) by expressing the wave function as

$$\psi(r,\theta,\phi)=R(r)\,Y_{lm}(\theta,\phi) \tag{14.21}$$

where $Y_{lm}(\theta,\phi)$ is a spherical harmonic, satisfying the equation

$$\frac{1}{\sin\theta}\frac{\partial}{\partial\theta}\left(\sin\theta\frac{\partial Y_{lm}}{\partial\theta}\right)+\frac{1}{\sin^2\theta}\frac{\partial^2Y_{lm}}{\partial\phi^2}+l(l+1)Y_{lm}=0$$

The radial part of equation (14.20) is

$$\frac{1}{r^2}\frac{\mathrm{d}}{\mathrm{d}r}\left[r^2\frac{\mathrm{d}R}{\mathrm{d}r}\right]-\frac{l(l+1)}{r^2}R=\frac{1}{c^2\hbar^2}\left[m_0^2c^4-\left(E+\frac{Ze^2}{r}\right)^2\right]R \tag{14.22}$$

Obviously, when we put $E = E' + m_0c^2$ and assume that E' and Ze^2/r are very small in comparison to m_0c^2, equation (14.22) reduces to the nonrelativistic case.[2] On defining

$$\rho = \alpha r \quad \text{where} \quad \alpha^2 = \frac{4(m_0^2c^4 - E^2)}{c^2\hbar^2}, \quad \gamma = \frac{Ze^2}{c\hbar}, \quad \lambda = \frac{2E\gamma}{c\hbar\alpha} \tag{14.23}$$

Equation (14.22) can be rearranged as

$$\frac{1}{\rho^2}\frac{\mathrm{d}}{\mathrm{d}\rho}\left[\rho^2\frac{\mathrm{d}R}{\mathrm{d}\rho}\right] + \left[\frac{\lambda}{\rho} - \frac{1}{4} - \frac{l(l+1) - \gamma^2}{\rho^2}\right]R = 0 \tag{14.24}$$

This equation (14.24) is similar to the corresponding equation for the nonrelativistic case if we replace the factor $l(l+1) - \gamma^2$ by $l(l+1)$. Expressing

$$u(u+1) = l(l+1) - \gamma^2 \tag{14.25}$$

equation (14.24) reduces to

$$\frac{1}{\rho^2}\frac{\mathrm{d}}{\mathrm{d}\rho}\left[\rho^2\frac{\mathrm{d}R}{\mathrm{d}\rho}\right] + \left[\frac{\lambda}{\rho} - \frac{1}{4} - \frac{u(u+1)}{\rho^2}\right]R = 0 \tag{14.26}$$

As we discussed for non-relativistic case, when

$$\lambda = n' + u + 1 \qquad n' = 0, 1, 2, \ldots \tag{14.27}$$

equation (14.26) has a well-behaved solution

$$R_{nu}(r) = -\left\{\left(\frac{2Z}{na_H}\right)^3\frac{(n-u-1)!}{2n(n+1)!}\right\}^{\frac{1}{2}} e^{-\rho/2}\rho^u L_{n-u-1}^{2u+1}(\rho) \tag{14.28}$$

---

[2]The quantity in the square brackets on right side of equation (14.22) for nonrelativistic case can be simplified as

$$m_0^2c^4 - \left(E + \frac{Ze^2}{r}\right)^2 = m_0^2c^4 - \left(E' + m_0c^2 + \frac{Ze^2}{r}\right)^2$$

$$= m_0^2c^4 - m_0^2c^4\left(1 + \frac{E' + Ze^2/r}{m_0c^2}\right)^2 = m_0^2c^4 - m_0^2c^4\left(1 + 2\frac{E' + Ze^2/r}{m_0c^2}\right)$$

$$= -2m_0c^2\left(E' + \frac{Ze^2}{r}\right)$$

Using this relation in equation (14.22), we get

$$\frac{1}{r^2}\frac{\mathrm{d}}{\mathrm{d}r}\left[r^2\frac{\mathrm{d}R}{\mathrm{d}r}\right] - \frac{l(l+1)}{r^2}R + \frac{2m_0}{\hbar^2}\left(E' + \frac{Ze^2}{r}\right) = 0$$

This is the corresponding equation for nonrelativistic case when we take $m = m_0$ and $E = E'$.

and u should have nonnegative value to make the radial factor $R(r)$ finite at the origin. From equation (14.25), the value u is

$$u = \frac{-1 \pm (1 + 4l^2 + 4l - 4\gamma^2)^{1/2}}{2}$$

$$= -\frac{1}{2} \pm \left[\left(l + \frac{1}{2}\right)^2 - \gamma^2\right]^{1/2} \tag{14.29}$$

For $l = 0$, both values of u are are negative and the function $R(r)$ diverges at the origin. However, for $l > 0$, the plus sign in this equation (14.29) gives a positive value for u whereas the minus sign gives a negative value for r. Further, as the value of γ is very small in most of the cases of interest (Z-small) the plus sign (with $l = 0$) gives the value of u very close to zero whereas the minus sign gives the value of u close to -1. For heavy nuclei, the interaction between nuclei and π-mesons at the short distances is no longer Coulombic and the divergence of $R(r)$ at the origin is of no interest. From equations (14.23) and (14.27), we have

$$\frac{2E\gamma}{c\hbar\alpha} = n' + u + 1$$

Using the values of α and u (with + sign) here, we have

$$\frac{E\gamma}{(m_0^2c^4 - E^2)^{1/2}} = n' + \frac{1}{2} + \left[\left(l + \frac{1}{2}\right)^2 - \gamma^2\right]^{1/2}$$

$$= n' + \frac{1}{2} + \left(l + \frac{1}{2}\right)\left[1 - \frac{\gamma^2}{2(l + 1/2)^2}\right]$$

$$= n' + l + 1 - \frac{\gamma^2}{2(l + 1/2)} \tag{14.30}$$

Expressing

$$n = n' + l + 1 \qquad \text{and} \qquad A = \left[n - \frac{\gamma^2}{2(l + \frac{1}{2})}\right]^2$$

equation (14.30) gives

$$\frac{E\gamma}{(m_0^2c^4 - E^2)^{1/2}} = A^{1/2}$$

$$\frac{E^2\gamma^2}{A} = m_0^2c^4 - E^2$$

This equation can be rearranged as

$$E = m_0c^2\left[1 + \frac{\gamma^2}{A}\right]^{-1/2}$$

On expanding the expression on the right side and neglecting the higher order terms, we have

$$E = m_0c^2\left[1 - \frac{\gamma^2}{2A} + \frac{3\gamma^4}{8A^2}\right] = m_0c^2\left[1 - \frac{\gamma^2 A^{-1}}{2} + \frac{3\gamma^4 A^{-2}}{8}\right] \quad (14.31)$$

On expanding A^{-1} and A^{-2}, we get

$$A^{-1} = \frac{1}{n^2}\left[1 - \frac{\gamma^2}{2n(l+1/2)}\right]^{-2} = \frac{1}{n^2}\left[1 + \frac{\gamma^2}{n(l+1/2)} + \ldots\right]$$

and

$$A^{-2} = \frac{1}{n^4}\left[1 - \frac{\gamma^2}{2n(l+1/2)}\right]^{-4} = \frac{1}{n^4}\left[1 + \frac{2\gamma^2}{n(l+1/2)} + \ldots\right]$$

Using these expressions for A^{-1} and A^{-2} in equation (14.31), we have

$$E = m_0c^2\left[1 - \frac{\gamma^2}{2n^2}\left\{1 + \frac{\gamma^2}{n(l+1/2)} + \ldots\right\} + \frac{3\gamma^4}{8n^4}\left\{1 + \frac{2\gamma^2}{n(l+1/2)} + \ldots\right\}\right]$$

On retaining terms up to γ^4, we get

$$\begin{aligned} E &= m_0c^2\left\{1 - \frac{\gamma^2}{2n^2}\left[1 + \frac{\gamma^2}{n(l+1/2)}\right] + \frac{3\gamma^4}{8n^4}\right\} \\ &= m_0c^2\left[1 - \frac{\gamma^2}{2n^2} - \frac{\gamma^4}{2n^4}\left(\frac{n}{l+1/2} - \frac{3}{4}\right)\right] \end{aligned} \quad (14.32)$$

The first term in equation (14.32) is the 'rest energy' of the particle and the second term $-m_0c^2\gamma^2/(2n^2)\left[\equiv -Z^2m_0e^4/(2\hbar^2n^2)\right]$ is the nonrelativistic energy for hydrogen atom. The last term gives the relativistic correction and is generally known as the 'fine structure energy'. It removes the l-degeneracy. For the transition between the ground state $l = 0$ and $l = (n-1)$, the splitting is

$$\begin{aligned} &-\frac{m_0c^2\gamma^4}{2n^4}\left\{\left[\frac{n}{n-1+1/2} - \frac{3}{4}\right] - \left[\frac{n}{1/2} - \frac{3}{4}\right]\right\} \\ &= \frac{2m_0c^2\gamma^4}{n^3}\,\frac{n-1}{2n-1} \end{aligned}$$

For hydrogen atom, the observed experimental split is only about half of that obtained from this expression. However, for the π-mesonic atoms, the agreement is found better.

4. Dirac equation for a free particle

In the preceding sections, it has been seen that the probability density $P(\vec{r}, t)$ could be negative as well as positive. In order to avoid negative values of the probability density, Dirac introduced an idea that when the equation is of the first order in time, it must be first order in space also. Probably he got this idea from the Maxwell's equations as they are the first order equations in both space as well as time. Let us start from the basic energy equation (14.1)

$$E = \pm(c^2p^2 + m_0^2c^4)^{1/2}$$

Replacing E and $\vec{p}$ by their operators $i\hbar\frac{\partial}{\partial t}$ and $-i\hbar\nabla$, respectively, and allowing the resulting operator equation to operate on the wave function $\Psi(\vec{r}, t)$, we get

$$i\hbar\frac{\partial}{\partial t}\Psi(\vec{r}, t) = \pm(-c^2\hbar^2\nabla^2 + m_0^2c^4)^{1/2}\Psi(\vec{r}, t) \qquad (14.33)$$

For proceeding further, there was an obvious need to express the square root of the operator. Dirac boldly expressed the operator as

$$c^2p^2 + m_0^2c^4 = \left[c(\alpha_xp_x + \alpha_yp_y + \alpha_zp_z) + \beta m_0c^2\right]^2 \qquad (14.34)$$

and then searched for the relations between $\vec{\alpha}$ and β so that this equation (14.34) is valid. (α_x, α_y and α_z are Cartesian components of $\vec{\alpha}$.) He obtained the following relations between α_x, α_y, α_z and β.

$$\begin{aligned}
&\alpha_x^2 = \alpha_y^2 = \alpha_z^2 = \beta^2 = 1\\
&\alpha_x\alpha_y + \alpha_y\alpha_x = \alpha_y\alpha_z + \alpha_z\alpha_y = \alpha_z\alpha_x + \alpha_x\alpha_z = 0\\
&\alpha_x\beta + \beta\alpha_x = \alpha_y\beta + \beta\alpha_y = \alpha_z\beta + \beta\alpha_z = 0
\end{aligned} \qquad (14.35)$$

It shows that the components of $\vec{\alpha}$ along with β anticommute in pairs with each other and their squares are unity. Such relations immediately suggest that they cannot be numbers, but square matrices of the same order. At present, we have knowledge of Pauli's spin matrices σ_x, σ_y, σ_z which anticommute with each other. Hence, it may be convenient to express α_x, α_y, α_z, β in terms of these matrices. As the Pauli's matrices are three in number and we have four parameters, one have to search for one more matrix which anti-commutes with the Pauli's matrices. Finding such matrix was not possible. Discussion for finding the matrices for α_x, α_y, α_z, β is given in the next section. From equations (14.1) and (14.34), the energy can be expressed as

$$E = \pm[c(\alpha_xp_x + \alpha_yp_y + \alpha_zp_z) + \beta m_0c^2]$$

We can take positive or negative sign. Further, the replacement of $\vec{\alpha}$ by $-\vec{\alpha}$ and β by $-\beta$ does not change the relationships between the components of $\vec{\alpha}$ and β. The relativistic Hamiltonian of a particle can be taken as

$$H = c\,\vec{\alpha}\,.\,\vec{p} + \beta m_0 c^2 \tag{14.36}$$

Replacing E and $\vec{p}$ by their operators and allowing the resulting operator equation to operate on $\Psi(r,t)$, we have

$$i\hbar\frac{\partial}{\partial t}\Psi(\vec{r},t) = -ic\hbar\left[\alpha_x\frac{\partial}{\partial x} + \alpha_y\frac{\partial}{\partial y} + \alpha_z\frac{\partial}{\partial z}\right]\Psi(r,t) + \beta m_0 c^2\Psi(\vec{r},t) \tag{14.37}$$

It is the Dirac's relativistic equation for a free particle.

5. Dirac matrices

Since the Hamiltonian in equation (14.36) should be Hermitian, the matrices α_x, α_y, α_z and β must be Hermitian. Hence, the α_x, α_y, α_z and β should be square matrices. Since the squares of all these four matrices are ± 1, the eigenvalues of them are ± 1. We have

$$\alpha_x = \alpha_x\beta^2 = \alpha_x\beta\beta = -\beta\alpha_x\beta$$

where in the last step we have used the anticommuting property $\alpha_x\beta + \beta\alpha_x = 0$. Now, we have

$$\mathrm{tr}(\alpha_x) = \mathrm{tr}(-\beta\alpha_x\beta) = -\mathrm{tr}(\alpha_x\beta^2) = -\mathrm{tr}(\alpha_x) \tag{14.38}$$

where the third step is obtained by using the cyclic property of trace of the product of matrices, *i.e.*, $\mathrm{tr}(ABC) = \mathrm{tr}(BCA)$. It is evident from equation (14.38) that $\mathrm{tr}(\alpha_x) = 0$ which is possible only when the number of $+1$ eigenvalues of the matrix is equal to the number of -1 eigenvalues. This restricts the dimension n of α_x to be an even number. Similar conclusions can be drawn for each of the α_y, α_z and β matrices also.

We have already three anticommuting 2×2 Pauli's spin matrices:

$$\sigma_x = \begin{bmatrix} 0 & 1 \\ 1 & 0 \end{bmatrix} \qquad \sigma_y = \begin{bmatrix} 0 & -i \\ i & 0 \end{bmatrix} \qquad \sigma_z = \begin{bmatrix} 1 & 0 \\ 0 & -1 \end{bmatrix}$$

Though the square of each of these matrices is unity, it is not possible find a fourth matrix that anticommutes with these three matrices. The next possibility is to take $n = 4$. It is customary to take β matrix having diagonal with two +1 eigenvalues and two −1 eigenvalues.

$$\beta = \begin{bmatrix} 1 & 0 & 0 & 0 \\ 0 & 1 & 0 & 0 \\ 0 & 0 & -1 & 0 \\ 0 & 0 & 0 & -1 \end{bmatrix} = \begin{bmatrix} I & 0 \\ 0 & -I \end{bmatrix}$$

Then three other 4×4 matrices anticommuting with β and satisfying the other conditions specified by the equations (14.35) are

$$\alpha_x = \begin{bmatrix} 0 & 0 & 0 & 1 \\ 0 & 0 & 1 & 0 \\ 0 & 1 & 0 & 0 \\ 1 & 0 & 0 & 0 \end{bmatrix} = \begin{bmatrix} 0 & \sigma_x \\ \sigma_x & 0 \end{bmatrix}$$

$$\alpha_y = \begin{bmatrix} 0 & 0 & 0 & -i \\ 0 & 0 & i & 0 \\ 0 & -i & 0 & 0 \\ i & 0 & 0 & 0 \end{bmatrix} = \begin{bmatrix} 0 & \sigma_y \\ \sigma_y & 0 \end{bmatrix}$$

$$\alpha_z = \begin{bmatrix} 0 & 0 & 1 & 0 \\ 0 & 0 & 0 & -1 \\ 1 & 0 & 0 & 0 \\ 0 & -1 & 0 & 0 \end{bmatrix} = \begin{bmatrix} 0 & \sigma_z \\ \sigma_z & 0 \end{bmatrix}$$

Obviously these matrices are Hermitian since σ_x, σ_y and σ_z are Hermitian. These matrices are the standard representation. Other representations are also possible. Since α's and β are 4×4 matrices, the Dirac wave function $\Psi(\vec{r}, t)$ must be a four-column vector:

$$\Psi(\vec{r}, t) = \begin{bmatrix} \Psi_1 \\ \Psi_2 \\ \Psi_3 \\ \Psi_4 \end{bmatrix} \quad \text{and} \quad \Psi^*(\vec{r}, t) = \begin{bmatrix} \Psi_1^* & \Psi_2^* & \Psi_3^* & \Psi_4^* \end{bmatrix} \tag{14.39}$$

These wave functions are sometimes referred to as the Dirac spinors.

Exercise 2. Show that the following matrices form a representation of Dirac's matrices:

$$\alpha_x = \begin{bmatrix} \sigma_x & 0 \\ 0 & -\sigma_x \end{bmatrix}, \qquad \alpha_y = \begin{bmatrix} 0 & I \\ I & 0 \end{bmatrix}$$

$$\alpha_z = \begin{bmatrix} \sigma_z & 0 \\ 0 & -\sigma_z \end{bmatrix}, \qquad \beta = \begin{bmatrix} 0 & iI \\ -iI & 0 \end{bmatrix}$$

where σ_x and σ_z are the Pauli's spin matrices and I is a unit matrix of order 2×2.

Solution: Here σ_x and σ_z are the Pauli's spin matrices. These matrices α_x, α_y, α_z and β are said to from Dirac's matrices when they satisfy the following relations.

$$\alpha_x^2 = \alpha_y^2 = \alpha_z^2 = \beta^2 = I$$

$$\alpha_x\alpha_y + \alpha_y\alpha_x = \alpha_y\alpha_z + \alpha_z\alpha_y = \alpha_z\alpha_x + \alpha_x\alpha_z = 0$$

$$\alpha_x\beta + \beta\alpha_x = \alpha_y\beta + \beta\alpha_y = \alpha_z\beta + \beta\alpha_z = 0$$

Here, I and 0 are the unit ans null matrices of order 4×4. We have

$$\alpha_x^2 = \begin{bmatrix} \sigma_x & 0 \\ 0 & -\sigma_x \end{bmatrix}\begin{bmatrix} \sigma_x & 0 \\ 0 & -\sigma_x \end{bmatrix} = \begin{bmatrix} \sigma_x^2 & 0 \\ 0 & \sigma_x^2 \end{bmatrix} = \begin{bmatrix} I & 0 \\ 0 & I \end{bmatrix} = I$$

$$\alpha_y^2 = \begin{bmatrix} 0 & I \\ I & 0 \end{bmatrix}\begin{bmatrix} 0 & I \\ I & 0 \end{bmatrix} = \begin{bmatrix} I & 0 \\ 0 & I \end{bmatrix} = I$$

$$\alpha_z^2 = \begin{bmatrix} \sigma_z & 0 \\ 0 & -\sigma_z \end{bmatrix}\begin{bmatrix} \sigma_z & 0 \\ 0 & -\sigma_z \end{bmatrix} = \begin{bmatrix} \sigma_z^2 & 0 \\ 0 & \sigma_z^2 \end{bmatrix} = \begin{bmatrix} 1 & 0 \\ 0 & 1 \end{bmatrix} = 1$$

$$\beta^2 = \begin{bmatrix} 0 & iI \\ -iI & 0 \end{bmatrix}\begin{bmatrix} 0 & iI \\ -iI & 0 \end{bmatrix} = \begin{bmatrix} I & 0 \\ 0 & I \end{bmatrix} = I$$

and

$$\alpha_x\alpha_y + \alpha_y\alpha_x = \begin{bmatrix} \sigma_x & 0 \\ 0 & -\sigma_x \end{bmatrix}\begin{bmatrix} 0 & I \\ I & 0 \end{bmatrix} + \begin{bmatrix} 0 & I \\ I & 0 \end{bmatrix}\begin{bmatrix} \sigma_x & 0 \\ 0 & -\sigma_x \end{bmatrix}$$

$$= \begin{bmatrix} 0 & \sigma_x \\ -\sigma_x & 0 \end{bmatrix} + \begin{bmatrix} 0 & -\sigma_x \\ \sigma_x & 0 \end{bmatrix} = \begin{bmatrix} 0 & 0 \\ 0 & 0 \end{bmatrix} = 0$$

$$\alpha_y\alpha_z + \alpha_z\alpha_y = \begin{bmatrix} 0 & I \\ I & 0 \end{bmatrix}\begin{bmatrix} \sigma_z & 0 \\ 0 & -\sigma_z \end{bmatrix} + \begin{bmatrix} \sigma_z & 0 \\ 0 & -\sigma_z \end{bmatrix}\begin{bmatrix} 0 & I \\ I & 0 \end{bmatrix}$$

$$= \begin{bmatrix} 0 & -\sigma_z \\ \sigma_z & 0 \end{bmatrix} + \begin{bmatrix} 0 & \sigma_z \\ -\sigma_z & 0 \end{bmatrix} = \begin{bmatrix} 0 & 0 \\ 0 & 0 \end{bmatrix} = 0$$

$$\alpha_z\alpha_x + \alpha_x\alpha_z = \begin{bmatrix} \sigma_z & 0 \\ 0 & -\sigma_z \end{bmatrix}\begin{bmatrix} \sigma_x & 0 \\ 0 & -\sigma_x \end{bmatrix} + \begin{bmatrix} \sigma_x & 0 \\ 0 & -\sigma_x \end{bmatrix}\begin{bmatrix} \sigma_z & 0 \\ 0 & -\sigma_z \end{bmatrix}$$

$$= \begin{bmatrix} \sigma_z\sigma_x & 0 \\ 0 & \sigma_z\sigma_x \end{bmatrix} + \begin{bmatrix} \sigma_x\sigma_z & 0 \\ 0 & \sigma_x\sigma_z \end{bmatrix}$$

$$= \begin{bmatrix} \sigma_z\sigma_x + \sigma_x\sigma_z & 0 \\ 0 & \sigma_z\sigma_x + \sigma_x\sigma_z \end{bmatrix} = \begin{bmatrix} 0 & 0 \\ 0 & 0 \end{bmatrix} = 0$$

$$\alpha_x\beta + \beta\alpha_x = \begin{bmatrix} \sigma_x & 0 \\ 0 & -\sigma_x \end{bmatrix}\begin{bmatrix} 0 & iI \\ -iI & 0 \end{bmatrix} + \begin{bmatrix} 0 & iI \\ -iI & 0 \end{bmatrix}\begin{bmatrix} \sigma_x & 0 \\ 0 & -\sigma_x \end{bmatrix}$$

$$= \begin{bmatrix} 0 & i\sigma_x \\ i\sigma_x & 0 \end{bmatrix} + \begin{bmatrix} 0 & -i\sigma_x \\ -i\sigma_x & 0 \end{bmatrix} = \begin{bmatrix} 0 & 0 \\ 0 & 0 \end{bmatrix} = 0$$

$$\alpha_y\beta + \beta\alpha_y = \begin{bmatrix} 0 & I \\ I & 0 \end{bmatrix}\begin{bmatrix} 0 & iI \\ -iI & 0 \end{bmatrix} + \begin{bmatrix} 0 & iI \\ -iI & 0 \end{bmatrix}\begin{bmatrix} 0 & I \\ I & 0 \end{bmatrix}$$

$$= \begin{bmatrix} -iI & 0 \\ 0 & iI \end{bmatrix} + \begin{bmatrix} iI & 0 \\ 0 & -iI \end{bmatrix} = \begin{bmatrix} 0 & 0 \\ 0 & 0 \end{bmatrix} = 0$$

$$\alpha_z\beta + \beta\alpha_z = \begin{bmatrix} \sigma_z & 0 \\ 0 & -\sigma_z \end{bmatrix}\begin{bmatrix} 0 & iI \\ -iI & 0 \end{bmatrix} + \begin{bmatrix} 0 & iI \\ -iI & 0 \end{bmatrix}\begin{bmatrix} \sigma_z & 0 \\ 0 & -\sigma_z \end{bmatrix}$$

$$= \begin{bmatrix} 0 & i\sigma_z \\ i\sigma_z & 0 \end{bmatrix} + \begin{bmatrix} 0 & -i\sigma_z \\ -i\sigma_z & 0 \end{bmatrix} = \begin{bmatrix} 0 & 0 \\ 0 & 0 \end{bmatrix} = 0$$

Since all the required relations are satisfied by α_x, α_y, α_z, and β, the set of given matrices represent the Dirac matrices.

6. Continuity equation

Dirac wave equation and its Hermitian conjugate equation are

$$i\hbar\frac{\partial\Psi}{\partial t} = -ic\hbar\,\vec{\alpha}\,.\nabla\Psi + \beta m_0c^2\Psi \tag{14.40}$$

and

$$-i\hbar\frac{\partial\Psi^\dagger}{\partial t} = ic\hbar\nabla\Psi^\dagger.\,\vec{\alpha} + \Psi^\dagger\beta m_0c^2 \tag{14.41}$$

The Hermitian property of $\vec{\alpha}$ and β, and the rule that the order of factors in any matrix product is reversed on taking the Hermitian conjugate are used in writing equation (14.41). Now, multiplying equation (14.40) by $\Psi^\dagger$ on left, equation (14.41) by Ψ on right and subtracting the latter from the former, we get

$$i\hbar\left[\Psi^\dagger\frac{\partial\Psi}{\partial t}+\frac{\partial\Psi^\dagger}{\partial t}\Psi\right]=-ic\hbar(\Psi^\dagger\,\vec{\alpha}\,.\nabla\Psi+\nabla\Psi^\dagger.\,\vec{\alpha}\,\Psi)$$

$$\frac{\partial}{\partial t}(\Psi^\dagger\Psi)+\nabla.(c\Psi^\dagger\,\vec{\alpha}\,\Psi)=0$$

$$\frac{\partial P(\vec{r},t)}{\partial t}+\nabla.\,\vec{j}\,(\vec{r},t)=0 \tag{14.42}$$

where

$$P(\vec{r},t)=\Psi^\dagger\Psi \qquad \text{and} \qquad \vec{j}\,(\vec{r},t)=c\Psi^\dagger\,\vec{\alpha}\,\Psi$$

Equation (14.42) is the equation of continuity and the quantities $P(\vec{r},t)$ and $\vec{j}\,(\vec{r},t)$ can be interpreted as the probability density and probability current density. The expression for P is the same as in the nonrelativistic case. In the nonrelativistic limit, $\vec{j}\,(\vec{r},t)$ here reduces to the nonrelativistic value. The normalization integral in Dirac's formalism takes the form

$$\int\Psi^\dagger(\vec{r},t)\Psi(\vec{r},t)\;d\tau=1$$

The expectation value of any operator A is defined by

$$\langle A\rangle=\int\Psi^\dagger(\vec{r},t)\;A\;\Psi(\vec{r},t)\;d\tau$$

where A in the integral is the corresponding operator and the wave function Ψ is normalized.

7. Plane wavc equation

As we did for the non-relativistic case and for the Klein-Gordon equation, let us try the plane wave solution for the Dirac equation also. With 4×4 matrices for $\vec{\alpha}$ and β, and column vector with four rows for the wave function Ψ, Dirac equation is equivalent to four simultaneous first-order differential equations which are linear and homogeneous in the four components Ψ_1, Ψ_2, Ψ_3 and Ψ_4. Let us now consider plane wave solutions of the form

$$\Psi_j(\vec{r},t)=u_j\;e^{i(\vec{k}.\vec{r}-\omega t)} \qquad j=1,2,3,4 \tag{14.43}$$

These Ψ_j's are eigenfunctions of energy and momentum with eigenvalues $\hbar\omega$ and $k\hbar$, respectively. Substitution of equations (14.43) and (14.39) in (14.37) gives[3]

$$Eu = (c\,\vec{\alpha}\,.\,\vec{p} + \beta m_0 c^2)u \tag{14.44}$$

where u is a four column vector partitioned into two components as the following.

$$u = \begin{bmatrix} u_1 \\ u_2 \\ u_3 \\ u_4 \end{bmatrix} = \begin{bmatrix} v \\ w \end{bmatrix} \quad \text{where} \quad v = \begin{bmatrix} u_1 \\ u_2 \end{bmatrix} \quad w = \begin{bmatrix} u_3 \\ u_4 \end{bmatrix} \tag{14.45}$$

Substituting equation (14.45) in (14.44), we get[4]

$$E\begin{bmatrix} v \\ w \end{bmatrix} = c\begin{bmatrix} 0 & \vec{\sigma}\,.\,\vec{p} \\ \vec{\sigma}\,.\,\vec{p} & 0 \end{bmatrix}\begin{bmatrix} v \\ w \end{bmatrix} + m_0c^2\begin{bmatrix} I & 0 \\ 0 & -I \end{bmatrix}\begin{bmatrix} v \\ w \end{bmatrix}$$

On simplification, we get

$$(E - m_0c^2)v = c(\vec{\sigma}\,.\,\vec{p})w \tag{14.46}$$

$$(E + m_0c^2)w = c(\vec{\sigma}\,.\,\vec{p})v \tag{14.47}$$

---

[3]We have

$$i\hbar\frac{\partial}{\partial t}\Psi_j = -ic\hbar\left[\alpha_x\frac{\partial}{\partial x} + \alpha_x\frac{\partial}{\partial x} + \alpha_y\frac{\partial}{\partial z}\Psi_j + \beta m_0c^2\Psi_j\right.$$

Using

$$\Psi_j(\vec{r},t) = u_j\; e^{i(k_x x + k_y y + k_z z - \omega t)}$$

in this equation, we get

$$\hbar\omega\Psi_j = c(\alpha_x\hbar k_x + \alpha_y\hbar k_y + \alpha_z\hbar k_z)\Psi_j + \beta m_0c^2\Psi_j = (c\,\vec{\alpha}\,.\,\vec{p} + \beta m_0c^2)\Psi_j$$

[4]We have

$$\vec{\alpha}.\,\vec{p} = \alpha_x p_x + \alpha_y p_y + \alpha_z p_z = \begin{bmatrix} 0 & \sigma_x \\ \sigma_x & 0 \end{bmatrix} p_x + \begin{bmatrix} 0 & \sigma_y \\ \sigma_y & 0 \end{bmatrix} p_y + \begin{bmatrix} 0 & \sigma_z \\ \sigma_z & 0 \end{bmatrix} p_z$$

$$= \begin{bmatrix} 0 & \sigma_x p_x + \sigma_y p_y + \sigma_z p_z \\ \sigma_x p_x + \sigma_y p_y + \sigma_z p_z & 0 \end{bmatrix} = \begin{bmatrix} 0 & \vec{\alpha}\,.\,\vec{p} \\ \vec{\alpha}\,.\,\vec{p} & 0 \end{bmatrix}$$

Here, each one of the equations (14.46) as well as (14.47) represents two equations, as the σ is a 2×2 matrix and v as well as w are two-component column vectors. Multiplying equation (14.46) by $(E+m_0c^2)$ and substituting the value of $(E+m_0c^2)w$ from equation (14.47), we get

$$(E^2 - m_0^2c^4)v = c(\vec{\sigma} \cdot \vec{p})c(\vec{\sigma} \cdot \vec{p})v = c^2(\vec{\sigma} \cdot \vec{p})^2 v$$

Since[5] $(\vec{\sigma} \cdot \vec{p})^2 = p^2$, we have

$$(E^2 - m_0^2c^4)v = c^2p^2v$$

$$(E^2 - c^2p^2 - m_0^2c^4)v = 0$$

As $v \neq 0$, we have

$$E^2 - c^2p^2 - m_0^2c^4 = 0$$

$$E = \pm(c^2p^2 + m_0^2c^4)^{1/2}$$

Thus, the energy can have two values

$$E_+ = +(c^2p^2 + m_0^2c^4)^{1/2}$$

$$E_- = -(c^2p^2 + m_0^2c^4)^{1/2}$$

For E_+, we shall use equation (14.47) and for E_-, we shall use (14.46). Let us discuss the cases for E_+ and E_- values of energy.

Exercise 3. Find out energy-momentum relation for a photon. Obtain expression for momentum and mass of a photon.

Solution: Photon is a relativistic particle. For a relativistic particle, energy-momentum relation is

$$E^2 - c^2p^2 - m_0^2c^4 = 0$$

[5]We have

$$\begin{aligned}(\vec{\sigma} \cdot \vec{p})^2 &= (\sigma_xp_x + \sigma_yp_y + \sigma_zp_z)^2 = \sigma_x^2p_x^2 + \sigma_y^2p_y^2 + \sigma_z^2p_z^2 + \sigma_x\sigma_yp_xp_y \\ &+\sigma_y\sigma_xp_yp_x + \sigma_y\sigma_zp_yp_z + \sigma_z\sigma_yp_zp_y + \sigma_z\sigma_xp_zp_x + \sigma_x\sigma_zp_xp_z = \sigma_x^2p_x^2 \\ &+\sigma_y^2p_y^2 + \sigma_z^2p_z^2 + (\sigma_x\sigma_y + \sigma_y\sigma_x)p_xp_y + (\sigma_y\sigma_z + \sigma_z\sigma_y)p_yp_z \\ &+(\sigma_z\sigma_x + \sigma_x\sigma_z)p_zp_x = p_x^2 + p_y^2 + p_z^2 = p^2\end{aligned}$$

For photon, we have $m_0 = 0$, and therefore, we have $E = pc$. For a photon of frequency ν, we have $E = h\nu$. Therefore, the momentum of a photon is

$$p = \frac{E}{c} = \frac{h\nu}{c}$$

Mass m of a photon is related to the momentum p as $p = mc$. Thus, the mass of a photon is

$$m = \frac{p}{c} = \frac{h\nu}{c^2}$$

7.1 E_+ eigenvalue

For the simplest case, let us take

$$v = \begin{bmatrix} 1 \\ 0 \end{bmatrix} \qquad \text{and} \qquad \begin{bmatrix} 0 \\ 1 \end{bmatrix}$$

We have

$$(\vec{\sigma} \, . \, \vec{p}) = \sigma_x p_x + \sigma_y p_y + \sigma_z p_z$$

$$= \begin{bmatrix} 0 & 1 \\ 1 & 0 \end{bmatrix} p_x + \begin{bmatrix} 0 & -i \\ i & 0 \end{bmatrix} p_y + \begin{bmatrix} 1 & 0 \\ 0 & -1 \end{bmatrix} p_z = \begin{bmatrix} p_z & p_- \\ p_+ & -p_z \end{bmatrix}$$

Here, we have defined $p_+ = p_x + ip_y$ and $p_- = p_x - ip_y$. From equation (14.47), with

$$v = \begin{bmatrix} 1 \\ 0 \end{bmatrix}$$

we get

$$w = \frac{c(\vec{\sigma} \, . \, \vec{p})v}{E_+ + m_0c^2} = \frac{c}{E_+ + m_0c^2} \begin{bmatrix} p_z & p_- \\ p_+ & -p_z \end{bmatrix} \begin{bmatrix} 1 \\ 0 \end{bmatrix}$$

$$= \begin{bmatrix} cp_z/(E_+ + m_0c^2) \\ cp_+/(E_+ + m_0c^2) \end{bmatrix}$$

For the second solution

$$v = \begin{bmatrix} 0 \\ 1 \end{bmatrix}$$

we get

$$w = \frac{c(\vec{\sigma}\,.\,\vec{p})v}{E_+ + m_0c^2} = \frac{c}{E_+ + m_0c^2}\begin{bmatrix} p_z & p_- \\ p_+ & -p_z \end{bmatrix}\begin{bmatrix} 0 \\ 1 \end{bmatrix}$$

$$= \begin{bmatrix} cp_-/(E_+ + m_0c^2) \\ -cp_z/(E_+ + m_0c^2) \end{bmatrix}$$

Thus, for the eigenvalue E_+, we have two independent solutions as the following.

$$u^{(1)} = \begin{bmatrix} 1 \\ 0 \\ cp_z/(E_+ + m_0c^2) \\ cp_+/(E_+ + m_0c^2) \end{bmatrix}, \qquad u^{(2)} = \begin{bmatrix} 0 \\ 1 \\ cp_-/(E_+ + m_0c^2) \\ -cp_z/(E_+ + m_0c^2) \end{bmatrix}$$

7.2 E_- eigenvalue

For the simplest case, let us take

$$w = \begin{bmatrix} 1 \\ 0 \end{bmatrix} \qquad \text{and} \qquad \begin{bmatrix} 0 \\ 1 \end{bmatrix}$$

From equation (14.46), with

$$w = \begin{bmatrix} 1 \\ 0 \end{bmatrix}$$

we get

$$v = \frac{c(\vec{\sigma}\,.\,\vec{p})w}{E_- - m_0c^2} = \frac{c}{E_- - m_0c^2}\begin{bmatrix} p_z & p_- \\ p_+ & -p_z \end{bmatrix}\begin{bmatrix} 1 \\ 0 \end{bmatrix}$$

$$= \begin{bmatrix} cp_z/(E_- - m_0c^2) \\ cp_+/(E_- - m_0c^2) \end{bmatrix}$$

For the second solution

$$w = \begin{bmatrix} 0 \\ 1 \end{bmatrix}$$

we get

$$v = \frac{c(\vec{\sigma}\,.\,\vec{p})w}{E_- - m_0c^2} = \frac{c}{E_- - m_0c^2}\begin{bmatrix} p_z & p_- \\ p_+ & -p_z \end{bmatrix}\begin{bmatrix} 0 \\ 1 \end{bmatrix}$$

$$= \begin{bmatrix} cp_-/(E_- - m_0c^2) \\ -cp_z/(E_- - m_0c^2) \end{bmatrix}$$

Thus, for the eigenvalue E_-, we have two independent solutions as the following.

$$u^{(3)} = \begin{bmatrix} cp_z/(E_- - m_0c^2) \\ cp_+/(E_- - m_0c^2) \\ 1 \\ 0 \end{bmatrix}, \quad u^{(4)} = \begin{bmatrix} cp_-/(E_- - m_0c^2) \\ -cp_z/(E_- - m_0c^2) \\ 0 \\ 1 \end{bmatrix}$$

In the nonrelativistic limit $|p| << m_0c$, $|p_x| << m_0c$, $|p_y| << m_0c$, $|p_z| << m_0c$ and therefore, we have $E_+ \equiv m_0c^2$, $E_- \equiv -m_0c^2$ and

$$u^{(1)} = \begin{bmatrix} 1 \\ 0 \\ 0 \\ 0 \end{bmatrix}, \quad u^{(2)} = \begin{bmatrix} 0 \\ 1 \\ 0 \\ 0 \end{bmatrix}, \quad u^{(3)} = \begin{bmatrix} 0 \\ 0 \\ 1 \\ 0 \end{bmatrix}, \quad u^{(4)} = \begin{bmatrix} 0 \\ 0 \\ 0 \\ 1 \end{bmatrix}$$

Obviously with the decrease of $\vec{p}$, the third and fourth components of $u^{(1)}$ and $u^{(2)}$ decrease, and finally become zero whereas the first and second components remain unaffected. Hence, the first and second components are generally known as the large components, and the third and fourth are generally known as the small components. On the other side, for $u^{(3)}$ and $u^{(4)}$, the first and the second components are known as the small components whereas the first and the second components are known as the large components.

8. Negative energy states

Each of the equations (14.46) and (14.47) has two equations, since $\vec{\sigma}$ is a 2×2 matrix. Therefore, each of the E_+ as well as E_- occur twice. When $p = 0$, we have $E_+ = m_0c^2$ and $E_- = -m_0c^2$. The value of m_0c^2 for electron is 0.511 MeV. This is known as the rest-energy of electron. The energy spectrum of a free particle thus has two branches corresponding to E_+ and E_-, one starting at m_0c^2 and extending to ∞ as $|p| = \infty$ and the other starting at $-m_0c^2$ and extending to $-\infty$ as $|p| = \infty$ (Figure 1). The two branches are separated by a forbidden gap of the width $2m_0c^2$, where no energy levels exist.

It is very difficult to imagine such negative energy states because even a small perturbation could cause a transition of an electron from a positive state to a state of negative energy and as a result of such

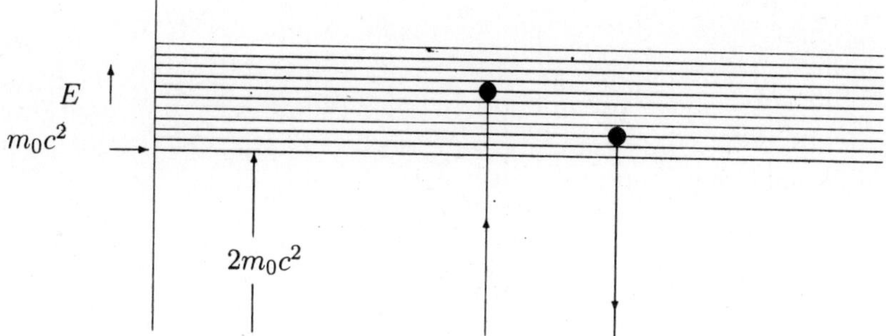

Figure 1: Energy levels of a free Dirac particle. Hollow circles represent the negative energy states of electrons which have no physically observable effects. Solid circles represent the positive energy states which correspond to the observable electrons.

transition an enormous amount of energy ($\geq m_0c^2$) is released. In a real situation, no such thing takes place. In order to resolve this problem, Dirac assumed that all the negative energy states are ordinarily occupied by the electrons and this sea of negative energy electrons have no physically observable effects. He further considered that when an electron occupying a negative energy state picks up energy and goes to a positive energy state, it takes its place as an ordinary observable electron. The empty space in the negative energy states behaves as if it is a particle having a positive charge. It responds to electric as well as magnetic fields. This empty space is known as a hole. This explanation was accepted in 1932 after the discovery of positron. A positron is a particle having mass equal to that of an electron and charge in equal amount to that of an electron, but opposite.

Exercise 4. The probability current density $\vec{j}(\vec{r},t)$ is expressed through the relation $\vec{j}(\vec{r},t) = c\Psi^\dagger \vec{\alpha}\, \Psi$, where Ψ is a four-component wave vector. Find out expressions for j_x, j_y and j_z in terms of the components of Ψ.

Solution: When $\vec{j}(\vec{r},t) = c\Psi^\dagger \vec{\alpha}\, \Psi$ then we have

$$j_x = c\Psi^\dagger \alpha_x \Psi = c\begin{bmatrix}\Psi_1^* & \Psi_2^* & \Psi_3^* & \Psi_4^*\end{bmatrix}\begin{bmatrix}0 & 0 & 0 & 1\\ 0 & 0 & 1 & 0\\ 0 & 1 & 0 & 0\\ 1 & 0 & 0 & 0\end{bmatrix}\begin{bmatrix}\Psi_1\\ \Psi_2\\ \Psi_3\\ \Psi_4\end{bmatrix}$$

$$= c\begin{bmatrix}\Psi_1^* & \Psi_2^* & \Psi_3^* & \Psi_4^*\end{bmatrix}\begin{bmatrix}\Psi_4\\ \Psi_3\\ \Psi_2\\ \Psi_1\end{bmatrix} = c(\Psi_1^*\Psi_4 + \Psi_2^*\Psi_3 + \Psi_3^*\Psi_2 + \Psi_4^*\Psi_1)$$

$$j_y = c\Psi^\dagger \alpha_y \Psi = c\begin{bmatrix}\Psi_1^* & \Psi_2^* & \Psi_3^* & \Psi_4^*\end{bmatrix}\begin{bmatrix}0 & 0 & 0 & -i \\ 0 & 0 & i & 0 \\ 0 & -i & 0 & 0 \\ i & 0 & 0 & 0\end{bmatrix}\begin{bmatrix}\Psi_1 \\ \Psi_2 \\ \Psi_3 \\ \Psi_4\end{bmatrix}$$

$$= c\begin{bmatrix}\Psi_1^* & \Psi_2^* & \Psi_3^* & \Psi_4^*\end{bmatrix}\begin{bmatrix}-i\Psi_4 \\ i\Psi_3 \\ -i\Psi_2 \\ i\Psi_1\end{bmatrix} = ic(-\Psi_1^*\Psi_4 + \Psi_2^*\Psi_3 - \Psi_3^*\Psi_2 + \Psi_4^*\Psi_1)$$

$$j_z = c\Psi^\dagger \alpha_z \Psi = c\begin{bmatrix}\Psi_1^* & \Psi_2^* & \Psi_3^* & \Psi_4^*\end{bmatrix}\begin{bmatrix}0 & 0 & 1 & 0 \\ 0 & 0 & 0 & -1 \\ 1 & 0 & 0 & 0 \\ 0 & -1 & 0 & 0\end{bmatrix}\begin{bmatrix}\Psi_1 \\ \Psi_2 \\ \Psi_3 \\ \Psi_4\end{bmatrix}$$

$$= c\begin{bmatrix}\Psi_1^* & \Psi_2^* & \Psi_3^* & \Psi_4^*\end{bmatrix}\begin{bmatrix}\Psi_3 \\ -\Psi_4 \\ \Psi_1 \\ -\Psi_2\end{bmatrix} = c(\Psi_1^*\Psi_3 - \Psi_2^*\Psi_4 + \Psi_3^*\Psi_1 - \Psi_4^*\Psi_2)$$

9. Spin of a Dirac particle

Since the spin of an electron carries no energy, it can only be observed through its coupling with the orbital motion of the electron. This can be demonstrated through the conservation of total angular momentum and through the spin-orbit coupling energy.

We know that a dynamical variable which is conserved should commute with the Hamiltonian. Consider the orbital angular momentum $\vec{l} = \vec{r} \times \vec{p}$. The time rate of change of l_x for a particle, whose Hamiltonian is given by equation (14.36), is

$$\begin{aligned} i\hbar\frac{dl_x}{dt} &= [l_x, H] \\ &= [yp_z - zp_y, c\,\vec{\alpha}\,.\,\vec{p} + \beta m_0c^2] \\ &= [yp_z, c\alpha_x p_x] - [zp_y, c\alpha_x p_x] + [yp_z, c\alpha_y p_y] - [zp_y, c\alpha_y p_y] \\ &\quad + [yp_z, c\alpha_z p_z] - [zp_y, c\alpha_z p_z] + [yp_z, \beta m_0c^2] - [zp_y, \beta m_0c^2] \end{aligned}$$

Since α_x, α_y, α_z and β commute with $\vec{r}$ and $\vec{p}$, all the commutators except the following two vanish and we have.

$$i\hbar\frac{dl_x}{dt} = [yp_z, c\alpha_y p_y] - [zp_y, c\alpha_z p_z] \tag{14.48}$$

Now, equations (14.48) takes the form

$$\begin{aligned} i\hbar\frac{dl_x}{dt} &= c[y, p_y]p_z\alpha_y - c[z, p_z]p_y\alpha_z \\ &= ic\hbar[\alpha_y p_z - \alpha_z p_y] \end{aligned} \tag{14.49}$$

This equation(14.49) shows that l_x is not a constant of motion. Similar relations can be found for l_y and l_z. It obviously shows that the orbital angular momentum $\vec{l}$ is not conserved. On the physical grounds, the total angular momentum of a system is supposed to be conserved. Therefore, we must have some other contribution to angular momentum, such that the commutator of its x-component with H must be negative of the right side of equation (14.49).

Now, we would like to show that the required contribution to angular momentum comes from the operator $\frac{1}{2}\hbar\,\vec{\sigma}'$, where

$$\vec{\sigma}' = \begin{bmatrix} \vec{\sigma} & 0 \\ 0 & \vec{\sigma} \end{bmatrix}$$

The equation of motion of its x-component is

$$\begin{aligned} i\hbar\frac{d\sigma_x'}{dt} &= [\sigma_x', c\,\vec{\sigma}\,.\,\vec{p} + \beta m_0 c^2] \\ &= [\sigma_x', c\sigma_x p_x + c\sigma_y p_y + c\sigma_z p_z + \beta m_0 c^2] \\ &= [\sigma_x', c\alpha_x p_x] + [\sigma_x', c\alpha_y p_y] + [\sigma_x', c\alpha_z p_z] + [\sigma_x', \beta m_0 c^2] \end{aligned} \tag{14.50}$$

We now have[6]

$$[\sigma'_x, \alpha_x] = 0; \quad [\sigma'_x, \alpha_y] = 2i\alpha_z; \quad [\sigma'_x, \alpha_z] = -2i\alpha_y; \quad [\sigma'_x, \beta m_0 c^2] = 0$$

Substituting these commutators in equations (14.50), we have

$$\begin{aligned} i\hbar\frac{d\sigma'_x}{dt} &= c[\sigma'_x, \alpha_x]p_x + c[\sigma'_y, \alpha_y]p_y + c[\sigma'_z, \alpha_z]p_z \\ &= -2ic(\alpha_y p_z - \alpha_z p_y) \\ i\hbar\frac{d}{dt}\left[\frac{1}{2}\hbar\sigma'_x\right] &= -ic\hbar(\alpha_y p_z - \alpha_z p_y) \end{aligned} \tag{14.51}$$

which is the negative of the right side of equation (14.49). Similar expressions can be obtained y and z components of $\vec{\sigma}'$. Combining equations

[6]We have

$$[\sigma'_x, \alpha_x] = \sigma'_x\alpha_x - \alpha_x\sigma'_x = \begin{bmatrix} \sigma_x & 0 \\ 0 & \sigma_x \end{bmatrix}\begin{bmatrix} 0 & \sigma_x \\ \sigma_x & 0 \end{bmatrix} - \begin{bmatrix} 0 & \sigma_x \\ \sigma_x & 0 \end{bmatrix}\begin{bmatrix} \sigma_x & 0 \\ 0 & \sigma_x \end{bmatrix}.$$

$$= \begin{bmatrix} 0 & \sigma_x^2 \\ \sigma_x^2 & 0 \end{bmatrix} - \begin{bmatrix} 0 & \sigma_x^2 \\ \sigma_x^2 & 0 \end{bmatrix} = 0$$

$$[\sigma'_x, \alpha_y] = \sigma'_x\alpha_y - \alpha_y\sigma'_x = \begin{bmatrix} \sigma_x & 0 \\ 0 & \sigma_x \end{bmatrix}\begin{bmatrix} 0 & \sigma_y \\ \sigma_y & 0 \end{bmatrix} - \begin{bmatrix} 0 & \sigma_y \\ \sigma_y & 0 \end{bmatrix}\begin{bmatrix} \sigma_x & 0 \\ 0 & \sigma_x \end{bmatrix}$$

$$= \begin{bmatrix} 0 & \sigma_x\sigma_y \\ \sigma_x\sigma_y & 0 \end{bmatrix} - \begin{bmatrix} 0 & \sigma_y\sigma_x \\ \sigma_y\sigma_x & 0 \end{bmatrix} = \begin{bmatrix} 0 & \sigma_x\sigma_y - \sigma_y\sigma_x \\ \sigma_x\sigma_y - \sigma_y\sigma_x & 0 \end{bmatrix}$$

$$= \begin{bmatrix} 0 & 2i\sigma_z \\ 2i\sigma_z & 0 \end{bmatrix} = 2i\alpha_z$$

$$[\sigma'_x, \alpha_z] = \sigma'_x\alpha_z - \alpha_z\sigma'_x = \begin{bmatrix} \sigma_x & 0 \\ 0 & \sigma_x \end{bmatrix}\begin{bmatrix} 0 & \sigma_z \\ \sigma_z & 0 \end{bmatrix} - \begin{bmatrix} 0 & \sigma_z \\ \sigma_z & 0 \end{bmatrix}\begin{bmatrix} \sigma_x & 0 \\ 0 & \sigma_x \end{bmatrix}$$

$$= \begin{bmatrix} 0 & \sigma_x\sigma_z \\ \sigma_x\sigma_z & 0 \end{bmatrix} - \begin{bmatrix} 0 & \sigma_z\sigma_x \\ \sigma_z\sigma_x & 0 \end{bmatrix} = \begin{bmatrix} 0 & \sigma_x\sigma_z - \sigma_z\sigma_x \\ \sigma_x\sigma_z - \sigma_z\sigma_x & 0 \end{bmatrix}$$

$$= \begin{bmatrix} 0 & -2i\sigma_y \\ -2i\sigma_y & 0 \end{bmatrix} = -2i\alpha_y$$

$$[\sigma'_x, \beta] = \sigma'_x\beta - \beta\sigma'_x = m_0c^2 \begin{bmatrix} \sigma_x & 0 \\ 0 & \sigma_x \end{bmatrix}\begin{bmatrix} I & 0 \\ 0 & I \end{bmatrix} - \begin{bmatrix} I & 0 \\ 0 & I \end{bmatrix}\begin{bmatrix} \sigma_x & 0 \\ 0 & \sigma_x \end{bmatrix}$$

$$= \begin{bmatrix} \sigma_x & 0 \\ 0 & -\sigma_x \end{bmatrix} - \begin{bmatrix} \sigma_x & 0 \\ 0 & -\sigma_x \end{bmatrix} = 0$$

(14.49) and (14.51), we have

$$\frac{d}{dt}\left[l_x + \frac{1}{2}\hbar\sigma'_x\right] = 0 \qquad \text{or} \qquad l_x + \frac{1}{2}\hbar\sigma'_x = \text{constant}$$

It shows that $\vec{l} + \frac{1}{2}\hbar\,\vec{\sigma}'$ commutes with the Hamiltonian and therefore it can be taken as the total angular momentum. From the structure of the $\vec{\sigma}'$ matrix we have

$$\sigma_x'^2 = \sigma_y'^2 = \sigma_z'^2 = 1$$

Hence, the eigenvalue of $\frac{1}{2}\hbar\,\vec{\sigma}'$ is either $+\frac{1}{2}\hbar$ or $-\frac{1}{2}\hbar$. This additional part

$$s = \frac{\hbar}{2}\,\vec{\sigma}' = \frac{\hbar}{2}\begin{bmatrix} \vec{\sigma} & 0 \\ 0 & \vec{\sigma} \end{bmatrix}$$

is interpreted as the spin angular momentum of electron. In this manner, the concept of spin evolved automatically from the Dirac's Hamiltonian.

Exercise 5. For

$$\vec{\sigma}' = \begin{bmatrix} \vec{\sigma} & 0 \\ 0 & \vec{\sigma} \end{bmatrix}$$

where $\vec{\sigma}$ are the Pauli's spin matrices, show that

$$\begin{aligned}
&\sigma_x'^2 = \sigma_y'^2 = \sigma_z'^2 = 1 \\
&\sigma'_x\sigma'_y = i\sigma'_z; \qquad \sigma'_y\sigma'_z = i\sigma'_x; \qquad \sigma'_z\sigma'_x = i\sigma'_y \\
&\sigma'_y\sigma'_x = -i\sigma'_z; \qquad \sigma'_z\sigma'_y = i\sigma'_x, \qquad \sigma'_x\sigma'_z = -i\sigma'_y \\
&\sigma'_x\sigma'_y + \sigma'_y\sigma'_x = \sigma'_y\sigma'_z + \sigma'_z\sigma'_y = \sigma'_z\sigma'_x + \sigma'_x\sigma'_z = 0
\end{aligned}$$

Solution: We have

$$\sigma_x'^2 = \begin{bmatrix} \sigma_x & 0 \\ 0 & \sigma_x \end{bmatrix}\begin{bmatrix} \sigma_x & 0 \\ 0 & \sigma_x \end{bmatrix} = \begin{bmatrix} \sigma_x^2 & 0 \\ 0 & \sigma_x^2 \end{bmatrix} = \begin{bmatrix} I & 0 \\ 0 & I \end{bmatrix} = 1$$

$$\sigma_y'^2 = \begin{bmatrix} \sigma_y & 0 \\ 0 & \sigma_y \end{bmatrix}\begin{bmatrix} \sigma_y & 0 \\ 0 & \sigma_y \end{bmatrix} = \begin{bmatrix} \sigma_y^2 & 0 \\ 0 & \sigma_y^2 \end{bmatrix} = \begin{bmatrix} I & 0 \\ 0 & I \end{bmatrix} = 1$$

$$\sigma_z'^2 = \begin{bmatrix} \sigma_z & 0 \\ 0 & \sigma_z \end{bmatrix}\begin{bmatrix} \sigma_z & 0 \\ 0 & \sigma_z \end{bmatrix} = \begin{bmatrix} \sigma_z^2 & 0 \\ 0 & \sigma_z^2 \end{bmatrix} = \begin{bmatrix} I & 0 \\ 0 & I \end{bmatrix} = 1$$

Now, we have

$$\sigma'_x\sigma'_y = \begin{bmatrix} \sigma_x & 0 \\ 0 & \sigma_x \end{bmatrix}\begin{bmatrix} \sigma_y & 0 \\ 0 & \sigma_y \end{bmatrix} = \begin{bmatrix} \sigma_x\sigma_y & 0 \\ 0 & \sigma_x\sigma_y \end{bmatrix} = \begin{bmatrix} i\sigma_z & 0 \\ 0 & i\sigma_z \end{bmatrix} = i\sigma'_z$$

Similarly, we have $\sigma'_y\sigma'_z = i\sigma'_x$ and $\sigma'_z\sigma'_x = i\sigma'_y$.
Now, we have

$$\sigma'_y\sigma'_x = \begin{bmatrix} \sigma_y & 0 \\ 0 & \sigma_y \end{bmatrix}\begin{bmatrix} \sigma_x & 0 \\ 0 & \sigma_x \end{bmatrix} = \begin{bmatrix} \sigma_y\sigma_x & 0 \\ 0 & \sigma_y\sigma_x \end{bmatrix} = \begin{bmatrix} -i\sigma_z & 0 \\ 0 & -i\sigma_z \end{bmatrix} = -i\sigma'_z$$

Similarly, we get $\sigma'_z\sigma'_y = -i\sigma'_x$ and $\sigma'_x\sigma'_z = -i\sigma'_y$. Obviously, we have

$$\sigma'_x\sigma'_y + \sigma'_y\sigma'_x = i\sigma'_z - i\sigma'_z = 0$$
$$\sigma'_y\sigma'_z + \sigma'_z\sigma'_y = i\sigma'_x - i\sigma'_x = 0$$
$$\sigma'_z\sigma'_x + \sigma'_x\sigma'_z = i\sigma'_y - i\sigma'_y = 0$$

Exercise 6. Show that $(\vec{\sigma'} \cdot \vec{A})(\vec{\sigma'} \cdot \vec{B}) = (\vec{A} \cdot \vec{B}) + i\,\vec{\sigma'} \cdot (\vec{A} \times \vec{B})$, where

$$\vec{\sigma'} = \begin{bmatrix} \vec{\sigma} & 0 \\ 0 & \vec{\sigma} \end{bmatrix}$$

where $\vec{\sigma}$ are the Pauli's spin matrices.

Solution: We have

$$(\vec{\sigma'} \cdot \vec{A})(\vec{\sigma'} \cdot \vec{B}) = (\sigma'_x A_x + \sigma'_y A_y + \sigma'_z A_z)(\sigma'_x B_x + \sigma'_y B_y + \sigma'_z B_z)$$
$$\sigma'^2_x A_x B_x + \sigma'^2_y A_y B_y + \sigma'^2_z A_z B_z + \sigma'_x\sigma'_y A_x B_y + \sigma'_x\sigma'_z A_x B_z$$
$$+\sigma'_y\sigma'_x A_y B_x + \sigma'_y\sigma'_z A_y B_z + \sigma'_z\sigma'_x A_z B_x + \sigma'_z\sigma'_y A_z B_y$$

Since $\sigma'^2_x = \sigma'^2_y = \sigma'^2_z = 1$, $\sigma'_x\sigma'_y = -\sigma'_y\sigma'_x$, $\sigma'_y\sigma'_z = -\sigma'_z\sigma'_y$, $\sigma'_z\sigma'_x = -\sigma'_x\sigma'_z$, as shown in the preceding exercise, we have

$$(\vec{\sigma'} \cdot \vec{A})(\vec{\sigma'} \cdot \vec{B}) = (\vec{A} \cdot \vec{B}) + \sigma'_x\sigma'_y(A_x B_y - A_y B_x)$$
$$+\sigma'_y\sigma'_z(A_y B_z - A_z B_y) + \sigma'_z\sigma'_x(A_z B_x - A_x B_z)$$

Now,

$$\sigma'_x\sigma'_y = \begin{bmatrix} \sigma_x & 0 \\ 0 & \sigma_x \end{bmatrix}\begin{bmatrix} \sigma_y & 0 \\ 0 & \sigma_y \end{bmatrix} = \begin{bmatrix} \sigma_x\sigma_y & 0 \\ 0 & \sigma_x\sigma_y \end{bmatrix} = i\begin{bmatrix} \sigma_z & 0 \\ 0 & \sigma_z \end{bmatrix} = i\sigma'_z$$

Similarly, we have $\sigma'_y\sigma'_z = i\sigma'_x$ and $\sigma'_z\sigma'_x = i\sigma'_y$. Using this result along with the cyclic relations, we have

$$(\vec{\sigma'} \,.\, \vec{A})(\vec{\sigma'} \,.\, \vec{B}) = (\vec{A} \cdot \vec{B}) + i\sigma'_z(A_xB_y - A_yB_x) + i\sigma'_x(A_yB_z - A_zB_y) + i\sigma'_y(A_zB_x - A_xB_z)$$

$$= (\vec{A} \cdot \vec{B}) + i\,\vec{\sigma'} \,.\, (\vec{A} \times \vec{B})$$

Exercise 7. Show that $(\vec{\alpha} \,.\, \vec{A})(\vec{\alpha} \,.\, \vec{B}) = (\vec{A} \cdot \vec{B}) + i\,\vec{\sigma'} \,.(\vec{A} \times \vec{B})$, where $\vec{A}$ and $\vec{B}$ commute with $\vec{\alpha}$ and

$$\vec{\sigma'} = \begin{bmatrix} \vec{\sigma} & 0 \\ 0 & \vec{\sigma} \end{bmatrix}$$

where $\vec{\sigma}$ are the Pauli's spin matrices.

Solution: We have

$$(\vec{\alpha} \,.\, \vec{A})(\vec{\alpha} \,.\, \vec{B}) = (\alpha_xA_x + \alpha_yA_y + \alpha_zA_z)(\alpha_xB_x + \alpha_yB_y + \alpha_zB_z)$$
$$\alpha_x^2A_xB_x + \alpha_y^2A_yB_y + \alpha_z^2A_zB_z + \alpha_x\alpha_yA_xB_y + \alpha_x\alpha_zA_xB_z + \alpha_y\alpha_xA_yB_x + \alpha_y\alpha_zA_yB_z + \alpha_z\alpha_xA_zB_x + \alpha_z\alpha_yA_zB_y$$

Since $\alpha_x^2 = \alpha_y^2 = \alpha_z^2 = 1$, $\alpha_x\alpha_y = -\alpha_y\alpha_x$, $\alpha_y\alpha_z = -\alpha_z\alpha_y$, $\alpha_z\alpha_x = -\alpha_x\alpha_z$, as shown earlier, we have

$$(\vec{\alpha} \,.\, \vec{A})(\vec{\alpha} \,.\, \vec{B}) = (\vec{A} \cdot \vec{B}) + \alpha_x\alpha_y(A_xB_y - A_yB_x) + \alpha_y\alpha_z(A_yB_z - A_zB_y) + \alpha_z\alpha_x(A_zB_x - A_xB_z)$$

Now,

$$\alpha_x\alpha_y = \begin{bmatrix} 0 & \sigma_x \\ \sigma_x & 0 \end{bmatrix}\begin{bmatrix} 0 & \sigma_y \\ \sigma_y & 0 \end{bmatrix} = \begin{bmatrix} \sigma_x\sigma_y & 0 \\ 0 & \sigma_x\sigma_y \end{bmatrix} = i\begin{bmatrix} \sigma_z & 0 \\ 0 & \sigma_z \end{bmatrix} = i\sigma'_z$$

Similarly, we have $\alpha_y\alpha_z = i\sigma'_x$ and $\alpha_z\alpha_x = i\sigma'_y$. Using this result along with the cyclic relations, we have

$$(\vec{\alpha} \,.\, \vec{A})(\vec{\alpha} \,.\, \vec{B}) = (\vec{A} \cdot \vec{B}) + i\sigma'_z(A_xB_y - A_yB_x) + i\sigma'_x(A_yB_z - A_zB_y) + i\sigma'_y(A_zB_x - A_xB_z)$$

$$= (\vec{A} \cdot \vec{B}) + i\,\vec{\sigma'} \,.\, (\vec{A} \times \vec{B})$$

10. Magnetic moment of electron

Let us now discuss about the intrinsic magnetic moment associated with the spin of an electron. For that, we consider an electron in an electromagnetic field having a vector potential $\vec{A}$ and a scalar potential ϕ. On replacing E by $E - e\phi$ and $\vec{p}$ by $\vec{p} - (e\vec{A}/c)$ in the Dirac equation, we have

$$\left[(E - e\phi) - c\,\vec{\alpha} \,.\, (\vec{p} - \frac{e\vec{A}}{c}) - \beta m_0 c^2\right]\Psi(\vec{r}, t) = 0 \qquad (14.52)$$

Operating equation (14.52) on the left by

$$(E - e\phi) + c\,\vec{\alpha} \,.\, (\vec{p} - \frac{e\vec{A}}{c}) + \beta m_0 c^2$$

we get

$$\left[(E - e\phi) + c\,\vec{\alpha} \,.\, \left(\vec{p} - \frac{e\vec{A}}{c}\right) + \beta m_0 c^2\right]$$

$$\left[(E - e\phi) - c\,\vec{\alpha} \,.\, \left(\vec{p} - \frac{e\vec{A}}{c}\right) - \beta m_0 c^2\right]\Psi(\vec{r}, t) = 0$$

$$\left[(E - e\phi)^2 - c(E - e\phi)\,\vec{\alpha} \,.\, \left(\vec{p} - \frac{e\vec{A}}{c}\right) - m_0 c^2 (E - e\phi)\beta\right.$$

$$+ c\,\vec{\alpha} \,.\, \left(\vec{p} - \frac{e\vec{A}}{c}\right)(E - e\phi) - c^2\left[\vec{\alpha} \,.\, \left(\vec{p} - \frac{e\vec{A}}{c}\right)\right]^2 - m_0 c^3\,\vec{\alpha} \,.\, \left(\vec{p} - \frac{e\vec{A}}{c}\right)$$

$$\left. + m_0 c^2 \beta (E - e\phi) - m_0 c^3 \beta\,\vec{\alpha} \,.\, \left(\vec{p} - \frac{e\vec{A}}{c}\right) - \beta^2 m_0^2 c^4\right]\Psi(\vec{r}, t) = 0 \qquad (14.53)$$

Since β commutes with E, and ϕ is a scalar, we have

$$\beta(E - e\phi) = (E - e\phi)\beta \qquad (14.54)$$

Since β commutes with $\vec{p}$ as well as with $\vec{A}$, we have

$$\vec{\alpha} \,.\, (c\,\vec{p} - e\,\vec{A})\beta = \vec{\alpha} \,.\, \beta(c\,\vec{p} - e\,\vec{A})$$

Using anticommuting property of β with $\vec{\alpha}$, we have

$$\vec{\alpha} \,.\, (c\,\vec{p} - e\,\vec{A})\beta = -\beta\,\vec{\alpha} \,.\, (c\,\vec{p} - e\,\vec{A}) \qquad (14.55)$$

Using equations (14.54) and (14.55) in (14.53), we get

$$\Big\{(E - e\phi)^2 - \Big[\vec{\alpha}.(c\,\vec{p} - e\,\vec{A})\Big]^2 - m_0^2c^4 - (E - e\phi)\,\vec{\alpha}.(c\,\vec{p} - e\,\vec{A}) + \vec{\alpha}.(c\,\vec{p} - e\,\vec{A})(E - e\phi)\Big\}\Psi = 0 \quad (14.56)$$

The second term in equation (14.56) can be simplified using the identity

$$(\vec{\alpha}\,.\,\vec{P})(\vec{\alpha}\,.\,\vec{Q}) = (\vec{P}\,.\,\vec{Q}) + i\,\vec{\sigma}'\,.\,(\vec{P}\times\vec{Q}) \quad (14.57)$$

where $\vec{P}$ and $\vec{Q}$ are operators commuting with $\vec{\alpha}$. Thus, we have

$$\Big[\vec{\alpha}\,.\,(c\,\vec{p} - e\,\vec{A})\Big]^2 = \Big[\vec{\alpha}\,.\,(c\,\vec{p} - e\,\vec{A})\Big].\Big[\vec{\alpha}\,.\,(c\,\vec{p} - e\,\vec{A})\Big]$$

$$= (c\,\vec{p} - e\,\vec{A}).(c\,\vec{p} - e\,\vec{A}) + i\sigma'\Big[(c\,\vec{p} - e\,\vec{A})\times(c\,\vec{p} - e\,\vec{A})\Big] \quad (14.58)$$

Further, we have

$$\begin{aligned}(c\,\vec{p} - e\,\vec{A})\times(c\,\vec{p} - e\,\vec{A})\Psi &= (-c\,e\,\vec{P}\times\vec{A} - ce\,\vec{A}\times\vec{P})\Psi\\ &= ice\hbar(\nabla\times\vec{A} + \vec{A}\times\nabla)\Psi\\ &= ice\hbar[(\nabla\times\vec{A})\Psi + \nabla\Psi\times\vec{A} + \vec{A}\times\nabla\Psi]\\ &= ice\hbar[(\nabla\times\vec{A})\Psi + \nabla\Psi\times\vec{A} - \nabla\Psi\times\vec{A}]\\ &= ice\hbar[(\nabla\times\vec{A})\Psi\end{aligned}$$

Expressing the magnetic field $\vec{B}$ as $\vec{B} = \nabla\times\vec{A}$, we get

$$(c\,\vec{p} - e\,\vec{A})\times(c\,\vec{p} - e\,\vec{A})\Psi = ice\hbar\,\vec{B}\,\Psi$$

From equation (14.58), we get

$$\Big[\vec{\alpha}.(c\,\vec{p} - e\,\vec{A})\Big]^2\Psi = (c\,\vec{p} - e\,\vec{A})^2\Psi - ce\hbar\,\vec{\sigma}'\,.\,\vec{B}\,\Psi$$

For the last two terms in equation (14.56), we have

$$\begin{aligned}&\Big[-(E - e\phi)\,\vec{\alpha}\,.\,(c\,\vec{p} - e\,\vec{A}) + \vec{\alpha}\,.\,(c\,\vec{p} - e\,\vec{A})(E - e\phi)\Big]\Psi\\ &= \Big[e\,\vec{\alpha}\,.\,(E\,\vec{A} - \vec{A}\,E) - c\,\vec{\alpha}\,.\,(E\,\vec{p} - \vec{p}\,E)\\ &\quad + ce\,\vec{\alpha}\,.\,(\phi\,\vec{p} - \vec{p}\,\phi) - e^2\,\vec{\alpha}\,.\,(\phi\,\vec{A} - \vec{A}\,\phi)\Big]\Psi\end{aligned}$$

where we considered that E commutes with $\vec{\alpha}$. As E commutes with $\vec{p}$, we have ($E\ \vec{p}{=}\vec{p}\ E$). Further, for the scalar nature of ϕ, we have $\phi\ \vec{A}{=}\vec{A}\ \phi$. Thus, we have

$$
\begin{aligned}
(\phi\ \vec{p}-\vec{p}\ \phi)\Psi &= -i\hbar\phi\nabla\Psi + i\hbar\nabla(\phi\Psi) \\
&= -i\hbar\phi\nabla\Psi + i\hbar\phi\nabla\Psi + i\hbar(\nabla\phi)\Psi = i\hbar(\nabla\phi)\Psi
\end{aligned}
$$

and

$$
(E\ \vec{A}-\vec{A}\ E)\Psi = i\hbar\frac{\partial}{\partial t}(\vec{A}\ \Psi) - i\hbar\ \vec{A}\ \frac{\partial\Psi}{\partial t}
$$

$$
= i\hbar\ \vec{A}\ \frac{\partial\Psi}{\partial t} + i\hbar\frac{\partial\vec{A}}{\partial t}\Psi - i\hbar\ \vec{A}\ \frac{\partial\Psi}{\partial t} = i\hbar\frac{\partial\vec{A}}{\partial t}\Psi
$$

Therefore, the last two terms in equation(14.56) together give

$$
\Big[ie\hbar\ \vec{\alpha}\ .\frac{\partial\vec{A}}{\partial t} + ice\hbar\ \vec{\alpha}\ .\nabla\phi\Big]\Psi = ice\hbar\ \vec{\alpha}\ .\Big[\frac{1}{c}\frac{\partial\vec{A}}{\partial t} + \nabla\phi\Big]\Psi
$$

$$
= -ice\hbar\ \vec{\alpha}\ .\ \vec{\varepsilon}\ \Psi
$$

where $\vec{\varepsilon}$ is the electric field, associated with the electromagnetic field and defined as

$$
\vec{\varepsilon} = -\frac{1}{c}\frac{\partial\vec{A}}{\partial t} - \nabla\phi
$$

Combining all the contributions, equation (14.56) can now be expressed as

$$
\Big[(E-e\phi)^2 - (c\vec{p}\ -e\vec{A})^2 - m_0^2c^4 + ce\hbar\ \vec{\sigma}'.\ \vec{B}\ -ice\hbar\ \vec{\alpha}\ .\ \ \vec{\varepsilon}\ \Big]\Psi = 0 \quad (14.59)
$$

The first-three terms here are precisely the same as we get in relativity for a charged particle in an electromagnetic field. In order to understand the physical significance of the last two terms in equation (14.59), let us consider the nonrelativistic limit of this equation. The total energy E includes the rest energy m_0c^2 and we have $e\phi << m_0c^2$. Expressing E' as

$$
E' - E - m_0c^2 \qquad (14.60)
$$

and considering that in the nonrelativistic limit, $E' << m_0c^2$, we have

$$
(E-e\phi)^2 - m_0^2c^4 = (E-e\phi+m_0c^2)(E-e\phi-m_0c^2)
$$

$$
= (E'+2m_0c^2-e\phi)(E'-e\phi) \cong 2m_0c^2(E'-e\phi) \quad (14.61)
$$

Using equation (14.61) in (14.59), we get

$$\left[2m_0c^2(E - e\phi) - (c\vec{p} - e\vec{A})^2 + ce\hbar\,\vec{\sigma}'\,.\,\vec{B} - ice\hbar\,\vec{\alpha}\,.\,\vec{\varepsilon}\,\right]\Psi = 0$$

This equation can be rearranged as

$$E'\Psi = \left[\frac{1}{2m_0}\left(\vec{p} - \frac{e\vec{A}}{c}\right)^2 + e\phi - \frac{e\hbar}{2m_0c}\,\vec{\sigma}'\,.\,\vec{B} + \frac{ie\hbar}{2m_0c}\,\vec{\alpha}\,.\,\vec{\varepsilon}\,\right]\Psi \quad (14.62)$$

The third term on the right side of equation (14.62) has the form of a magnetic moment $e\hbar\vec{\sigma}'/(2m_0c)$ interacting with a magnetic field $\vec{B}$. The corresponding magnetic moment of the electron is

$$\mu = \frac{e\hbar\vec{\sigma}'}{2m_0c} = \frac{e\,\vec{s}}{m_0c}$$

where $\vec{s} = \hbar\vec{\sigma}'/2$. This value was postulated by Uhlenbeck and Goudsmit. The last term in equation (14.62) in the nonrelativistic limit reduces to zero.

11. Spin-orbit interaction

Another achievement of the Dirac equation is that the spin-orbit interaction comes out automatically. The Dirac Hamiltonian for an electron in a central field $V(r)$ is

$$H = c\,\vec{\alpha}\,.\,\vec{p} + \beta m_0c^2 + V(r)$$

This Hamiltonian satisfies the wave equation

$$i\hbar\frac{\partial}{\partial t}\Psi(\vec{r}, t) = H\Psi(\vec{r}, t)$$

Since the Hamiltonian is independent of time, the wave function Ψ can be expressed as the product of two parts where one depends on the space and the other on the time

$$\Psi(\vec{r}, t) = \psi(\vec{r})\,\exp\left[-\frac{iEt}{\hbar}\right]$$

It gives that Ψ is an eigenfunction of $i\hbar(\partial/\partial t)$ with eigenvalue E. Now, we express E as in equation(14.60) and the time independent $\psi(\vec{r})$ as a two component vector as

$$\psi(\vec{r}) = \begin{bmatrix} \psi_1 \\ \psi_2 \end{bmatrix}$$

Here, the ψ_1 represents the first-two components of the vector ψ and ψ_2 the last-two components. Expressing the Dirac matrices in the form

$$\vec{\alpha} = \begin{bmatrix} 0 & \vec{\sigma} \\ \vec{\sigma} & 0 \end{bmatrix}, \qquad \beta = \begin{bmatrix} I & 0 \\ 0 & -I \end{bmatrix}$$

where $\vec{\sigma}$ are the Pauli's spin matrices. The Dirac equation is

$$\begin{bmatrix} 0 & c\,\vec{\sigma}\,.\,\vec{p} \\ c\,\vec{\sigma}\,.\,\vec{p} & 0 \end{bmatrix}\begin{bmatrix} \psi_1 \\ \psi_2 \end{bmatrix} + \begin{bmatrix} m_0c^2 & 0 \\ 0 & -m_0c^2 \end{bmatrix}\begin{bmatrix} \psi_1 \\ \psi_2 \end{bmatrix} + V\begin{bmatrix} \psi_1 \\ \psi_2 \end{bmatrix}$$
$$= (E' + m_0c^2)\begin{bmatrix} \psi_1 \\ \psi_2 \end{bmatrix}$$

On simplification, we get the following pair of the coupled equations.

$$(E' - V)\psi_1 - c(\vec{\sigma}\,.\,\vec{p})\psi_2 = 0 \tag{14.63}$$

and

$$(E' + 2m_0c^2 - V)\psi_2 - c(\vec{\sigma}\,.\,\vec{p})\psi_1 = 0 \tag{14.64}$$

From equation (14.64), we have

$$\psi_2 = \frac{(\vec{\sigma}\,.\,\vec{p})}{2m_0c\left[1 + \frac{E'-V}{2m_0c^2}\right]}\psi_1$$
$$= \left[1 + \frac{E' - V}{2m_0c^2}\right]^{-1}\frac{(\vec{\sigma}\,.\,\vec{p})}{2m_0c}\psi_1$$
$$\cong \left[1 - \frac{E' - V}{2m_0c^2}\right]\frac{(\vec{\sigma}\,.\,\vec{p})\psi_1}{2m_0c}$$

Here, we used the fact that $(E' - V) << 2m_0c^2$. Using this value of ψ_2 in equation (14.63), we have

$$E'\psi_1 = V\psi_1 + \frac{1}{2m_0}(\vec{\sigma}\,.\,\vec{p})(\vec{\sigma}\,.\,\vec{p})\psi_1$$
$$-\frac{1}{4m_0^2c^2}(\vec{\sigma}\,.\,\vec{p})(E' - V)(\vec{\sigma}\,.\,\vec{p})\psi_1 \tag{14.65}$$

We have

$$(\vec{\sigma}\,.\,\vec{p})(\vec{\sigma}\,.\,\vec{p}) = (\sigma_xp_x + \sigma_yp_y + \sigma_zp_z)(\sigma_xp_x + \sigma_yp_y + \sigma_zp_z)$$
$$= (\sigma_x^2p_x^2 + \sigma_y^2p_y^2 + \sigma_z^2p_z^2) = p_x^2 + p_y^2 + p_z^2 = p^2 \tag{14.66}$$

As the E' commutes with $\vec{\sigma}$ as well as $\vec{p}$, we have

$$(\vec{\sigma} \cdot \vec{p})E'(\vec{\sigma} \cdot \vec{p}) = E'(\vec{\sigma} \cdot \vec{p})(\vec{\sigma} \cdot \vec{p}) = E'p^2 \tag{14.67}$$

We now have

$$(\vec{\sigma} \cdot \vec{p})V(\vec{\sigma} \cdot \vec{p}) = (\vec{\sigma} \cdot \vec{p}\ V)(\vec{\sigma} \cdot \vec{p}) \tag{14.68}$$

Using the relation

$$(\vec{\sigma} \cdot \vec{B})(\vec{\sigma} \cdot \vec{C}) = \vec{B} \cdot \vec{C} + i\, \vec{\sigma} \cdot (\vec{B} \times \vec{C})$$

where $\vec{B}$ and $\vec{C}$ are operators, we have

$$(\vec{\sigma} \cdot \vec{p}\ V)(\vec{\sigma} \cdot \vec{p}) = \left\{(\vec{p}V) \cdot \vec{p}\right\} + i\, \vec{\sigma} \cdot \left\{(\vec{p}V) \times \vec{p}\right\} \tag{14.69}$$

From equations (14.68) and (14.69), we have

$$(\vec{\sigma} \cdot \vec{p})V(\vec{\sigma} \cdot \vec{p}) = \{(\vec{p}V) \cdot \vec{p}\} + i\, \vec{\sigma} \cdot \left\{(\vec{p}V) \times \vec{p}\right\} \tag{14.70}$$

Using[7]

$$\vec{p}V = V\vec{p} - i\hbar(\nabla V)$$

We have

$$(\vec{\sigma} \cdot \vec{p})V(\vec{\sigma} \cdot \vec{p}) = \left\{V\vec{p} - i\hbar(\nabla V)\right\} \cdot \vec{p} + i\, \vec{\sigma} \cdot \left\{V\vec{p} - i\hbar(\nabla V)\right\} \times \vec{p}$$

$$= Vp^2 - i\hbar(\nabla V) \cdot \vec{p} + \vec{\sigma}\, \hbar(\nabla V) \times \vec{p} \tag{14.71}$$

Using equations (14.66), (14.67) and (14.71) in (14.65) we get

$$E'\psi_1 = \left[V + \frac{p^2}{2m_0} - \frac{E'p^2}{4m^2c^2} + \frac{Vp^2}{4m_0^2c^2} - \frac{i\hbar}{4m_0^2c^2}(\nabla V \cdot \vec{p}) + \frac{\hbar}{4m_0^2c^2}\, \vec{\sigma} \cdot (\nabla V \times \vec{p})\right]\psi_1 \tag{14.72}$$

---

[7] We have

$$(\vec{p}\ V - V\ \vec{p})\psi = -i\hbar\nabla(V\psi) + i\hbar V\nabla\psi$$
$$= -i\hbar V\nabla\psi + i\hbar V\nabla\psi - i\hbar(\nabla V)\psi = -i\hbar(\nabla V)\psi$$

Therefore, we have

$$\vec{p}\ V - V\ \vec{p} = -i\hbar(\nabla V)$$
$$\vec{p}\ V = V\ \vec{p} - i\hbar(\nabla V)$$

For the spherically symmetric potential, we have

$$(\nabla V).\vec{p} = -i\hbar(\nabla V).\nabla = i\hbar\frac{dV}{dr}\frac{\partial}{\partial r} \tag{14.73}$$

and

$$\begin{aligned}\frac{\hbar}{4m_0^2c^2}\vec{\sigma}.\left\{(\nabla V)\times\vec{p}\right\} &= \frac{1}{2m_0^2c^2}\frac{\hbar}{2}\vec{\sigma}.\frac{1}{r}\frac{dV}{dr}(\vec{r}\times\vec{p}) \\ &= \frac{1}{2m_0^2c^2}\frac{1}{r}\frac{dV}{dr}\vec{s}.\vec{l}\end{aligned} \tag{14.74}$$

where $\vec{l} = \vec{r}\times\vec{p}$ is angular momentum. Using equations (14.73) and (14.74) in (14.72) we have

$$\begin{aligned}E'\psi_1 = \Bigg[V &+ \frac{p^2}{2m_0} - \frac{E'-V}{4m_0^2c^2}p^2 - \frac{\hbar^2}{4m_0^2c^2}\frac{dV}{dr}\frac{\partial}{\partial r} \\ &+ \frac{1}{2m_0^2c^2}\frac{1}{r}\frac{dV}{dr}\vec{l}.\vec{s}\Bigg]\psi_1\end{aligned} \tag{14.75}$$

The first and second terms on the right side of this equation correspond to the nonrelativistic Schrödinger equation. In the third term, we may replace $(E'-V)$ by $p^2/2m_0$. This is acceptable as it is only a second-order term. By doing so, the term becomes $p^4/(8m_0^3c^2)$. It can be interpreted as the classical relativistic mass correction in the following manner.

$$\begin{aligned}E' &= E - m_0c^2 \\ &= (c^2p^2 + m_0c^2c^4)^{1/2} - m_0c^2 \\ &= m_0c^2\left[1 + \frac{p^2}{2m_0^2c^2} - \frac{p^4}{8c^4m_0c^4}\right] - m_0c^2 \\ &= \frac{p^2}{2m_0} - \frac{p^4}{8c^2m_0^3}\end{aligned}$$

The last term in equation (14.75) has no classical analogue. It is the spin-orbit interaction energy which appears here as an automatic consequence of the Dirac equation.

12. Radial equation for an electron in a central potential

The relativistic Hamiltonian for an electron in a central potential $V(r)$ is

$$H = c\,\vec{\alpha}\,.\,\vec{p} + \beta m_0 c^2 + V(r) \tag{14.76}$$

Considering $\vec{P}=\vec{r}$ and $\vec{Q}=\vec{p}$ in equation (14.57), we have

$$\begin{aligned}(\vec{\alpha}\,.\,\vec{r})(\vec{\alpha}\,.\,\vec{p}) &= \vec{r}\,.\,\vec{p} + i\,\vec{\sigma}'\,.\,(\vec{r}\times\vec{p})\\ &= \vec{r}\,.\,\vec{p} + i\,\vec{\sigma}'\,.\,\vec{l}\end{aligned} \tag{14.77}$$

Let us introduce a new operator k by expressed by the relation

$$k\hbar = \beta(\vec{\sigma}'\,.\,\vec{l} + \hbar) \tag{14.78}$$

so that

$$\vec{\sigma}'\,.\,\vec{l} = \frac{(k-\beta)\hbar}{\beta}$$

Using this value of $\vec{\sigma}'\,.\,\vec{l}$ and multiplying both sides by $(\vec{\alpha}\,.\,\vec{r})$, equation (14.77) can be expressed as

$$(\vec{\alpha}\,.\,\vec{r})^2(\vec{\alpha}\,.\,\vec{p}) = (\vec{\alpha}\,.\,\vec{r})\Big[\vec{r}\,.\,\vec{p} + \frac{i(k-\beta)\hbar}{\beta}\Big]$$

Since[8] $(\vec{\alpha}\,.\,\vec{r})^2 = r^2$, we have

$$\vec{\alpha}\,.\,\vec{p} = \frac{\vec{\alpha}\,.\,\vec{r}}{r^2}\Big[\vec{r}\,.\,\vec{p} - i\hbar + \frac{ik\hbar}{\beta}\Big] \tag{14.79}$$

where we used $(\vec{\alpha}.\vec{r})^2 = r^2$. For convenience, the expression for $(\vec{\alpha}\,.\,\vec{p})$ can be expressed in terms of radial momentum operator p_r and the radial velocity operator α_r defined as

$$p_r = \frac{\vec{r}\,.\,\vec{p} - i\hbar}{r} \quad \text{and} \quad \alpha_r = \frac{\vec{\alpha}\,.\,\vec{r}}{r} \tag{14.80}$$

In terms of p_r and α_r, equation (14.79) can be expressed as

$$\begin{aligned}\vec{\alpha}\,.\,\vec{p} &= \frac{\vec{\alpha}\,.\,\vec{r}}{r}\Big[\frac{\vec{r}\,.\,\vec{p} - i\hbar}{r} + \frac{ik\hbar\beta}{r\beta^2}\Big]\\ &= \alpha_r\Big[p_r + \frac{ik\beta\hbar}{r}\Big]\end{aligned}$$

---

[8] We have

$$(\vec{\alpha}\,.\,\vec{r})(\vec{\alpha}\,.\,\vec{r}) = \vec{r}\,.\,\vec{r} + i\,\vec{\sigma}'\,.\,(\vec{r}\times\vec{r}) = r^2$$

where we used $\beta^2 = 1$. Using this values of $\vec{\alpha} \, . \, \vec{p}$ in equation(14.76), we get

$$H = c\alpha_r p_r + \frac{ic\hbar\alpha_r k\beta}{r} + \beta m_0 c^2 + V(r) \tag{14.81}$$

Here, the operator k commutes with this Hamiltonian given in equation (14.81).

Let us now evaluate the eigenvalues of the operator k. For this, we square equation (14.78) and have ($\beta^2 = 1$)

$$k^2\hbar^2 = \left(\vec{\sigma}' \, . \, \vec{l}\,\right)^2 + 2\hbar\left(\vec{\sigma}' . \vec{l}\,\right) + \hbar^2 \tag{14.82}$$

From equation (14.57) we have

$$\begin{aligned}(\vec{\sigma}' . \vec{l}\,)(\vec{\sigma}' . \vec{l}\,) &= l^2 + i\,\vec{\sigma}' .(\vec{l} \times \vec{l}\,) \\ &= l^2 + i\,\vec{\sigma}' . \, i\hbar\,\vec{l} \\ &= l^2 - \hbar(\vec{\sigma}' . \vec{l}\,)\end{aligned} \tag{14.83}$$

where we have used the relation $\vec{l} \times \vec{l} = i\hbar\,\vec{l}$, derived earlier. Using equation (14.83) in (14.82), we have

$$k^2\hbar^2 = l^2 + \hbar(\vec{\sigma}' \, . \, \vec{l}\,) + \hbar^2$$

Since

$$\vec{s} = \frac{1}{2}\,\hbar\,\vec{\sigma}'$$

and eigen value of s^2 is $s(s+1)\hbar^2$ which is equal to $3\hbar^2/4$ for $s = 1/2$ for electron. Using these values, we have

$$\begin{aligned}k^2\hbar^2 &= l^2 + 2\,\vec{s} \, . \, \vec{l} + \frac{3\hbar^2}{4} + \frac{\hbar^2}{4} \\ &= l^2 + +2\,\vec{s} \, . \, \vec{l} + s^2 + \frac{\hbar^2}{4} \\ &= (\vec{l} + \vec{s}\,)^2 + \frac{\hbar^2}{4} = j^2 + \frac{\hbar^2}{4}\end{aligned} \tag{14.84}$$

where $\vec{j} = \vec{l} + \vec{s}$, is the total angular momentum. Since the eigenvalue of the operator j^2 is $j(j+1)\hbar^2$, the above equation (14.84) can be expressed as

$$k^2 = j(j+1) + \frac{1}{4} = \left[j + \frac{1}{2}\right]^2$$

Thus, we have

$$k = \pm\left[j + \frac{1}{2}\right]$$

For an electron in the orbitals s, p, d, ..., we have $j = 1/2,\ 3/2,\ 5/2$ ..., respectively and therefore, the possible values of k are ± 1, ± 2, ± 3, ... Since the operators H and k commute, it is possible to have a representation in which they are diagonal. It follows from equation (14.80) that $\alpha_r^2 = 1$, $\beta^2 = 1$ and $\alpha_r\beta + \beta\alpha_r = 0$. Hermitian matrices satisfying these relations can be 2×2 matrices and we take them as the following.

$$\alpha_r = \begin{bmatrix} 0 & -i \\ i & 0 \end{bmatrix} \qquad \beta = \begin{bmatrix} 1 & 0 \\ 0 & -1 \end{bmatrix}$$

For 2×2 matrices α_r and β, the radial part of the wave function has the two components

$$\begin{bmatrix} F/r \\ G/r \end{bmatrix} \tag{14.85}$$

Now, we have[9]

$$p_r = -i\hbar\frac{\partial}{\partial r} - \frac{i\hbar}{r}$$

Using the expressions for α_r, β and p_r in equation (14.81) and operating the resultant radial function in equation (14.85), we get the following two-component equation.

$$(E - V)\begin{bmatrix} F/r \\ G/r \end{bmatrix} = -ic\hbar\begin{bmatrix} 0 & -i \\ i & 0 \end{bmatrix}\left(\frac{\partial}{\partial r} + \frac{1}{r}\right)\begin{bmatrix} F/r \\ G/r \end{bmatrix}$$

---

[9]We have

$$\nabla \equiv \hat{r}\frac{\partial}{\partial r} + \frac{\hat{\theta}}{r}\frac{\partial}{\partial\theta} + \frac{\hat{\phi}}{r\sin\theta}\frac{\partial}{\partial\phi}$$

Thus, we have

$$\vec{p} = -i\hbar\nabla = -i\hbar\left[\hat{r}\frac{\partial}{\partial r} + \frac{\hat{\theta}}{r}\frac{\partial}{\partial\theta} + \frac{\hat{\phi}}{r\sin\theta}\frac{\partial}{\partial\phi}\right]$$

Hence,

$$\vec{p}\,.\,\vec{r} = -i\hbar r\frac{\partial}{\partial r}$$

and therefore, we have

$$p_r = \frac{\vec{r}\,.\,\vec{p}}{r} - \frac{i\hbar}{r} = -i\hbar\frac{\partial}{\partial r} - \frac{i\hbar}{r}$$

$$+\frac{ic\hbar k}{r}\begin{bmatrix} 0 & -i \\ i & 0 \end{bmatrix}\begin{bmatrix} 1 & 0 \\ 0 & -1 \end{bmatrix}\begin{bmatrix} F/r \\ G/r \end{bmatrix} + m_0c^2\begin{bmatrix} 1 & 0 \\ 0 & -1 \end{bmatrix}\begin{bmatrix} F/r \\ G/r \end{bmatrix} \tag{14.86}$$

Since

$$\left(\frac{\partial}{\partial r} + \frac{1}{r}\right)\frac{F}{r} = \frac{1}{r}\frac{\mathrm{d}F}{\mathrm{d}r} \qquad \text{and} \qquad \left(\frac{\partial}{\partial r} + \frac{1}{r}\right)\frac{G}{r} = \frac{1}{r}\frac{\mathrm{d}G}{\mathrm{d}r}$$

equation (14.86) reduces to

$$\begin{bmatrix} 0 & -\hbar c \\ \hbar c & 0 \end{bmatrix}\begin{bmatrix} \frac{1}{r}\frac{\mathrm{d}F}{\mathrm{d}r} \\ \frac{1}{r}\frac{\mathrm{d}G}{\mathrm{d}r} \end{bmatrix} - \frac{c\hbar k}{r}\begin{bmatrix} 0 & 1 \\ 1 & 0 \end{bmatrix}\begin{bmatrix} F/r \\ G/r \end{bmatrix} + m_0c^2\begin{bmatrix} 1 & 0 \\ 0 & -1 \end{bmatrix}\begin{bmatrix} F/r \\ G/r \end{bmatrix}$$

$$+(V-E)\begin{bmatrix} F/r \\ G/r \end{bmatrix} = 0$$

This equation can be expressed in the form of the following coupled equations.

$$\frac{\mathrm{d}G}{\mathrm{d}r} + \frac{kG}{r} + \frac{1}{c\hbar}(E - V - m_0c^2)F = 0 \tag{14.87}$$

and

$$\frac{\mathrm{d}F}{\mathrm{d}r} - \frac{kF}{r} - \frac{1}{c\hbar}(E - V + m_0c^2)G = 0 \tag{14.88}$$

These are radial equations for an electron in a central field. For the known form of the potential $V(r)$, these equations can be solved for the energy eigenvalues.

Exercise 8. Show that the following 2×2 matrices α_r and β satisfy the relations $\alpha_r^2 = 1$, $\beta^2 = 1$ and $\alpha_r\beta + \beta\alpha_r = 0$.

$$\alpha_r = \begin{bmatrix} 0 & -i \\ i & 0 \end{bmatrix} \qquad \beta = \begin{bmatrix} 1 & 0 \\ 0 & -1 \end{bmatrix}$$

Solution: We have

$$\alpha_r^2 = \begin{bmatrix} 0 & -i \\ i & 0 \end{bmatrix}\begin{bmatrix} 0 & -i \\ i & 0 \end{bmatrix} = \begin{bmatrix} 1 & 0 \\ 0 & 1 \end{bmatrix} = 1$$

$$\beta^2 = \begin{bmatrix} 1 & 0 \\ 0 & -1 \end{bmatrix}\begin{bmatrix} 1 & 0 \\ 0 & -1 \end{bmatrix} = \begin{bmatrix} 1 & 0 \\ 0 & 1 \end{bmatrix} = 1$$

$$\alpha_r\beta + \beta\alpha_r = \begin{bmatrix} 0 & -i \\ i & 0 \end{bmatrix}\begin{bmatrix} 1 & 0 \\ 0 & -1 \end{bmatrix} + \begin{bmatrix} 1 & 0 \\ 0 & -1 \end{bmatrix}\begin{bmatrix} 0 & -i \\ i & 0 \end{bmatrix}$$

$$= \begin{bmatrix} 0 & i \\ i & 0 \end{bmatrix} + \begin{bmatrix} 0 & -i \\ -i & 0 \end{bmatrix} = \begin{bmatrix} 0 & 0 \\ 0 & 0 \end{bmatrix} = 0$$

13. Hydrogen atom

Let us now use equations (14.87) and (14.88) to find out energy eigenvalues of hydrogen atom. For the hydrogen atom, the potential energy is $V(r) = -Ze^2/r$. For convenience, let us introduce new parameters as

$$\alpha = \frac{(m_0^2c^4 - E^2)^{1/2}}{c\hbar}$$

$$= \left(\frac{m_0c^2 + E}{c\hbar}\right)^{1/2}\left(\frac{m_0c^2 - E}{c\hbar}\right)^{1/2} = \alpha_1^{1/2}\alpha_2^{1/2}$$

where

$$\alpha_1 = \frac{m_0c^2 + E}{c\hbar} \qquad \text{and} \qquad \alpha_2 = \frac{m_0c^2 - E}{c\hbar}$$

Now, equations (14.87) and (14.88) can be expressed as

$$\frac{dG}{dr} + \frac{kG}{r} - \left(\alpha_2 - \frac{Ze^2}{c\hbar r}\right)F = 0 \tag{14.89}$$

and

$$\frac{dF}{dr} - \frac{kF}{r} - \left(\alpha_1 + \frac{Ze^2}{c\hbar r}\right)G = 0 \tag{14.90}$$

Defining $\gamma = Ze^2/c\hbar$ and introducing a new variable $\rho = \alpha r$ equations (14.89) and (14.90) can be expressed as the following.

$$\frac{dG}{d\rho} + \frac{kG}{\rho} - \left(\frac{\alpha_2}{\alpha} - \frac{\gamma}{\rho}\right)F = 0 \tag{14.91}$$

and

$$\frac{dF}{d\rho} - \frac{kF}{\rho} - \left(\frac{\alpha_1}{\alpha} + \frac{\gamma}{\rho}\right)G = 0 \tag{14.92}$$

As in the nonrelativistic case, let us look for a solution of the form

$$G(\rho) = \sum_{n=0}^{\infty} a_n\, \rho^{s+n}\, e^{-\rho}, \qquad a_0 \neq 0 \tag{14.93}$$

and

$$F(\rho) = \sum_{n=0}^{\infty} b_n \, \rho^{s+n} \, e^{-\rho}, \qquad b_0 \neq 0 \tag{14.94}$$

Using equations (14.93) and (14.94) in (14.91) and (14.92) and equating the coefficients of $\rho^{s+n+1} \, e^{-\rho}$ to zero, we get

$$(s+n+k)a_n - a_{n-1} + \gamma b_n - \frac{\alpha_2}{\alpha} b_{n-1} = 0 \tag{14.95}$$

and

$$(s+n-k)b_n - b_{n-1} - \gamma a_n - \frac{\alpha_1}{\alpha} a_{n-1} = 0 \tag{14.96}$$

For $n = 0$, equations (14.95) and (14.96) give

$$(s+k)a_0 + \gamma b_0 = 0 \qquad \text{and} \qquad (s-k)b_0 - \gamma a_0 = 0 \tag{14.97}$$

For nonvanishing solution of equation (14.97), we have the determinant

$$\begin{vmatrix} s+k & \gamma \\ -\gamma & s-k \end{vmatrix} = 0$$

or

$$s = \pm(k^2 - \gamma^2)^{1/2}$$

Here, the negative solution is not acceptable as it gives diverging nature of F and G at the origin. Thus, the acceptable solution is with the positive sign. For $n > 0$, a relation between a_n and b_n can be obtained by multiplying equation (14.95) by α and (14.96) by α_2 and subtracting the products, we get

$$a_n \Big[(s+n+k)\alpha + \alpha_2 \gamma\Big] = b_n \Big[\alpha_2(s+n-k) - \alpha\gamma\Big] \tag{14.98}$$

here we have used the relation $\alpha^2 = \alpha_1 \alpha_2$. As we found in the non-relativistic case, here also regular solutions are possible only when the series (14.93) and (14.94) terminate. Let the series terminate at $n = n'$ so that $a_{n'+1} = b_{n'+1} = 0$. Replacing n by $(n'+1)$ in equation (14.95), we get

$$(s+n'+1+k)a_{n'+1} - a_{n'} + \gamma b_{n'+1} - \frac{\alpha_2}{\alpha} b_{n'} = 0 \tag{14.99}$$

For $a_{n'+1} = b_{n'+1} = 0$, equation (14.99) reduces to

$$a_{n'} = -\frac{\alpha_2}{\alpha} \, b_{n'}, \qquad n' = 0, 1, 2, \ldots \tag{14.100}$$

Equation (14.96) also gives the same relation between $b_{n'}$ and $a_{n'}$. For getting energy levels, we put $n = n'$ in equation (14.98) and use equation (14.100) so that

$$-\frac{\alpha_2}{\alpha} b_{n'}[(s+n'+k)\alpha + \alpha_2\gamma] = b_{n'}[\alpha_2(s+n'-k) - \alpha\gamma]$$

Since $b_{n'} \neq 0$, this equation gives

$$2\alpha(s+n') = \gamma(\alpha_1 - \alpha_2)$$

Using the values of α_1, α_2, α in this equation, we get

$$(m_0^2c^4 - E^2)^{1/2}(s+n') = E\gamma$$

On squaring, we have

$$(m_0^2c^4 - E^2)(s+n')^2 = E^2\gamma^2$$

This equation can be rearranged as

$$E = m_0c^2\left[1 + \frac{\gamma^2}{(s+n')^2}\right]^{-1/2}$$

Using the value of s with positive sign and expanding with the help of Binomial series and retaining terms up to γ^4, we have

$$E = m_0c^2\left[1 + \gamma^2\left\{(k^2-\gamma^2)^{1/2} + n'\right\}^{-2}\right]^{-1/2}$$

$$= m_0c^2\left[1 + \gamma^2\left\{(n'+|k|) - \frac{1}{2}\frac{\gamma^2}{|k|}\right\}^{-2}\right]^{-1/2}$$

Expressing $n = n' + |k|$, we get

$$E = m_0c^2\left[1 + \frac{\gamma^2}{n^2}\left(1 - \frac{1}{2}\frac{\gamma^2}{|k|n}\right)^{-2}\right]^{-1/2}$$

$$= m_0c^2\left[1 + \frac{\gamma^2}{n^2}\left(1 + \frac{\gamma^2}{|k|n}\right)\right]^{-1/2}$$

$$= m_0c^2\left[1 - \frac{1}{2}\frac{\gamma^2}{n^2}\left(1 + \frac{\gamma^2}{|k|n}\right) + \frac{3}{8}\frac{\gamma^4}{n^4}\right]$$

$$= m_0c^2\left[1 - \frac{\gamma^2}{2n^2} - \frac{\gamma^4}{2n^4}\left(\frac{n}{|k|} - \frac{3}{4}\right)\right] \tag{14.101}$$

Now, $n = n' + |k| = n' + j + \frac{1}{2}$ is the total quantum number of hydrogen atom and $|k|$ can take the values $1, 2, \ldots, n$. We extend the discussion to nonrelativistic limit for labeling the energy levels by l, the orbital angular momentum quantum number. Then the values of l in terms of k are

$$l = |k| = j + \frac{1}{2} \qquad \text{for} \quad k < 0$$

$$l = k - 1 = j - \frac{1}{2} \qquad \text{for} \quad k > 0$$

The relativistic effects partly remove the degeneracy of the energy level for a given n in the nonrelativistic case. The structure energy levels is determined by the quantum number n whereas the fine structure is obtained with the help of $|k|$ or j. The total spread in the energy of the fine structure levels for a given value of n can easily be found by taking the difference in energy for the states $|k| = 1$ and $|k| = n$ is

$$E = -\frac{\gamma^4 m_0 c^2}{2n^4}\left[\left(\frac{n}{n} - \frac{3}{4}\right) - \left(\frac{n}{1} - \frac{3}{4}\right)\right]$$

$$= \frac{\gamma^4 m_0 c^2 (n-1)}{2n^4}$$

This value is in agreement with the experiment.

Exercise 9. Find out the energy of the levels of the hydrogen atom for $n = 1$, 2 and 3.

Solution: The energy of the level with quantum number n of hydrogen atom is

$$E = m_0 c^2 \left[1 - \frac{\gamma^2}{2n^2} - \frac{\gamma^4}{2n^4}\left(\frac{n}{|k|} - \frac{3}{4}\right)\right]$$

where $\gamma = Ze^2/c\hbar$, $|k|$ can take the values $1, 2, \ldots, n$ and the orbital quantum number l is expressed as

$$l = |k| = j + \frac{1}{2} \qquad \text{for} \quad k < 0$$

$$l = k - 1 = j - \frac{1}{2} \qquad \text{for} \quad k > 0$$

For $n = 1$, 2 and 3 of hydrogen atom, the values of n, k, l, j, term of the level and its energy are given in the following table:

n	k	l	j	Term	E/m_0c^2
1	-1	1	$\frac{1}{2}$	$1^2P_{1/2}$	$1-\frac{\gamma^2}{2}-\frac{\gamma^4}{8}$
	1	0	$\frac{1}{2}$	$1^2S_{1/2}$	$1-\frac{\gamma^2}{2}-\frac{\gamma^4}{8}$
2	-1	1	$\frac{1}{2}$	$2^2P_{1/2}$	$1-\frac{\gamma^2}{8}-\frac{5\gamma^4}{128}$
	1	0	$\frac{1}{2}$	$2^2S_{1/2}$	$1-\frac{\gamma^2}{8}-\frac{5\gamma^4}{128}$
	-2	2	$\frac{3}{2}$	$2^2D_{3/2}$	$1-\frac{\gamma^2}{8}-\frac{\gamma^4}{128}$
	2	1	$\frac{3}{2}$	$2^2P_{3/2}$	$1-\frac{\gamma^2}{8}-\frac{\gamma^4}{128}$
3	-1	1	$\frac{1}{2}$	$3^2P_{1/2}$	$1-\frac{\gamma^2}{18}-\frac{\gamma^4}{72}$
	1	0	$\frac{1}{2}$	$3^2S_{1/2}$	$1-\frac{\gamma^2}{18}-\frac{\gamma^4}{72}$
	-2	2	$\frac{3}{2}$	$3^2D_{3/2}$	$1-\frac{\gamma^2}{18}-\frac{\gamma^4}{216}$
	2	1	$\frac{3}{2}$	$3^2P_{3/2}$	$1-\frac{\gamma^2}{18}-\frac{\gamma^4}{216}$
	-3	3	$\frac{5}{2}$	$3^2F_{5/2}$	$1-\frac{\gamma^2}{18}-\frac{\gamma^4}{648}$
	3	2	$\frac{5}{2}$	$3^2D_{5/2}$	$1-\frac{\gamma^2}{18}-\frac{\gamma^4}{648}$

14. Problems and questions

1. Discuss how Klein-Gordon equation leads to positive as well as negative probability density values.

2. Starting from the Klein-Gordon equation, obtain the equation of continuity.

3. Obtain an expression for the energy of a charged particle obeying Klein-Gordon equation in a Coulomb potential. Discuss the significance of various terms.

4. Discuss the reasons: (i) Why Dirac preferred a first-order equation both in time and space coordinates. (ii) The dimension of the Dirac matrices has to be a even number.

5. Obtain expressions for the probability density and the probability current density in the Dirac theory.

6. What are negative energy states? Discuss about a hole.

7. For a Dirac particle moving in a central potential, show that the orbital angular momentum is not a constant of motion.

8. Show that the following matrices form a representation of Dirac's matrices:

$$\alpha_x = \begin{bmatrix} \sigma_x & 0 \\ 0 & -\sigma_x \end{bmatrix}, \qquad \alpha_y = \begin{bmatrix} 0 & I \\ I & 0 \end{bmatrix}$$

$$\alpha_z = \begin{bmatrix} \sigma_z & 0 \\ 0 & -\sigma_z \end{bmatrix}, \qquad \beta = \begin{bmatrix} 0 & iI \\ -iI & 0 \end{bmatrix}$$

where σ_x and σ_z are the Pauli's spin matrices.

9. For

$$\vec{\sigma'} = \begin{bmatrix} \vec{\sigma} & 0 \\ 0 & \vec{\sigma} \end{bmatrix}$$

where $\vec{\sigma}$ are the Pauli's spin matrices, show that

$$\sigma_x'^2 = \sigma_y'^2 = \sigma_z'^2 = 1$$
$$\sigma_x'\sigma_y' = i\sigma_z'; \qquad \sigma_y'\sigma_z' = i\sigma_x'; \qquad \sigma_z'\sigma_x' = i\sigma_y'$$
$$\sigma_y'\sigma_x' = -i\sigma_z'; \qquad \sigma_z'\sigma_y' = -i\sigma_x'; \qquad \sigma_x'\sigma_z' = -i\sigma_y'$$
$$\sigma_x'\sigma_y' + \sigma_y'\sigma_x' = \sigma_y'\sigma_z' + \sigma_z'\sigma_y' = \sigma_z'\sigma_x' + \sigma_x'\sigma_z' = 0$$

10. Show that $(\vec{\sigma'} \,.\, \vec{A})(\vec{\sigma'} \,.\, \vec{B}) = (\vec{A} \,.\, \vec{B}) + i\,\vec{\sigma'} \,.(\vec{A} \times \vec{B})$, where

$$\vec{\sigma'} = \begin{bmatrix} \vec{\sigma} & 0 \\ 0 & \vec{\sigma} \end{bmatrix}$$

and $\vec{\sigma}$ are the Pauli's spin matrices

11. Show that $(\vec{\alpha} \,.\, \vec{A})(\vec{\alpha} \,.\, \vec{B}) = (\vec{A} \,.\, \vec{B}) + i\,\vec{\sigma'} \,.(\vec{A} \times \vec{B})$, where $\vec{A}$ and $\vec{B}$ commute with $\vec{\alpha}$ and

$$\vec{\sigma'} = \begin{bmatrix} \vec{\sigma} & 0 \\ 0 & \vec{\sigma} \end{bmatrix}$$

and $\vec{\sigma}$ are the Pauli's spin matrices

12. Show that the following 2×2 matrices α_r and β satisfy the relations $\alpha_r^2 = 1$, $\beta^2 = 1$ and $\alpha_r\beta + \beta\alpha_r = 0$.

$$\alpha_r = \begin{bmatrix} 0 & -i \\ i & 0 \end{bmatrix} \qquad \beta = \begin{bmatrix} 1 & 0 \\ 0 & -1 \end{bmatrix}$$

13. If $\vec{\sigma}' = \begin{bmatrix} \vec{\sigma} & 0 \\ 0 & \vec{\sigma} \end{bmatrix}$

show that

(*i*) $\sigma'^2_x = \sigma'^2_y = \sigma'^2_z = 1$

(*ii*) $[\sigma'_x, \sigma_x] = 0, \quad [\sigma'_x, \sigma_y] = 2i\ \alpha_z, \quad [\sigma'_x, \sigma_z] = -2i\ \alpha_y$

where $\vec{\sigma}$ are the Pauli matrices and $\alpha_x, \alpha_y, \alpha_z$ are the Dirac matrices.

14. Find out the energy of the levels of the hydrogen atom for $n = 1$, 2 and 3.

15. Write notes on the following

(i) Magnetic moment

(ii) Spin-orbit interaction

(iii) Concept of hole

Some Useful Integrals

$$\int_0^\infty x^{2n+1}\,\mathrm{e}^{-ax^2}\,\mathrm{d}x = \frac{n!}{2a^{n+1}}$$ n is a positive integer, $a > 0$

$$\int_0^\infty x^{2n}\mathrm{e}^{-ax^2}\,\mathrm{d}x = \frac{\Gamma(n+1/2)}{2a^{n+1/2}}$$ n is a positive integer, $a > 0$

$$\int_{-\infty}^\infty x^n\mathrm{e}^{-ax^2}\,\mathrm{d}x = 0$$ n is an odd integer, $a > 0$

$$\int_0^\infty x^n\mathrm{e}^{-ax}\,\mathrm{d}x = \frac{n!}{a^{n+1}}$$ $n \geq 0 \quad a > 0$

$$\int_0^\infty \frac{x}{\mathrm{e}^x - 1}\,\mathrm{d}x = \frac{\pi^2}{6}$$

$$\int_0^\infty \frac{x^3}{\mathrm{e}^x - 1}\,\mathrm{d}x = \frac{\pi^4}{15}$$

$$\int_0^\infty \mathrm{e}^{-ax}\sin(bx)\,\mathrm{d}x = \frac{b}{a^2+b^2}$$ $a > 0$

$$\int_0^\infty \mathrm{e}^{-ax}\cos(bx)\,\mathrm{d}x = \frac{a}{a^2+b^2}$$ $a > 0$

$$\int x\,\mathrm{e}^{ax}\,\mathrm{d}x = \frac{ax-1}{a^2}\mathrm{e}^{ax}$$

$$\int x^2\mathrm{e}^{ax}\,\mathrm{d}x = \left(\frac{x^2}{a} - \frac{2x}{a^2} + \frac{2}{a^3}\right)\mathrm{e}^{ax}$$

Bibliography

G. Aruldhas, Quantum Mechanics (2002) Prentice-Hall of India Pvt. Ltd., New Delhi

V. Devanathan, Quantum Mechanics, (2005) Narosa Publishing House, New Delhi

Amit Goswami, Quantum Mechanics, 2nd ed., (1997) Wm.C. Brown Publisher, U.S.A

P.M. Mathews and K. Venkatesan, A Text Book of Quantum Mechanics, (1976) Tata McGraw-Hill, New Delhi

V.K. Thankappan, Quantum Mechanics, 3rd ed., (1993) Wiley Eastern, New Delhi

Values of Some Physical Quantities

Physical quantity	Symbol	Value
Speed of light in vacuum	c	2.998×10^{8} m s^{-1}
Permeability of vacuum	μ_0	$4\pi \times 10^{-7}$ H m^{-1}
Permittivity of vacuum	$\epsilon_0 = 1/\mu_0 c^2$	8.854×10^{-12} F m^{-1}
Planck constant	h	6.626×10^{-34} J s
Planck constant/2π	$\hbar$	1.055×10^{-34} J s
Boltzmann constant	k	1.381×10^{-23} J K^{-1}
Stefan-Boltzmann constant	σ	5.670×10^{-8} J/m^2 s K^4
Charge of electron	e	1.602×10^{-19} C
		4.803×10^{-10} e.s.u.
Electron rest mass	m_e	9.109×10^{-31} kg
		0.511 MeV/c^2
Proton rest mass	m_p	1.673×10^{-27} kg
		938.3 MeV/c^2
Neutron rest mass	m_n	1.675×10^{-27} kg
		939.6 MeV/c^2
Ratio m_p and m_e	m_p/m_e	1836.15
Atomic mass unit	a.m.u.	1.661×10^{-27} kg
Avogadro number	N	6.022×10^{23} mol^{-1}
Gas constant	R	8.314 J mol^{-1} K^{-1}
Bohr radius	a_0	5.292×10^{-11} m
Fine structure constant	α	7.297×10^{-3}
Reciprocal of α	$1/\alpha$	137.036
Rydberg constant	R_∞	1.097×10^{7} m^{-1}
Bohr magneton	μ_B	9.274×10^{-24} J T^{-1}
Nuclear magneton	μ_n	5.051×10^{-27} J T^{-1}
Compton wavelength of electron	$h/m_e c$	2.426×10^{-12} m

Index